Microbiology
Theory & Application

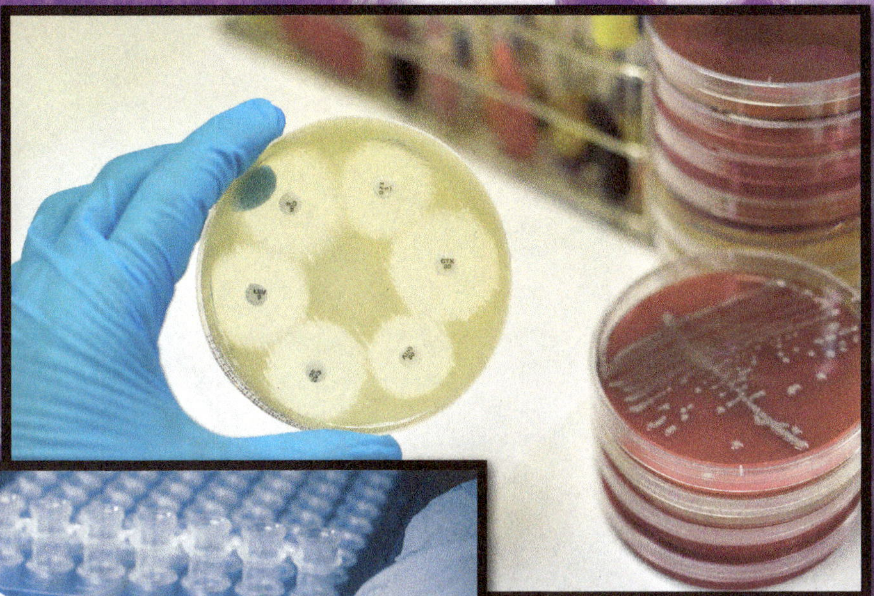

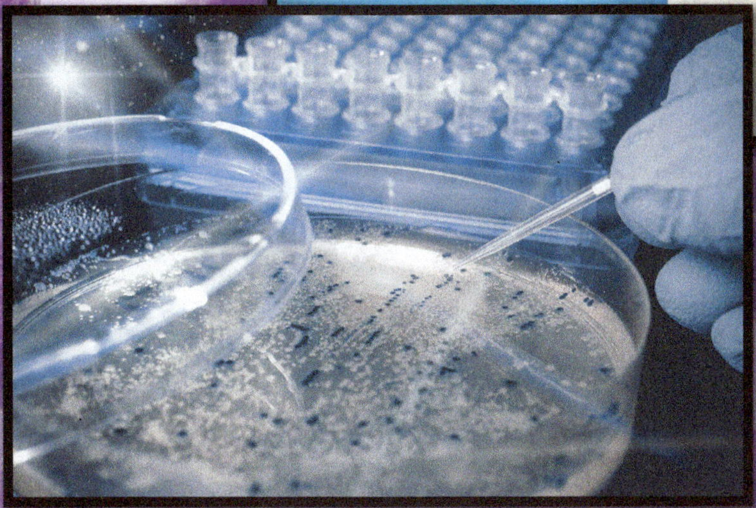

Audra Swarthout • Michael Pressler

Kendall Hunt
publishing company

Cover images © Shutterstock, Inc.

Kendall Hunt
publishing company

www.kendallhunt.com
Send all inquiries to:
4050 Westmark Drive
Dubuque, IA 52004-1840

ISBN 978-1-5249-2349-5

Printed in the United States of America

Contents

Introduction and History of Microbiology

A question we often ask our students on the first day of class is "What is microbiology and why do you need to study this subject?" The simplest definition of **microbiology** is that it is the study of living organisms called **microorganisms** that are generally too small to be seen by the unaided eye. Common microorganisms include bacteria, archaea, fungi, protozoa, algae, and helminths or parasitic worms. Nonliving microbes include viruses, viroids, and prions. The field of microbiology also encompasses other broad fields of science such as immunology, epidemiology, genetics, metabolism, and biotechnology. To answer why you need to study the subject of microbiology would require a little background information that many students are familiar with. Common answers as to the need to study microbiology revolve around how microbes can make us sick; there are many examples of **pathogenic (disease-causing) microorganisms** that do indeed make us visit the doctor's office. For students going into the healthcare profession, it is very important to understand the relationships that humans have with microbes and how we can prevent and treat disease. The pathogenic microbes only make up a tiny portion of the microbes in the world, many of which we encounter on a daily basis. In fact only about 1% of the microbes in the world can be considered harmful to humans. The other 99% of microbes either have no significant impact on our daily lives or are relatively benign. Let's take a quick look at how microbes impact our daily lives.

Microbes in Our Lives

Normal Microbiota

The focus of this course and this text will be on the single-celled microorganisms called bacteria. Bacteria were once classified or grouped as plants, which gave rise to the use of the term *flora* for microbes. The term *flora* has since been replaced by *microbiota*, as scientists have further classified bacteria into a group of their own. Bacteria and other microbes are normally present in and on the human body. These microbes are referred to as **normal microbiota**. These microbial populations inhabit body surfaces such as the skin, oral cavity, respiratory tract, gastrointestinal tract, and genitourinary tract. Our normal microbiota serve as one of the first line defenses against pathogens, by preventing the growth of pathogens on body surfaces. The abundance of microorganisms on and inside the body is quite astounding: there are approximately 90 trillion microbial cells that are constantly present inside and on the human body. These microbial cells far outnumber the cells that you are composed of, which is approximately 10 trillion cells.

ABUNDANCE OF MICROBES	
Skin Surface	2 million microbes/square inch
Feces	100 billion bacteria/gram
Air	50–100 microbes/cubic foot
Soil	1–10+ million microbes/gram

The microorganisms that inhabit the skin are divided into two populations: resident and transient microbiota. **Resident microbiota** is the regular, stable microbiota of the skin. They live in and colonize the deeper layers of the epidermis, hair follicles and glands. **Transient microbiota** is acquired by routine contact and their variety constantly changes. They are found only on skin surfaces and do not colonize. Since they are picked up by contact from infected persons or objects, they are highly influenced by personal hygiene. Oily skin, humidity, occupational exposure, and clothing are further influences to the composition of transient microbiota. Objects that are frequently in contact with our bodies carry populations of our microbiota on their surfaces. These items include jewelry, keys, and other items mentioned in the lab. In a healthcare setting, we transfer microbiota to bedding, chairs, eating utensils, and other objects which we come in contact with.

You might be asking how we develop our normal microbiota. Are we born with it? The developing fetus, protected inside the mother's womb, is normally free of all microorganisms. The colonization of the body will begin as the baby makes its wonderful debut into life. This will occur as the placental membranes rupture and during the birthing process. As the baby breathes and is touched and cared for by healthcare workers, parents and family members, other microorganisms are introduced. Within less than 12 hours of delivery, staphylococci, streptococci, and lactobacilli will colonize a newborn. If the mother breastfeeds the baby, then the microbiota of the baby's large intestine will consist mainly of *Bifidobacterium* species. Bottle-feeding, first teeth, weaning from the bottle, and beginning to eat foods, are major events by which more microorganisms are introduced. Newborns are more at risk of infections leading to complications and death because their normal microbiota have not fully developed and their immune system is not yet fully functional. Often it takes the first year of the baby's life for the normal microbiota to fully develop.

Common Resident Microbiota

Conjunctiva of Eye: *Staphylococci (Staphylococcus epidermidis, Staphylococcus aureus); Viridans Streptococci; Moraxella catarrhalis; Corynebacteria; Haemophilus; Neisseria.*

Outer Ear: *Staphylococcus epidermidis; Pseudomonas.*

Oral Cavity:

Bacteria: *Staphylococci (Staphylococcus epidermidis, Staphylococcus aureus); Viridans Streptococci (Streptococcus mutans, S. salivaricus, S. sanguis, S. mitis, S. oralis, etc); Streptococcus pneumoniae; Bacteroides; Fusobacterium.; Lactobacillus; Moraxella catarrhalis; Actinomycetes; Corynebacterium; Haemophilus influenzae.*

Fungi: *Candida albicans.*

Upper Respiratory (Nasal passages, Throat & Pharynx):
Bacteria: *Staphylococci (S. epidermidis, S. aureus); Viridans Streptococci; Streptococcus pneumoniae; Streptococcus pyogenes; Lactobacillus; Corynebacterium; Moraxella catarrhalis; Neisseria; Haemophilus influenzae; Bacteroides; Fusobacterium.*

Fungi: *Candida albicans.*

Skin:
Bacteria: *Staphylococci (S. epidermidis, S. aureus); Micrococcus; Viridans Streptococci; Corynebacterium; Propionibacterium acnes; Peptococcus; Peptostreptococcus; Pseudomonas.*

Fungi: *Pityrosporum ovale & orbiculare; Candida; Cryptococcus neoformans.*

Arthropods: *Demodex folliculorum* mite.

Esophagus: Usually little to no normal microbiota.

Bacteria: *Streptococcus viridans*

Lower Gastrointestinal (Lower portion of small intestine, large intestine & rectum):
The intestinal tract is a harsh environment for microorganisms, yet the bulk of our normal microbiota inhabits this region of the body. In fact, the colon may contain 10^{12} bacteria per gram of material. Most of these are anaerobes. These organisms inhibit the growth of other pathogens, but some can be opportunistic (e.g. *C. difficile* can produce pseudomembranous colitis).

Bacteria: *Bacteroides; Fusobacterium; Enterococcus faecalis; Escherichia coli; Lactobacillus; Staphylococcus aureus; Clostridium; Bifidobacterium; Enterobacter; Klebsiella; Eubacterium; Streptococci; Pseudomonas; Salmonella; Peptococcus; Peptostreptococcus; Ruminococcus; Proteus; Citrobacter; Shigella.*

Fungi: *Candida.*

Upper Gastrointestinal:
The upper gastrointestinal tract (the stomach, duodenum, jejunum, and upper ileum) normally contains limited microbiota; the bacterial concentrations is less than 10^4 organisms/ml of intestinal secretions. Most of these organisms are derived from the oropharynx and pass through the gut with each meal.

Stomach: *Helicobacter pylori* in up to 50% of the population.

Female Reproductive (External genitalia, vagina & cervix):

Bacteria: *Lactobacillus acidophilus; Streptococcus; Corynebacterium; Escherichia coli; Mycobacterium; Proteus; Staphylococci (S. epidermidis, S. aureus); Bacteroides; Enterobacter; Clostridium; Fusobacterium; Enterococcus.*

Fungi: *Candida.*

Protozoa: *Trichomonas.*

Lower Urinary (Lower portion of urethra): *Escherichia coli; Proteus mirabilis; Staphylococcus epidermidis; Lactobacillus; Klebsiella; Pseudomonas; Enterococcus.*

Microbe/Germ-Free Areas of the Body

Lower Respiratory: Trachea & bronchi have sparse microbiota; bronchioles & alveoli have no normal microbiota and are usually sterile.

Reproductive: No normal microbiota and usually sterile; upper reproductive of females is also sterile.

Upper Urinary: Kidneys, ureters, bladder & upper urethra of both males & females have no normal microbiota and are usually sterile.

Cardiovascular (Heart & blood vessels), Nervous (Brain & spinal cord), Muscular, Skeletal: No normal microbiota and usually sterile.

Liver, glands, bone marrow, middle & inner ear, internal eye and sinuses: No normal microbiota and usually sterile.

Opportunistic Pathogens

Some of our normal microbiota may be considered opportunistic pathogens, which are organisms that do not normally cause disease in a normal healthy individual. They cause disease only if the individual is immunocompromised or if the organism enters into a part of the body that it does not belong to. One example of an opportunistic bacterium found on the skin is *Staphylococcus aureus*. This organism is known to cause a wide variety of infections from mild infections of the skin to sometimes fatal infections such as toxic shock syndrome.

Infectious Disease

When a pathogen overcomes the host's immune system, disease may result. Infectious diseases may be classified in a variety of ways. We will briefly look at some organisms that cause what are referred to as **emerging infectious diseases**. An emerging infectious disease is one that is relatively new and has not been seen historically. New diseases emerge for a variety of reasons, such as mutation of microbes, human exposure to animal pathogens and sometimes changing technology and environments. Examples of emerging infectious diseases are infections caused by methicillin resistant *Staphylococcus aureus* (MRSA); note that MRSA is not a disease, but the organism that causes disease. MRSA is an opportunistic organism that causes the same types of infections as normal *Staphylococcus aureus*, the difference being that MRSA is more difficult to treat since the organism is resistant to the antibiotic methicillin. 'Resistant' means that the antibiotic will not effectively treat the infection and the patient will not get well. This organism is considered emerging since it was first seen in the 1980s; it is not historically prevalent like diseases such as the bubonic plague. Another example of an emerging organism that causes disease is *Escherichia coli* O157:H7; this organism is a toxin-producing strain of *E. coli* (normal microbiota) and was first seen in 1982. *Escherichia coli* O157:H7 is the leading cause of diarrhea (often bloody) worldwide and can be fatal in young children and elderly individuals. Other examples are Ebola, HIV virus, Avian influenza, and cat scratch fever.

Not only can infectious diseases be new, but some old diseases that were once controlled are rearing their ugly heads again; these are called **re-emerging diseases.** This resurgence of old diseases can be attributed to three contributing factors:

1. Increase in world travel.
2. Certain populations of humans have a lax attitude, are misinformed, or are afraid of vaccinations.
3. There is an increasing population of immunocompromised people.

Think about how these factors could lead to a re-emergence of a disease such as Pertussis (whooping cough) or measles. Both of these diseases have vaccinations readily available in the United States. If there is a population of people who have not been vaccinated, they are susceptible to contract the disease. A person traveling from overseas where vaccination is not as common can bring the causative organisms to the U.S. and infect susceptible populations. Since the 1980s, the number of pertussis cases has increased significantly, affecting adults and adolescents. Since immunity from the DTaP vaccine declines after a few years, there is no protection by the age of 12, and therefore booster shots are required. Due to the lax attitude some people have there is an increasing population susceptible to contract pertussis. Are you up to date on your vaccines?

Microbes in the Environment

Decomposition

Bacteria and fungi are considered decomposers and are involved in sewage treatment and even the breakdown of toxic waste such as petroleum products. These organisms are responsible for the loss of billions of pounds of food each year due to spoilage, bread mold, soured milk, rotted fruits and vegetables; we are all familiar with these household items and have all thrown food away due to spoilage. The result of this spoilage is the waste of money that we all spend just to throw food away. Don't just throw those rotted foods into the garbage, which will be shipped to the nearby landfill. Make a pile in your back yard and let the microbes turn that waste into compost, which can act as a natural fertilizer for your garden. The decomposition process takes the complex macromolecules in the food and converts them to simpler forms utilized by plants.

Element Cycling

Microbes are essential for life on earth—they are involved in the carbon, nitrogen, phosphorus, and sulfur cycles. You will learn in a later chapter how we as humans rely on microbes for our lives to exist.

Food Production

Not only do microbes spoil our food, but they are also involved in producing food such as cheese, yogurt, bread, sauerkraut, and vinegar, and let's not forget alcohol!

After reading this introduction, you should have a better appreciation for microbes and microbiology. As the semester progresses, you will gain more in-depth knowledge on how we as humans interact with microbes on a daily basis.

A Brief History of Microbiology

The 1600s–1800s was an instrumental time for the development of the field of microbiology. Perhaps the most influential events that led to the development of the field were the debates over spontaneous generation beginning in 1668, and the development of the microscope in the late 1600s. Prior to the 1670s, there were no microscopes and there was little to no knowledge of microorganisms. Early scientists were asking very basic questions about life and disease such as:

Is spontaneous generation of microbial life possible?
What causes disease?
How can disease be prevented?

Let's look at these questions one by one and answer these questions by observing early scientific experiments.

Is spontaneous generation of microbial life possible?

In the early history of human civilizations, there was very little scientific understanding of basic biological processes such as how life arises. It was Aristotle (384–322 BC) who proposed the theory of **spontaneous generation**, which is a hypothesis that living organisms arise from nonliving matter and that there is an unseen "life force" that causes life to arise. The conflicting theory of **biogenesis,** which we now know as true, states that living organisms arise from pre-existing life. We will briefly look some important experiments that disproved the theory of spontaneous generation.

Francesco Redi (1626–1697)

Francesco Redi in 1668 conducted some experiments using raw meat. Common knowledge at the time led people to believe that maggots arose spontaneously from rotting meat, therefore having a belief in spontaneous generation. By taking a flask with a piece of meat and leaving it open to the air, maggots will form on the rotting meat. Repeating this same experiment and covering the flask with a cork, no maggots form on the meat. This simple experiment disproves the common knowledge that maggots arise spontaneously from rotting meat! Right? Well Francesco anticipated criticism that by not allowing air to contact the meat the "life force" will not survive. He answered this criticism by covering a third flask with a fine mesh so that air could contact the rotting meat. Maggots did indeed form; however, they were on the top of the mesh and not in contact with the meat. Since the maggots had no food they quickly died. Observations of his experiment showed flies landing on the meat in the open flask and on top of the mesh on the third flask; Redi therefore concluded that the fly is leading to the development of the maggots, thereby supporting biogenesis. But why did the meat rot?

Anton van Leeuwenhoek (1632–1723)

A few years after Redi, in 1673, Anton van Leeuwenhoek developed the first simple microscopes and discovered the microbial world. For this reason Anton is commonly referred to as the Father of Microbiology. Upon examining a drop of pond water he saw numerous "animalcules" zipping around in his sample. Anton developed over 500 microscopes during his time which were more like a specialized magnifying glass by today's standards; yet he was able to take a glimpse into the microbial world, thus forever changing how humans viewed the world.

John Needham (1713–1781)

Following the discovery of the microbial world, scientists were wondering if spontaneous generation of microbial life was possible. In 1745, Needham used a flask of nutrient broth and heated it sufficiently to kill the "life force"; he then transferred the broth to a new flask and sealed the flask. After a day, he observed turbid microbial growth. At this time, there was no knowledge of heat-resistant bacterial structures and there was little to no aseptic techniques used. At the time, his results gave evidence for spontaneous generation. But some criticized his lack of a clean environment and Lazzaro Spallanzani suggested that microbes entered his flask from the air during his transfer technique.

Lazzaro Spallanzani (1729–1799)

As mentioned above Lazzaro criticized Needham's work, and in 1765, he repeated Needham's experiments; however, upon the completion of his boiling, he immediately and directly sealed the flasks. The result after one day, two days, three days . . . was no turbid microbial growth.

Spallanzani thus gave evidence against spontaneous generation and supported biogenesis. His work was criticized as well, with the argument that since no air was allowed in his sample the "life force" could not survive.

Louis Pasteur (1822–1895)

Almost 100 years after Spallanzani conducted his experiments, Pasteur demonstrated that microorganisms are indeed present in the air, and once and for all disproved the theory of spontaneous generation. Pasteur's swan-necked flasks allowed air to enter into his boiled nutrient broth and any microbes that entered would land in the neck of the flask and not contaminate his broth. The broth would remain sterile indefinitely; only if the flask was tipped and the broth came into contact with the contaminated neck, would it show growth.

Scientific Method

The process of disproving spontaneous generation helped the development of the scientific method, a process that every scientist today follows without even thinking about it. An outline of the scientific method is given below:

1. Observation leads to a question.
2. Question generates a hypothesis.
3. Hypothesis is tested through experimentation.
4. Results prove or disprove the hypothesis:
   a. Accept hypothesis leads to a theory or law.
   b. Reject hypothesis leads to modification of hypothesis and more experimentation.

What causes disease?

In the early 1860s, Louis Pasteur developed the **germ theory of disease,** the theory that states that microorganisms cause disease. Although he wasn't the first person to propose the theory, he was the first to develop experiments giving support to the theory.

Robert Koch (1843–1910) later used the germ theory and linked suspected causative agents of disease and developed experimental steps for directly linking a specific microbe to a specific disease. He used cows to demonstrate that *Bacillus anthracis* causes anthrax. In 1884, Koch developed postulates, which are the experimental steps that must be used to link an organism to the disease. Knowledge of causative agents led to antibiotic development and vaccine development, resulting in the eradication of smallpox worldwide and the significant decline in childhood diseases such as polio and measles.

Koch's Postulates

1. **The microorganism must be found in abundance in all individuals with the disease and absent from all healthy individuals.**
2. **The organism must be isolated from the diseased and grown in pure culture in the laboratory.**
3. **When introduced into a healthy host, the newly infected host must develop the same disease.**
4. **The organism must be re-isolated from the newly diseased host and grown in pure culture.**

There are exceptions to Koch's Postulates such as an organism's inability to be grown in pure culture in the lab. For instance, the organism that causes syphilis cannot be grown in pure culture. Some organisms cannot be grown in the lab at all, such as *Mycobacterium leprae* (causative agent of leprosy), which has never been grown in a lab. Other exceptions are ethical exceptions due to high mortality of the disease or the lack of a cure for human-specific viruses such as HIV or Ebola.

How can disease be prevented?

Knowledge of disease prevention did not come about until the mid-1800s, and even then, physicians either did not believe or did not practice preventive measures fully, until later on in the 1800s. Ignaz Semmelweis was a physician who noticed that mothers who delivered babies in the midwife ward of a hospital were less likely to develop childbed fever (causative agent *Streptococcus sp.*) than mothers who delivered in the physician ward. Semmelwies hypothesized that his medical students were transferring "cadaver particles" from autopsy studies into delivery rooms and the "cadaver particles" were causing the childbed fever. Semmelwies began requiring his medical students to wash their hands with chlorinated water, thereby removing the particles. His results were drastic: in the subsequent year, mortality from childbed fever dropped from 18% to 1%! Ignaz was unsuccessful in gaining support for this preventive measure, since doctors at the time did not believe that they could be the cause of the infections. Semmelweis later developed depression and was committed to a mental asylum, where he later died from an infection caused by *Streptococcus*! This is an interesting story about the development of hand-washing as a normal practice to prevent the spread of disease, and we cannot stress enough that just hand-washing alone is the number one way to prevent the spread of nosocomial infections.

Another instrumental figure who answered this question on how disease can be prevented is Joseph Lister (1827–1912). He is stated to have saved more lives in the 19th century than all the wars during that time had sacrificed. Lister introduced a system to prevent surgical wound infections called **aseptic techniques.** Aseptic techniques are used to prevent contamination from unwanted microorganisms. Lister applied Pasteur's developments in microbiology and promoted sterile techniques during surgeries. During the civil war, a field surgeon may have wiped his scalpel on the sole of his boot in between patients; this is far from sterile!

Must Read Before You Begin Lab
Laboratory Safety

There are always hazards in the laboratory due to the fact that we are working with living microorganisms. Microbes are "unseen" unless we grow them in large populations. The microorganisms we choose for our lab work are ones which normally do not cause infection in a healthy person, or have a moderate potential to infect. However, any microbe has the potential to harm, especially in larger quantities. Furthermore, there is always the potential for pathogenic mutations (the microbe mutates in such a fashion as to become a greater hazard). You must know the locations and instructions for use of all safety equipment provided in the lab: first aid kit, eye wash station, sharps container, biohazard container, and fire extinguisher. In case of an emergency, use the prep room phone to dial 9111.

Exposure Control Plan for Category a Students

A ***Category A student*** is a student who is involved in educational assignments and/or training that requires procedures or tasks which involve exposure or reasonably anticipated exposure to blood or other potentially infectious material (OPIM) or that involves a likelihood for spills or splashes of blood or other potentially infectious material. This includes procedures or tasks conducted in routine and non-routine situations. Within the college, there are certain programs and/or courses that require such procedures and tasks. Microbiology is one of those courses, as will be your clinical program. As a result, a big part of this Microbiology course will be learning all about **OSHA** (Occupational Safety and Health Administration), **MIOSHA** (Michigan Occupational Safety and Health Administration), and **CDC** (Centers for Disease Control and Prevention). OSHA and MIOSHA establish and enforce standards in the workplace to ensure employee safety. Delta College has an **Exposure Control manual** for all **Category A workers**, because this is required by OSHA and MIOSHA. CDC establishes guidelines that you are required to follow in this course, within your clinical program, and within your career. The entire *Delta College Exposure Control Manual For Category A Students* is at the end of the manual.

Exposure means reasonably anticipated skin, eye, mucous membrane, or parenteral contact with blood or other potentially infectious materials that may result from the performance of a student's educational or clinical responsibilities. This definition excludes incidental exposures that may take place, that are neither reasonably nor routinely expected, and that the student is not required to incur in the normal course of their training.

Exposure Incident means a specific exposure of the student's non-intact skin or mucous membranes (eye, nose, or mouth) to blood or other infectious body fluids, which results from the performance of a student's educational or clinical responsibilities. This includes exposure via the parenteral route.

Standard Operating Procedures (SOPs) means any of the following, which address the performance of the individual's responsibilities so as to reduce the risk of exposure to blood and other potentially infectious material:

- Written policies
- Written procedures
- Written directives
- Written standards of practice
- Written protocols
- Written systems of practice
- Elements of an infection control program

Work Practice Controls means controls that reduce exposure by altering the manner in which a task is performed. Category A students will use these controls to reduce transmission of pathogens regardless of route.

Engineering Controls means controls (e.g., sharps disposal containers, self-sheathing needles, safer medical devices, such as sharps with engineered sharps injury protections and needleless systems) that isolate or remove the blood-borne pathogens hazard from the workplace. Engineering controls deal with the physical environment, including buildings and equipment.

Standard/Universal Precautions

In the healthcare field, we refer to measures employed to protect yourself and your patients as **Isolation Precautions**. Since 1877, when the need for such precautions was first recognized, a series of precaution guidelines have evolved. In 1985, Universal Precautions (UP) was developed for all healthcare workers. These guidelines emphasized applying blood and body fluid precautions universally to all persons regardless of their infection status. In 1987, Body Substance Isolation (BSI) guidelines were proposed. BSI emphasized isolation of all moist and potentially infectious body substances from all patients. In 1989, OSHA (Occupational Safety and Health Administration) published the Blood-Borne Pathogen Regulations, governing occupation exposure to blood-borne pathogens in healthcare settings. By the 1990s, many were uncertain about which guidelines to follow!

At this time, most healthcare settings were using a mixed combination of all the guidelines, referring to them as universal precautions. As a result, a new set of guidelines was proposed and agreed upon by the CDC, HICPAC (Hospital Infection Control Practices Advisory Committee), Public Health Service, and U.S. Department of Health and Human Services. These are the **Standard/Universal Precautions**. They also include a set of **Transmission-Based Precautions**, for selected infections/diseases transmitted either by airborne routes, droplets, or by contact.

Standard/Universal Precautions means the use of barriers (protective clothing, eye wear, masks, and gloves) to control the transmission of infectious diseases. This is a method of infection control in which all human blood and body fluids are treated as if known to be infectious for HIV/AIDS, HBV (Hepatitis B Virus), HCV (Hepatitis C Virus), and other blood-borne pathogens.

You are to follow Standard/Universal Precautions each and every time you work with a patient. Standard Precautions reduce the risk of transmission of microorganisms from recognized and unrecognized sources of infection and are meant to bring about the control of infections. They apply to all patients regardless of their diagnosis or presumed infection status. Standard Precautions apply to (1) blood, (2) all body fluids, secretions and excretions (except sweat) regardless of whether they contain visible blood, (3) nonintact skin, and (4) mucous membranes. Bodily fluids include urine, feces, pus, saliva, spit, tears, mucus, vomit, sputum, vaginal or penal secretions, afterbirth and any other fluid-like substance which could come from a patient. The most obvious time you will be exposed directly to these bodily fluids is during specimen collection. Therefore, as you learn to collect specimens in the lab, you will learn how to employ isolation precautions.

Transmission-Based Precautions apply to patients known or suspected to be infected with (1) a pathogen which is highly transmissible, and (2) an epidemiologically important pathogen, such as a multidrug-resistant microorganism. Transmission-Based Precautions are used in addition to Standard Precautions. There are three types of Transmission-Based Precautions: Airborne, Droplet, and Contact.

Airborne Precautions are used in addition to Standard Precautions for patients known or suspected to have serious illnesses such as tuberculosis, measles, and chickenpox, transmitted by airborne droplet nuclei. These small, infective particles (less than 5 um in size) can be free-floating or combined with dust particles in the air.

Droplet Precautions are used in addition to Standard Precautions for patients known or suspected to have serious illnesses transmitted by large-particle droplets, 5 um in size, which can be spread by coughing, sneezing, or talking. Diseases fitting this category include: invasive *Haemophilus influenzae* type b (causing meningitis, pneumonia, epiglottitis, and sepsis), invasive *Neisseria meningitidis* (causing meningitis, pneumonia, and sepsis), diphtheria, *Mycoplasma pneumonia*, pertussis, pneumonic plague, streptococcal infections (pharyngitis, pneumonia, or scarlet fever), adenovirus, influenza, mumps, parvo virus B19, and rubella.

Contact Precautions are used in addition to Standard Precautions for patients known or suspected to have serious illnesses transmitted by direct patient contact or by contact with patient-care equipment and articles. These include (1) multidrug-resistant bacteria causing gastrointestinal, respiratory, skin, or wound infections, (2) enteric infections involving *Clostridium difficile*, *E. coli* O157:H7, *Shigella*, hepatitis A, or rotavirus, (3) RSV (respiratory syncytial virus), (4) viral hemorrhagic conjunctivitis, (5) viral hemorrhagic infections (Ebola, Lassa, or Marburg), and (6) skin infections involving cutaneous diphtheria, herpes simplex virus, impetigo, major abscesses or cellulitis, pediculosis, scabies, staphylococcal furunculosis, and disseminated zoster.

In the healthcare field you will constantly move from patient to patient and instrument to instrument. If you fail to wash your hands completely as you leave one patient or procedure, you will be taking contaminants with you to the next patient or procedure. This transmission is known as **cross contamination**. Cross contamination happens all too frequently in the healthcare setting. In fact, lack of proper hand-washing (or failing to wash at all) is the greatest cause of nosocomial infections.

A **nosocomial infection** is an infection that occurs within the hospital setting. The patient did not "arrive" with this infection, but gained it while in the healthcare setting due to lack of proper aseptic techniques on the part of the healthcare worker. An important part of this course will be learning to wash your hands so that you are not spreading germs!

Healthcare-Associated Infections (HAIs) are infections associated with healthcare delivery in any setting, such as hospitals, long-term care facilities, ambulatory settings, and home care.

Community-Acquired Infections (CAIs) are infections that have entered the community at large.

Laboratory Aseptic Techniques

Asepsis refers to the procedures that prevent contamination by microbes or their toxins. ***You must employ aseptic techniques when handling objects in this laboratory.***

Removal of gloves. There is a technique to removing gloves so that you do not contaminate your hands in the process.

- Grab one glove with your other gloved hand and peel off the glove by turning it inside out. Hold the removed glove in a rolled ball in the gloved hand.
- Reach inside the neck of second glove to grab hold of it. Do not touch the outside of the glove because it is considered to be contaminated.
- Peel off the second glove, turning it inside out as you go. The first glove ends up inside the second glove.
- Discard the gloves in the appropriate container. If they are contaminated, use the biohazard bag. If they are not contaminated, use the waste basket.

Test tube handling. Test tubes must be carried by the glass tube. They are never picked up or carried by the cap. The caps are not sealed and the test tube would fall and spill its contents. If you are carrying more than one test tube, place the tubes in a test tube rack.

Re-suspension of a culture. Before you take a sample from a test tube, the culture needs to be re-suspended. Gravity may have caused much of the sample to collect at the bottom of the test tube. **Test tubes are never shaken up and down!** This would cause the sample to leak out around the cap, down the sides of the tube and onto your hands (and perhaps all over the area!). To re-suspend, roll the tube between your hands. In some instances, the vortex mixer is used. These methods should re-suspend the sample without causing a spill.

Inoculating loops and needles. Loops and needles are not sterile when you take them from their container for use in the lab. They must be sterilized before and after each use. We use the flame of the Bunsen burner to sterilize them. You will be taught how to do this to ensure sterilization.

Test tube caps and flaming. Test tube caps must not be contaminated while taking a sample from a test tube. This means that test tube caps cannot be set down on the table. You will learn how to hold the cap within your fingers as you work with a tube. The lip of the tube must be kept free of contamination. This means that the top of the test tube must be run through the flame after the cap is removed and before the cap is replaced. You will also learn this skill.

Pouring media from flasks. Flasks must not be contaminated when used in the lab. The flask bottom must be held, while the lip of the flask is flamed.

Observing growth on Petri plates. In the second period of a lab exercise, after you have "grown" your microorganisms on Petri plates, you will observe them. Caution: You must be aware that moisture that has collected within the Petri plate may be contaminated with microorganisms. If that moisture were allowed to leak out, it could contain microorganisms.

Observation of plates: If you choose to elevate your plate for view, this must be done with utmost care. ***Never hold the plate over your lab manual, your lap or your eyes—hold it out away from yourself.*** You will learn that the best method for observing plate growth is by using the Quebec colony counter, which is designed to hold your plate and magnify it.

Lab Exercise 1.1

Introduction to the Microbiology Laboratory

Student Objectives

1. Define the following terms: microbe, clean, dirty, sterile, aseptic techniques, culture, bacteria colony, incubate, culture medium, disinfect, antiseptic, biohazard, personal protective equipment (PPE), contamination, infection, Petri plate and autoclave.
2. Evaluate the standard hand-washing procedure for working in the microbiology lab.
3. Understand the standard surface disinfection procedure for working in the microbiology lab.
4. Understand the standard spill clean-up procedure for working in the microbiology lab.
5. Know about the benefits and drawbacks of broth cultures, agar slants and agar plate cultures.
6. Understand the reasons why working with microorganisms can be a potentially dangerous situation.
7. Identify the safety precautions required in microbiology laboratories to protect workers.
8. Evaluate the isolation precautions (standard and transmission-based) which should be used by all healthcare workers.
9. Recognize the kinds of accidents that could occur in the lab and how they can be avoided.
10. Identify the types of equipment used for working with microorganisms in the lab.
11. Understand why students must inoculate and incubate as a part of their study.
12. Know what normal microbiota is, plus the difference between resident and transient microbiota.
13. Know what cross contamination is, how it can occur, and how to prevent it.
14. Know what nosocomial infections are, how they can occur, and how to prevent them.
15. Understand the three main types of hand care.
16. Identify and describe the main types of hand hygiene used in occupational settings.
17. Identify and evaluate the common health-care-associated and community-associated infections.
18. Evaluate the importance of the current Exposure Control manual for Category-A students.

Microbiology Laboratory Safety Rules

Laboratory Environment:

1. Alert your instructor if you meet one of the following criteria:
   a. Are immunocompromised
   b. Are pregnant
   c. Live with or care for an immunocompromised individual
2. Know the location and use of emergency safety equipment in the laboratory.
3. Never eat, drink, or chew gum in the laboratory.
4. Do not touch your face, apply cosmetics, adjust contact lenses or bite nails while in the lab.
5. Do not handle personal items while in the lab. Place **all** personal items (cell phones, book bags, purses, coats or other belongings) in the cubby-hole space provided. Do NOT place them on the laboratory tables.
   a. ***You may occasionally use your cell phone to document your results; however, do not handle your phone or other personal items while wearing gloves and wipe down personal items with an alcohol prep pad after use.
6. **Disinfect** your bench top **before** and **after** each laboratory session.
7. Treat each microorganism used in the laboratory as if it were potentially harmful.
8. **Wash your hands** when entering and before leaving the laboratory.
9. Do not enter the laboratory without permission. Do not work without an instructor present.

Personal Safety and Personal Protective Equipment (PPE):

1. Never place objects into your mouth while in the laboratory.
2. Wear a **laboratory coat** to cover your street clothes, wear slacks to cover your legs, wear socks and **closed-toe shoes** (no sandals!) to cover your feet.
3. Keep your laboratory coat inside the lab room; do NOT remove it from the lab area. Store your laboratory coat in the lab room during the entire semester.
4. Wear gloves during microbial experiments to protect your hands and prevent cross-contamination.
5. Your **eyes** must be **protected** in the laboratory. Wear safety glasses.
6. It is recommended that you do not wear contacts in the laboratory.
7. Do not wear artificial fingernails or have long nails in the laboratory. Tie long hair back and restrain fluffy or flyaway hair with a head band or scarf.
8. Report ALL accidents to your instructor. Place broken glass in the sharps container.
9. Use barrier precautions to prevent skin and mucous membrane exposure.

Laboratory Work:

1. Use Universal and Standard precautions.
2. Use the **autoclave cart** for media and cultures after the experiment is completed.
3. Use **biohazard bags** for contaminated paper toweling and materials only. Use regular wastebaskets for normal trash. Use Nalgene containers for contaminated pipettes.
4. Never take cultures out of the laboratory and do not perform unauthorized experiments.
5. Use caution when using the Bunsen burners, hot plates and electrical equipment.
6. You must sit while doing lab procedures.

Laboratory Routines

The following are basic **routines** for the microbiology student for each lab exercise:

1. Read the laboratory exercises *before* coming to class. Many students find it helpful to make a "to do" list. You need to be the judge if you are struggling to finish labs if these steps are necessary. People work at different speeds and the labs are planned so that the average student has plenty of time to finish the exercises.
2. Notice safety concerns and ensure safety precautions before beginning experiments.
3. Understand the required materials to complete the experiment.
4. Label all of your experiments with your name, table number, and course section number.
5. Be prepared to work independently as well as with your table partners.
6. Keep accurate notes and records of your procedures and results so you can refer to them for future work and tests. Your notes are essential to ensure that you perform all the necessary steps and observations.
7. Clean your work area before you leave the laboratory.
   - Stains and reagent bottles should be returned to their original location.
   - All tape on glassware should be removed before placing glassware into the autoclave cart.
   - Any potential contamination should be placed in marked biohazard containers.
   - Used paper towels should be discarded *only* in the regular trash cans (they are not biohazards, unless they are used to clean up a spill and are heavily contaminated).
   - Clean and properly store microscopes.
8. Use the laboratory disinfectant and paper towels to clean and disinfect the table surfaces using the '**APPLY, WIPE, REAPPLY**' method. Do this before and after each laboratory session.
   - **Apply** the disinfectant over the entire table surface.
   - Use paper toweling to **wipe** and dry the table surface. The toweling goes into the regular waste basket, not the biohazard bag.
   - **Reapply** the table surface with a heavy, soaking spray. Spray your table close enough so aerosols will not irritate the lungs. Allow the spray to air-dry for at least **10** minutes.

INTRODUCTION

Microbiology provides the first opportunity that undergraduate students have to experiment with microscopic living organisms. The microbiology laboratory is an opportunity to learn about microbes and their behavior. You will also learn about safety, aseptic techniques, and standard operating procedures. Microbes are relatively easy to grow and lend themselves to experimentation. The microbes used in this microbiology course are referred to by their **scientific names** (genus and species of the organism).

Microorganisms must be **cultured** (grown) to complete most of the exercises in this lab manual. Cultures will be set up during one laboratory period and will be examined for growth at the next laboratory period. Since there is variability in any population of living organisms, not all the laboratory experiments will "work" as the textbook or lab manual says. For example, during lab you will observe a very detailed description of an experiment or organism and probably find that this description does not match your reference exactly. Therefore, *observing*, *recording*, and *critically analyzing* your results carefully are important parts of each laboratory exercise. ***Accurate records* and *good organization* of lab work will enhance your enjoyment and facilitate your learning.**

DEFINITIONS

Culture: A method of growing microorganisms in large numbers in a culture medium in the laboratory so that we may observe and study them.

Bacterial colony: A visible cluster of bacteria growing on or in a medium (broth, agar) that descended from a single bacterial cell. Members of the colony are presumed to be genetically identical.

Inoculation: A method used in the laboratory for transferring microorganisms in an aseptic manner from one surface/medium to another using an inoculating loop, inoculating needle, swab, or other sterile transfer device. (Examples include from one test tube to another test tube, test tube to a microscope slide, or test tube to a Petri plate.)

Incubation: A method used in the laboratory to provide microorganisms with the appropriate temperature needed for growth. (We will incubate most of our cultures in a 37°C incubator in the prep room.)

Optimal temperature: The temperature a microorganism prefers and that will allow us to get the best growth for observation and testing.

HAND HYGIENE

It is important that you fully understand that your body, and the bodies of your patients, are covered with microorganisms both on the inside and outside. It is estimated that the typical body has over 100 trillion microorganisms living on or in it! These microbes are referred to as the **normal microbiota**.

To avoid accidental transfer of microorganisms between yourself and your patients, you must wash your hands and wear gloves. Then you must remove your gloves, wash your hands, and put on new gloves between patients.

The process of hand washing is the number one way to prevent nosocomial infections. *Every person's hand-washing habits are different. Each of us needs to identify those "tough" areas on our hands which are not being washed thoroughly. Typical areas of bacterial accumulation include around the nails, in the knuckles, in creases, and around jewelry (rings, watches, etc.). The following hand-washing activity is designed to identify areas of your hands that are not being thoroughly cleaned.*

HAND HYGIENE PROCEDURE

Current Recommendations for Hand Hygiene

There are three types of hand hygiene recognized in health care:

- Hand-washing using a plain soap along with an alcohol-based hand gel.
- Hand-washing using an antimicrobial soap.
- Surgical hand scrub.

Which level you employ at any particular time depends upon the activity in which you are involved.

Hand-Washing Using a Plain Soap Along With an Alcohol-Based Hand Gel

- This is the current recommendation procedure used in healthcare.
- Alcohol is not a good cleaning agent, as it loses its effectiveness in the presence of dirt and organic matter. Therefore, washing hands that are visibly dirty is required before applying an alcohol-based hand gel.

Step One: Wash hands with plain soap to remove dirt and debris.

- Soaps are detergent-based products that contain esterified fatty acids and sodium or potassium hydroxide. Their cleaning activity is attributed to their detergent properties, which result in removal of dirt, soil, and various organic substances from hands.
- Plain soaps have minimal, if any, antimicrobial activity. The purpose of hand-washing is to physically remove dirt, organic matter, and transient microbiota.
- Proper hand-washing includes mechanical friction, use of appropriate soap, proper rinsing and drying. The facility you work in will determine when hand-washing with plain soap is adequate or when hand-washing with an antimicrobial soap is required.
- Make sure you remove jewelry and roll out paper towels ahead of time. Skin underneath rings is more heavily colonized with bacteria than comparable areas on fingers without rings.
- *Antimicrobial soap* is a soap containing an antiseptic agent, usually triclosan. Triclosan enters bacterial cells and affects the cytoplasmic membrane, synthesis of RNA, fatty acids, and proteins.

Washing

- Use warm water. The water should be above body temperature, neither cold nor hot.
- Dampened hands should be thoroughly covered with either a plain soap or an antimicrobial soap (3–5 mL is recommended), then rubbed vigorously for at least 15 seconds, generating friction on all surfaces of the hands and fingers.
- Fingernails should be thoroughly cleaned. Washing should proceed from the tips of the fingers up to and including the wrists.

Rinsing

- Use warm water. The water should be above body temperature, neither cold nor hot.
- Rinsing should begin from the fingertips downward to the wrists.
- Hands should be thoroughly rinsed to remove soap and debris. (Note: Some hand washing procedures include the cleansing of the forearms.)

Drying

- Hands should be dried using paper towels.
- Drying should proceed from the tips of the fingers up to and including the wrists.
- A paper towel should be used to turn off the faucet, so as not to re-contaminate the hands.

Step Two: Disinfect hands using alcohol-based hand gels.

- 60% to 70% ethanol or isopropyl alcohol hand rubs, containing emollients to minimize skin drying, are considered the best.
- Alcohol is thought to work by denaturing proteins. It works well against many kinds of microorganisms, reducing the number of viable microorganisms on hands.
- Apply the product to the palm of one hand and rub hands together for one minute, covering all surfaces of the hands and fingers, generating friction on all surfaces.
- This technique is only effective if a sufficient amount of alcohol of appropriate concentration is used.
- Alcohol-based hand rubs may be used between several activities or patient contacts.
- Alcohol-based hand rubs may be used both before and after gloving to perform routine activities and procedures.
- Frequent use of alcohol-based hand gels can cause drying of skin unless emollients, humectants, or other skin-conditioning agents are added to the formulations.

Step Three:

Wash hands with plain soap whenever hands are visibly dirty or begin to feel gritty (as if there is a build-up of the gel on them). Also wash hands when contact with blood or other bodily fluids occurs. Use hand lotions to support good skin health.

Skin Hygiene: When Is Clean Too Clean?

Water content, humidity, pH, intracellular lipids, and rates of shedding help maintain the protective barrier properties of the skin. The barrier may be compromised through abrasion, skin dryness, irritation, or cracking. Among persons in occupations such as healthcare, in which frequent hand washing is required, long-term changes in the skin can result in chronic damage, irritant contact dermatitis and eczema, and changes in the microbiota. Irritant contact dermatitis, which is associated with frequent hand-washing, is an occupational risk for healthcare professionals. Damaged skin more often harbors increased numbers of pathogenic microbiota. The goal in healthcare should be to identify skin hygiene practices that provide adequate protection from transmission of infecting agents, while minimizing the risk for changing the make-up and condition of the healthcare worker's skin and increasing resistance of the bacterial microbiota.

Nail Length and Artificial Nails

The majority of normal microbiota are found around and under the fingernails. The current recommendation is that nails should be kept clean, short (less than ¼ inch), and smooth. Long nails are not allowed because they can harbor more transient microbiota. They also make cleaning (washing, rinsing and drying) more difficult. Further, gloves are harder to get on over long nails, and they lead to more tears. Many facilities do not allow any nail polish. If nail polish is allowed at all, it must be clear nail polish. Dark nail polish obscures the view, thereby interfering with the cleaning process. Artificial nails and extenders are not allowed in healthcare settings. Artificial nails have been found to harbor higher numbers of microorganisms. Due to the frequency of hand washing, the area underneath the artificial nails doesn't get sufficient time to dry. As a result, infections of the nail beds often occur. These infections are easily spread to patients while doing procedures, with or without gloves. Artificial nails are also subject to breaking loose, creating unnecessary problems in the midst of a procedure. Since artificial nails are often long, they also include the problems inherent with long nails.

Clean-Up of Biohazardous Material

You must either employ a disinfectant that has the ability to function as both a cleaning and disinfecting agent, **or** you must first use a cleaning agent and then employ a disinfectant, because many disinfectants are inactivated by the presence of organic matter.

In our laboratory, we employ a disinfecting agent that is also capable of functioning as a cleaning agent; thus, the same agent is used for both steps of the process. This is a common practice in healthcare settings.

Do not make incorrect assumptions! Because it is a common practice to employ one chemical that performs both steps, many healthcare workers and/or facilities never check. It is not uncommon to find out that cleaning and disinfection agents are being used incorrectly within a facility. There needs to be a process by which all chemicals are checked for quality and safety!

Biohazard Clean-Up Procedure

Blood or Body Fluid Containing Blood: Clean-up Protocol in Healthcare

Use the following method:

- 'APPLY, WIPE, REAPPLY' method

Remember, many disinfectants are inactivated by the presence of organic matter. This means that the active ingredient of the disinfectant combines with the proteins in the blood, mucus or pus. When this happens, it is no longer available to react with the microbial portion of the spill. This is the reason why we must **clean any organic matter first, before we begin our disinfection procedure**. The following are some disinfectants which are affected by the presence of organic material: alcohols, quaternary ammonium compounds (QUATS), chlorine compounds, phenolics, and hexachlorophene.

Secure the area, yourself, and the spill using the following procedures:

1. **Communicate:** Alert people in the immediate area of the spill.
2. **Protective equipment**: Wear a lab coat with long sleeves, back-fastening gown or jumpsuit, disposable gloves, disposable shoe covers, safety goggles, and mask or full-face shield.
3. **Limit and disposal:** Cover the spill area with paper towels or other absorbent materials. Carefully transport the contaminated paper towels and dispose them into a biohazard waste container along with the gloves you were wearing. Put on a new pair of gloves.
4. **Disinfect:** Carefully pour a freshly prepared 1 in 10 dilution of household bleach around the edges of the spill, and then onto the spill. Avoid splashing. Use paper towels to wipe up the spill, working from the edges into the center. Place paper towels and gloves you are wearing in biohazard waste for disposal. Put on a new pair of gloves.
5. **Reapply:** Reapply disinfectant and allow a 20-minute contact period. If after 20 minutes the disinfectant is still present, wipe up remaining liquid with a paper towel and dispose paper towel and gloves in biohazard waste.

Things to consider when cleaning biohazardous material:

Environmental Conditions

The disinfectant collects in surface defects, such as grooves and cracks. It collects in places where the smooth surface is broken up by screw tops, cubbyholes, joints and edges. All surfaces have defects. When you are trying to clean up bacterial contaminant and disinfect a surface, your cleaning agents and procedures must be able to deal with all aspects of the surface or object being cleaned. Pay special attention to defects in the surface, being certain to spray these completely during the soaking spray. When done properly, surfaces can be successfully disinfected.

Air Quality

Another important factor that must be considered in all healthcare settings is the quality of the air. When we spray chemicals for cleaning and disinfection purposes, a portion of those chemicals are released into the air we are breathing. Persons suffering from respiratory problems of any kind (from asthma to emphysema) are negatively affected by this process. That is why the second method involving the use of gauze pads with no spraying is becoming more commonplace in healthcare settings. This method avoids aerosolization of the chemicals.

Sterilization and Disinfection

Sterilization and disinfection are two processes carried out in the microbiology lab to diminish the possibility of contamination of a culture. Because microbes exist all around us, it is important that we develop laboratory skills that prevent contamination by unwanted microorganisms. We refer to these methods as **aseptic techniques.** The awareness that bacteria are everywhere must be constantly in our minds when handling cultures. In this microbiology lab, you will be handling tubes of bacteria and unless you learn how to handle them in such a way as to keep foreign organisms from them, you will always be working with contaminated cultures. Without **pure cultures** (cultures containing only one type of organism) the study of microbiology becomes incredibly difficult. **Sterilization** is a process that kills all organisms on or in a substance. **Disinfection** is a process that kills pathogenic microorganisms. Sterilization is an absolute state, whereas disinfection is not—some organisms can survive disinfection. In order to ensure that our cultures contain only the organism that we are trying to study, we must sterilize the media before inoculation. The most prevalent sterilization is the use of a moist heat mechanism called the **autoclave**. The autoclave is a device that permits the chamber to be filled with saturated steam (no air must be present; it would lower the temperature), and this saturated steam is maintained at a designated temperature and pressure for a given period of time. Generally, the temperature used is 121°C and the pressure is 15 lbs/in^2. The time is usually set for 15–20 minutes, although this may vary according to the nature and amount of material to be autoclaved. The pressure is used to raise the temperature of the steam above 100°C. Some microbes may form structures called endospores that enable the organism to survive boiling temperatures. However, endospores are killed at the temperatures and pressures reached during autoclaving.

Hand-Washing Activity

Materials

Laboratory materials: Antibacterial soap at hand-washing sinks, paper towels, Glitterbug UV lotion, UV view box, china marker, masking tape, and incubation tray.

Materials per student: Tryptic soy agar (TSA) plate.

NOTE: Do not wash hands until Step 5 of the procedure!

1. Use a china marker to divide your plate in half. Label one half of the bottom of a Petri plate "**pre-wash**", and the other half "**post-wash**", your initials, your table #, and your section #.
2. Lift the cover with one hand and lightly touch the surface of the agar on the "**pre-wash**" side, with the fingers or thumb of your other hand. Do not break the surface of the agar.
3. Replace the cover.
4. Apply the Glitterbug UV lotion all over your hands.
5. **Now wash your hands** for 15 to 20 seconds or as you "normally" wash your hands and dry them with paper toweling.
6. The instructor will gather the class to observe the results of each others' hands under the UV view box. Areas on the hands which were not cleaned will glow a bright bluish white!
7. If your hands had bacterial contaminant on them, this would mean that you did not do the best washing job possible. **What could have gone wrong? Answer question #1 in the lab report.**
8. Read the hand-washing technique described earlier in the chapter. **Wash your hands thoroughly following proper hand-washing techniques**.
9. Repeat Steps 2-3. Lift the cover with one hand and lightly touch the surface of the agar on the "**post-wash**" side, with the fingers or thumb of your other hand. Do not break the surface of the agar.
10. Place 2 pieces of tape along the edge of your plate and place the Petri dish on the tray, agar side up, for incubation.

Hand-Washing Activity Day 2

1. Observe the pre- and post-wash petri dish and answer the questions in your lab report.
2. Estimate the number of species on each side of the plate. To estimate the number of species, look closely at the different colonies. The colonies that share the same characteristics (morphology) are most likely the same species.

Analyze an isolated colony and identify the colony characteristics of each species. Record your observations in the lab report.

a. Size of the isolated colonies (small, medium, or large when compared to isolated colonies of the other microbes).
b. Color of the isolated colonies (clear, white, tan, bright yellow, pink/red, etc.).
c. Opacity of the isolated colonies (transparent/clear, translucent/partially clear, or opaque/not clear).
d. Configuration of the isolated colonies (see diagram on next page).
e. Margins of the isolated colonies (see diagram on next page).
f. Elevations of the isolated colonies (flat, raised, rounded/convex, drop-like/like a drop of water, bumpy/hilly, or crater-like with a sunken center.

See image below for example configurations and margins.

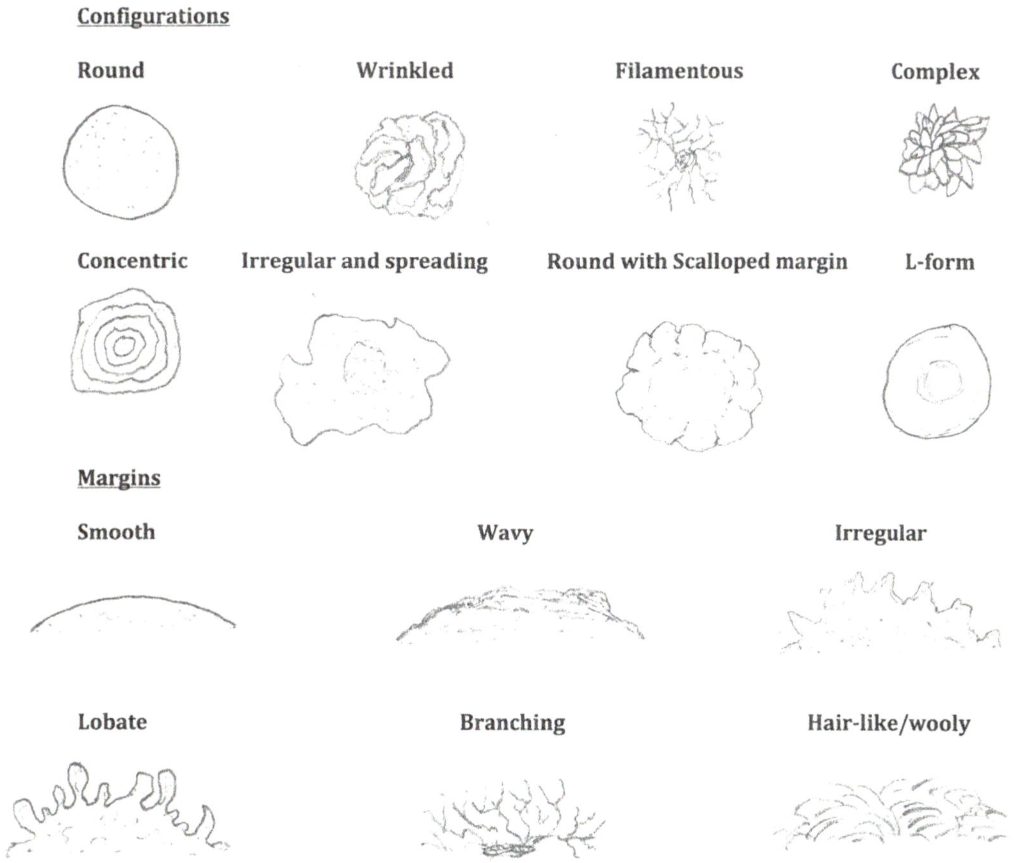

Courtesy of Authors

Spill Clean-Up of Organic Material Activity

Materials

Laboratory materials: Disinfectant, paper towels, and biohazard bags.

Materials per table: Test tubes with simulated bodily fluid and Glitterbug spray added and hand-held UV light.

Materials per student: PPE (Personal Protective Equipment): gloves, safety glasses, and lab coats. (Note: You may not have a lab coat with you today. Since our spill is not being done with a real bodily fluid, you need not be concerned.)

1. You must first practice how to remove your gloves without contaminating yourself. Your instructor will demonstrate this procedure.
2. Simulate a bodily fluid spill using your test tube. Spill some fluid on your table surface and some on the floor. (It is not necessary to spill on other objects or yourself!)
3. Use paper towels to remove the excess organic material to the biohazard bag. Your gloves should follow right behind the paper toweling into the bag. Put on new gloves before proceeding.
4. Use the standard **'APPLY, WIPE, REAPPLY'** method to disinfect the area of the spill.
5. Your instructor will turn out the lights and use the hand-held UV light to check for remaining contaminants.

Name ______________________

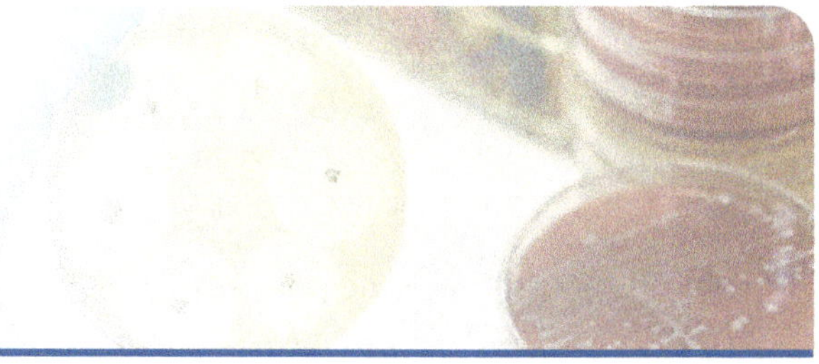

Exercise 1.1 Lab Report

Hand-Washing Activity

1. What does the presence of Glitterbug UV lotion on your hands tell you about the first hand-washing technique you performed? Describe what could have gone wrong.

   The Glitterbug UV lotion shows the presence of bacteria in my hand. I think that I do not use proper hand washing technique or contamination might have occur with exposure to the laboratory environment.

2. Compare the "pre-wash" plate to the "post-wash" plate. Describe the bacterial colonies present on each plate (color, size, etc.), and explain the differences between each plate.

3. Estimate the number of species on each side of your plate.
   Pre-wash number of species:______________________
   Post-wash number of species:______________________

4. Was there a reduction in the number of species? How can you tell if the colonies are different species?

5. What is cross contamination?

6. What is a nosocomial infection? What is the best way to prevent these types of infections?

7. How will this knowledge affect your behavior as a healthcare professional?

8. Describe the current recommended procedure for hand hygiene in the healthcare field.

Clean-Up and Disinfection of Table Surface or Spill

9. Consider the activity of cleaning the surface of an inanimate object, such as an examination table. Describe the correct method for standard clean-up and disinfection.

10. Explain why organic matter must be cleaned away before a disinfectant is applied.

11. In the lab, what should you do if you spill a bacterial culture on the lab bench? Explain the procedure.?

12. What are aseptic techniques?

13. What aseptic techniques did you use during the hand-washing procedure?

14. Describe the autoclave and why is it used?

Lab Exercise 1.2

Environmental Sampling and Normal Microbiota

Student Objectives

1. Take environmental samples, given the appropriate media and collection tools.
2. Understand that microorganisms are everywhere, including in and on your body.
3. Describe how microorganisms are collected, inoculated, cultured, incubated, and autoclaved.
4. Understand the use of various media in the laboratory.
5. Understand the factors which influence the growth of environmental samples.
6. Identify the normal, resident microbiota of the human body and the areas of the body that have no microbiota and are usually sterile.

INTRODUCTION

Microorganisms are all around us, and they play vital roles in the ecology of life on earth. In addition, some microorganisms provide important commercial benefits by producing chemicals (including antibiotics) and certain foods. Microorganisms are also major tools in basic research in the biological sciences. Finally, as we all know, some microorganisms cause disease. In this course, you will gain first-hand experience with a variety of microorganisms. You will learn techniques to identify, study, and work with them.

This exercise is designed to demonstrate that microorganisms, like bacteria, are everywhere. You will be using culture media to collect microorganisms in a variety of ways. To ensure that these exposures cover as wide a range as possible, specific assignments will be made for each student.

Ways to Begin a Laboratory Exercise using bacteria. There are two ways a lab exercise can begin. How we begin determines what we can do and how many lab periods are needed to complete a lab exercise. Let's take a look at each of these ways:

I—Beginning Lab by Collecting a Specimen

In the first method, you are given a growth medium and directions for collecting a specimen. The specimen may come from the lab room or it may come from you. Throughout the semester you will perform throat and urine cultures. Remember, to find out what microbe is infecting your patient there has to be some method of specimen collection employed. You need to know as much as you can about collecting, growing and testing specimens.

II—Beginning Lab with a Prepared Culture

In the second method, you are given a culture of microorganisms, which the lab technicians have already grown. These are usually grown in test tubes of liquid broth media. Sometimes they come to you on solid media. Solid media can be in test tubes or on flat, round, covered plates. Solid media in test tubes is called an "agar deep." If the solid media has been solidified at an angle in the test tube it is called an "agar slant." Hot liquid media is placed in a slant rack (so that the tube is slanted) and the medium hardens at an angle. This allows us to have a large surface area within a small, confined tube. The plates used for solid media are called Petri plates, named after the scientist attributed with their invention. Sometimes these plates are prepared with solid media ahead of time. Often you must pour the media in the plates yourself. This means you are given melted media (you'll find it in flasks in the hot water bath where it is being kept liquid) and empty sterile Petri plates. You will be taught the proper procedure for pouring the plates so that you do not contaminate them. Remember, everything we do in the lab must be done in a manner that prevents contamination of your work or yourselves! You must be able to work in an aseptic manner within the healthcare setting.

USE OF CULTURE MEDIA

A **culture medium** (plural, media) is food used for growing microorganisms. It contains nutrients that provide the organism with an appropriate physical and chemical environment for growth. The nutritional needs provided by a culture medium include:

- Water – necessary for biochemical reactions and as a solvent.
- Carbon – necessary for cell structure and energy.

- Nitrogen – necessary for protein synthesis, RNA, DNA, and ATP production. Many organisms obtain nitrogen (N) from amino acids and peptides. Beef extract and peptone found in nutrient broth or nutrient agar provide the nitrogen needed.
- Minerals – needs vary, but most require phosphate (P) for ATP. Many organisms require elements such as sodium (Na), potassium (K), calcium (Ca), magnesium (Mg), manganese (Mn), iron (Fe), zinc (Zn), copper (Cu), and cobalt (Co) for growth.
- Growth factors or vitamins – essential substances that an organism needs for growth.

Factors in the chemical and physical environment of a growth medium include:

- Hydrogen ion balance – many bacteria prefer a pH near 7.0, while fungi prefer a lower pH.
- Oxygen – some organisms require oxygen, while for others oxygen is toxic.
- Temperature – optimum temperature for growth varies with different microorganisms. Human pathogens grow best at our body temperature of 37°C.

Culture media may be classified by composition. If the exact composition of the medium is known, it is referred to as a **defined medium.** These media are made from chemical compounds that are highly purified and precisely defined, so that they are readily reproducible. **Nutrient broth** is a **complex** or **undefined** medium that contains beef extract and peptone, both ingredients for which the exact composition is unknown. **Peptone** can serve as a source of nitrogen and carbon, but what specific amino acids or carbohydrates are present is not precisely known. **Beef extract** provides amino acids, vitamins, and minerals for microbial growth. Most of the media used in this laboratory are complex. Media prepared as liquids are called broths. A liquid medium may be rendered solid (or semisolid) by the addition of a solidifying agent such as agar, silica gel, or gelatin. A solid medium is used for studying surface growth of microorganisms, to assist in obtaining a pure culture, and for quantitative studies. **Agar**, the preferred solidifying agent, is a complex carbohydrate derived from seaweed which most organisms cannot degrade and use as a carbon source, and which therefore remains solid. Also, agar liquefies at about 100°C and solidifies at around 40°C. For laboratory use, agar can be held in a molten state in water baths set at a temperature of 50°C. At this temperature, it does not injure most bacteria when it is poured over them; this property is useful in later labs when you perform food quality analysis and urinalysis. Once solidified, agar can be incubated at temperatures approaching 100°C; this property is particularly useful when thermophilic bacteria are being grown.

If a medium contains an excess of the nutrients required by one microorganism, such that it can outcompete the other organisms in the medium, it is referred to as an **enrichment medium**. A medium that inhibits the growth of undesired microorganisms without interfering with the growth of desired organisms is termed a **selective medium**. A medium that contains indicators that give color reactions enabling us to distinguish one type of organism from another is referred to as a **differential medium**. An example of a differential medium is a blood agar plate. You will learn about the blood agar plate in a later laboratory exercise. Some media do not fit into these categories and are referred to as **general purpose media**. This category includes nutrient broth, which can support the growth of many different organisms. General purpose media are used for propagating large numbers of bacteria because they are generally less expensive than most selective and differential media.

Environmental Sampling Activity

> **Materials**
>
> **Per student:** Sterile swab, one TSA plate, and one TS broth tube.

Agar Plate

1. Label the bottom of your plate with your initials, table number, and exposure number.
2. Expose your nutrient agar plate according to your assignment in the table.

Exposure Method for Nutrient Agar Plate	Table Number	
1. open to the air in the laboratory for 20 minutes	1	5
2. open to the air in a room other than lab for 20 minutes	2	4
3. cough on the agar surface	3	2
4. moist lips pressed against agar surface	4	3
5. tongue pressed against agar surface	5	1
6. several coins pressed lightly on agar surface and then removed	1	5
7. hair brushed against agar surface	2	4
8. touch agar surface to cell phone	3	2
9. touch agar surface to vending machine or drinking fountain	4	3
10. touch agar surface to jewelry	5	1

Broth Tube (Individual)

1. Moisten a sterile swab by immersing it into the TS broth tube and then press the swab against the inside wall of the tube. Immediately place the cap back on the broth tube.
2. Rub the moistened swab over a part of your body. Put the swab into the nutrient broth and gently swish it around to release any microorganisms. Remove the swab and dispose of it in the trash container. Immediately place the cap back on the broth tube.
3. Label the tube on the glass part with your table number and initials.

Incubation

Place your labeled **plate** and **tube** on the classroom tray and test tube rack to be incubated.

Name ________________________

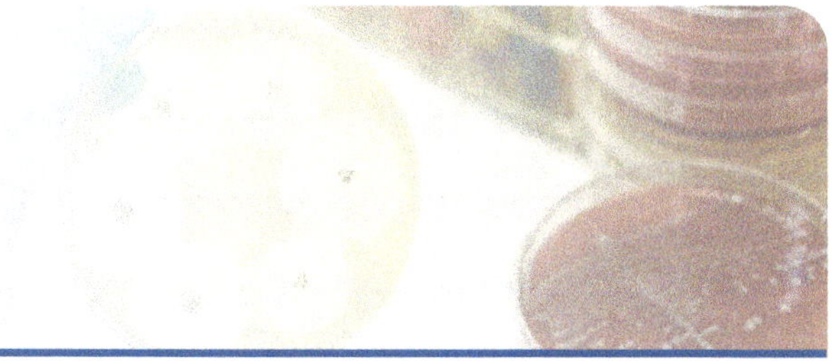

Exercise 1.2 Lab Report

1. Describe the variety of colony types seen on your **TSA plate** with reference to size, shape and color. Be sure to list where you obtained your sample.

____

____

____

2. How many species are present on your TSA plate? How can you tell they are different species?

____

____

____

3. Explain the bacterial growth in your **TS broth tube**. Use the diagram below. Be sure to list the body part you obtained your sample from.

____

____

____

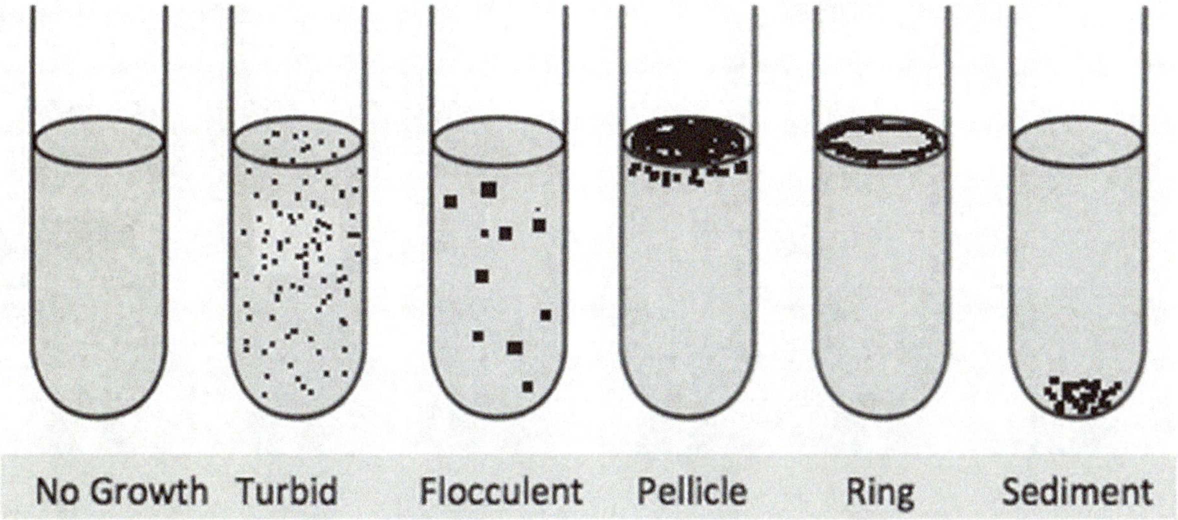

Courtesy of Authors

4. Can you tell how many species are in your TSB? Why or Why not?

5. **Compare your results** of the **agar** and **broth** with the results of other students in the lab. What **conclusions** can you draw based on where you took your samples and the outcomes of those samples? What factors would influence the different results you observe?

6. Differentiate between enriched media, selective media, and differential media.

7. Name the areas of the body that are normally considered to be "germ-free".

8. What did you learn that you will apply in your career?

Classification, Survey of the Microbial World, and Microscopy

Scientists, called taxonomists, classify all organisms as either prokaryotic or eukaryotic, which can be further classified in one of the three domains of life: **Bacteria**, **Archaea**, and **Eukarya**. Classification indicates that organisms can be grouped together into taxa with similar characteristics. By grouping organisms with similar characteristics, taxonomists can determine how organisms are related to one another and learn how organisms evolved from common ancestors. Bacteria and archaea are living cells that do not contain a nucleus enclosing their DNA; these microorganisms are characterized as **prokaryotes**. Living cells, including fungi, protozoa, plant and animal cells, which do contain a nucleus enclosing their DNA, are **eukaryotes**. Therefore, it is important that the microscope be used properly so that you can identify if the nucleus is present, as well as evaluate distinct shapes, sizes, and structures of these microorganisms.

Prokaryotes vs. Eukaryotes

Cells can be classified as **prokaryotic** or **eukaryotic**. Prokaryotic cells are exemplified by their lack of membrane nuclei and organelles. The word prokaryote comes from Greek, "pro" meaning 'before' and "karyon" meaning 'nucleus'—Literally, "before the nucleus". Eukaryotic cells have membrane-bound nuclei and organelles. The prefix "eu" means "true", so eukaryotic means "true nucleus".

Table 2.1 Differences Between Prokaryotes and Eukaryotes

	Prokaryotes	Eukaryotes
Size	Small: 0.3–2 um	Larger: 5–50 um
	Always unicellular	Can be unicellular or multicellular
Nuclei?	NO	YES
Membrane-bound organelles?	NO	YES
Where found?	Everywhere	Usually in environments that are not extreme

Classification and Nomenclature

In the 1700s, Carolus Linnaeus developed a taxonomic system in order to categorize living things. His system grouped organisms with similar characteristics, and had only two kingdoms: Animalia and Plantae. In the roughly 300 years since the development of his taxonomic scheme, we have learned that there are many other types of organisms, and the similarities and differences are not always obvious. For example, in Linnaeus' time, mushrooms were grouped with plants. Today, they are in their own kingdom, Fungi, and we know that they are actually more closely related to animals than to plants.

Today, microbiologists use the Domain system of classification, developed by Carl Woese. This system is based on gene sequences, and groups organisms into three domains: Bacteria, Archaea and Eukarya. The kingdom system fits within the domain system. (Figure 2.1)

Domains 'Bacteria' and 'Archaea' are both prokaryotic. It had previously been thought that archaea were a type of bacteria, but due to the work of Woese and others, it is now known that although they look like bacteria, Archaea are different at the molecular level. In addition, having a different type of ribosome than bacteria, archaea do not have peptidoglycan in their cell walls.

The domain 'Eukarya' contains all of the eukaryotic kingdoms, such as Animalia, Plantae, Fungi and Protista.

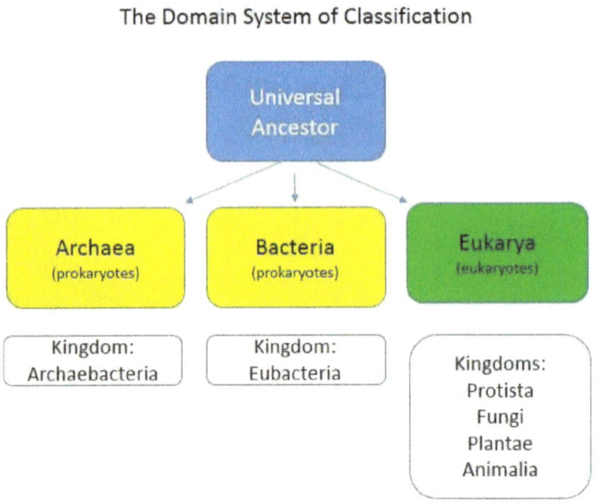

Fig. 2.1

Courtesy of Authors

Nomenclature

The simplest taxonomic classification is the specific epithet of the organism or the species. The species cannot be used alone, however, since there are often species that share the same name. Therefore, the species must be precluded with its genus; for example *Mycoplasma pneumoniae* and *Streptococcus pneumoniae* both share the same species name, but are not related since they are from different genera. This combination of an organism's genus and species is referred to as a scientific name. The scientific name of an organism can often describe characteristics common to the organism. In the example of *Mycoplasma pneumoniae* and *Streptococcus pneumoniae*, the species name of both of these organisms indicate that these organisms cause the disease pneumonia.

Streptococcus, the genus, describes the cell's morphology and arrangement. Take the example of *Staphylococcus aureus*; *aureus* is Latin for golden and describes the bacteria's appearance when grown on agar media—*Staphylococcus aureus* will grow as golden colored colonies. The scientific name might also honor a researcher such as *Escherichia coli*, named after Theodore Escherich or describe a habitat—for instance, the term *coli* would indicate that this organism would be found associated with fecal material in the colon.

Most bacteria are referred to by their genus and species names. Example: Genus: *Escherichia*, Species: *coli* = *Escherichia coli.* The genus can be shortened to just the capitalized first initial: *E. coli.* The genus and species should always be italicized, unless they are hand-written, in which case they must be underlined.

Strains are slight variations within the species. *E. coli* has many different strains, some of which are harmless and many of which are harmful. *E. coli* Nissle 1917 (aka EcN) is a non-pathogenic, potentially probiotic form of *E. coli. E. coli* O157:H7 is a pathogenic form of *E. coli* that is transmitted in food and causes life-threatening infections.

Bacteria

Bacteria are in their own domain; they are prokaryotic, single-celled living organisms. Bacteria are smaller than eukaryotic cells, lack a nucleus, lack most organelles (they require ribosomes though), and have a single chromosome. Bacteria contain a thick, protective cell wall composed of a molecule called Peptidoglycan. Bacteria can be round (**coccus**), rod-shaped (**bacillus**) or **spiral-shaped**; these shapes are referred to as the **cell morphology**. Bacteria can also take on a variety of shapes under the microscope; we call this **pleomorphic**. Bacteria can be found as single units (mono-), grouped in twos (diplo-), grouped in chains (strepto-), or grouped in bunches (staphylo-), these terms are referred to as the **cell arrangement**. If we apply those possible groupings to a round bacterium, we come up with these terms: monococcus, diplococcus, streptococcus, or staphylococcus. Shape and grouping arrangement is often employed in the naming of a bacterium. In this lab, we will use traditional methods to characterize, classify, and identify bacteria. These methods include the use of the Gram stain and cell morphology and arrangement, colony morphology, and the use of biochemical testing. The methods for characterizing bacteria will be discussed in a later chapter.

Bacterial Cell Morphology and Arrangement

The most common bacterial morphology are **coccus** (plural: cocci), which are round, and **bacillus** (plural: bacilli) which are rod-shaped.

Other morphologies include:

- **Vibrio**, which looks a rod with a slight bend
- **Spirillum**, an "S" or spiral shaped cell
- **Spirochete**, a corkscrew-shaped cell with flagella for motility

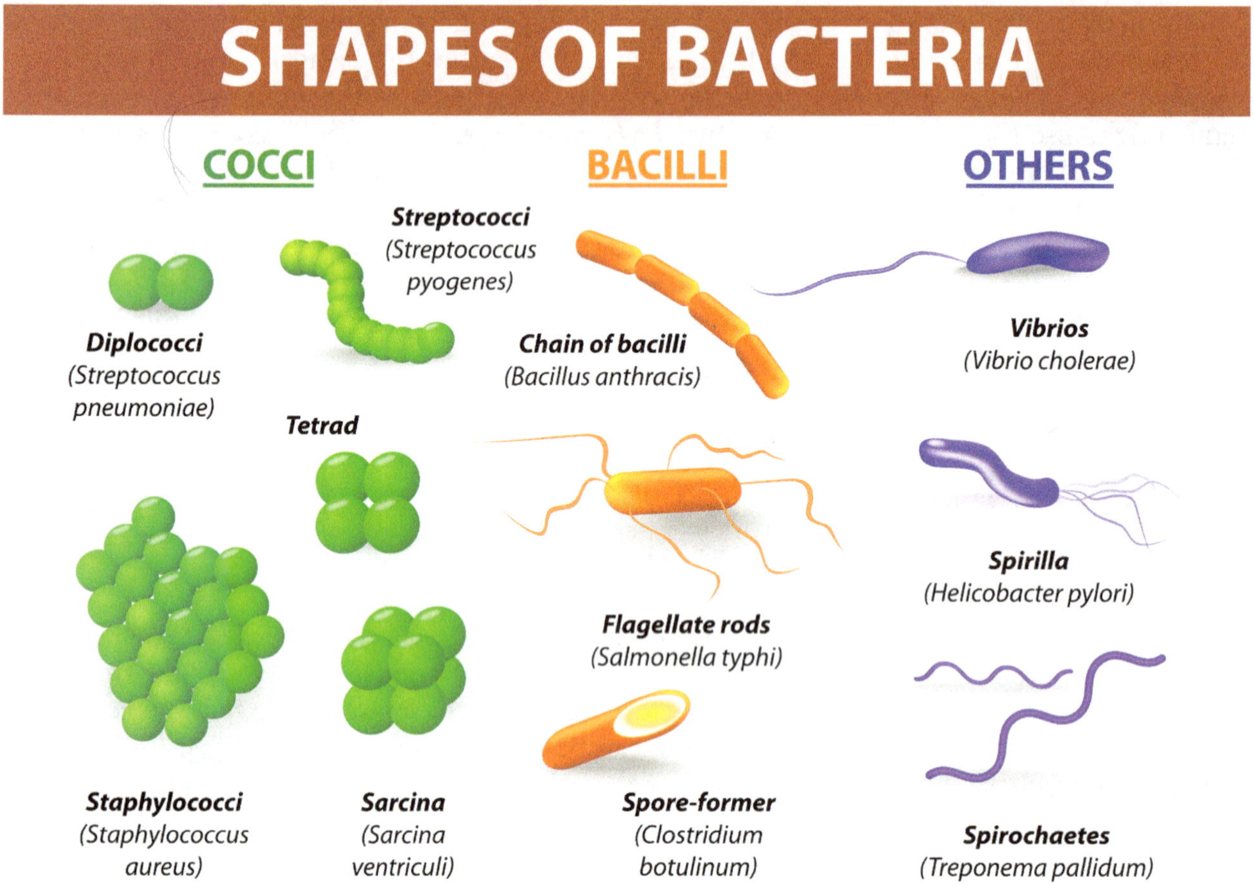

Fig. 2.2 Shapes of Bacteria
Designua/Shutterstock.com

Cell Arrangements

Bacteria may exist as single cells, or in arrangements involving multiple cells. These are not multicellular organisms, and the bacteria within them will survive on their own if removed from the arrangement.

The arrangements are determined by the way the bacteria reproduce, specifically, the plane on which the bacteria divide when going through cell division.

Colony Morphology

When grown on a semi solid media such as tryptic soy agar bacterial cells grow in clusters called colonies. Only one bacterial cell is needed in order to form a colony. The colony characteristics such as the size, color, configurations, and margins are referred to as the colony morphology. Organisms that share colony morphologies are sometimes but not always related. In Lab exercises such as the hand-washing activity in lab 1.1 the organisms with the same colony morphology are most likely the same species. Colony morphology is not to be confused with cell morphology, cell = microscopic and colony = macroscopic. Analysis of an organisms colony morphology helps in the classification and identification of unknown bacteria.

The Most Common Bacterial Arrangements

Diplo- refers to 2 cells, formed by both cocci and bacilli Example: *Neisseria gonorrhea*, a Gram negative diplococcus, causes gonorrhea	
Strepto- a chain of cells, like beads on a string, formed by both cocci and baciilli Examples: *Steptococcus mutans* (strep throat), *Streptobacillus moniliformis* (rat bite fever)	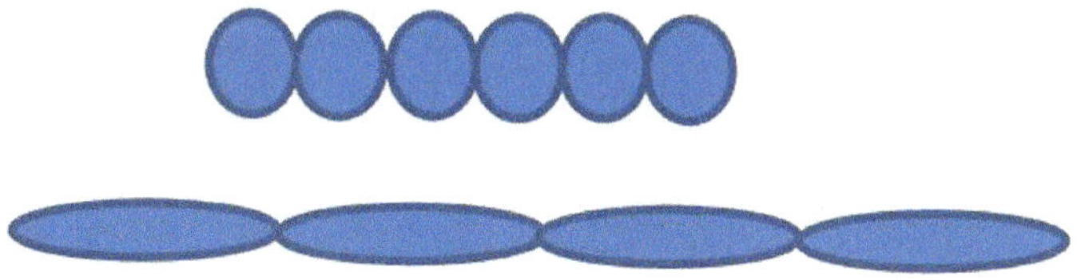
Staphylo- a cluster of cells, only cocci form this arrangement Example: *Staphylococcus aureus,* causes skin infections, MRSA	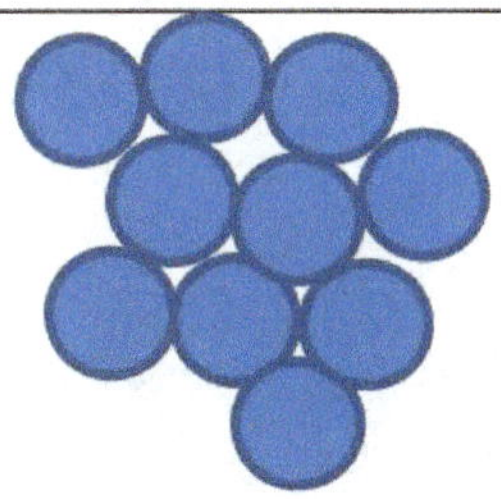
Tetrad: four cocci in a "packet" Example: *Micrococcus luteus*, non-pathogenic resident skin flora	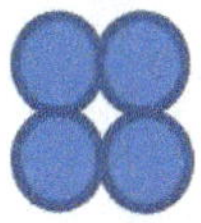
Octad: eight cocci in a packet Example: *Sarcina ventriculi,* a rare microbe	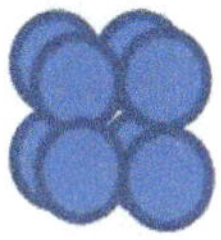
Palisade layer: formed only by bacilli, cells are lined up "side-by-side"	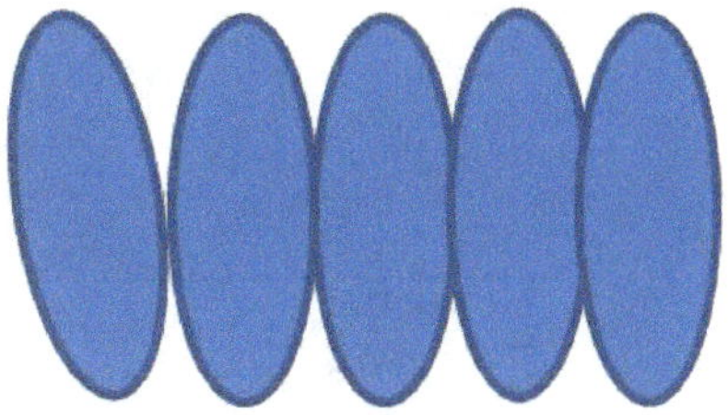

Fig. 2.3 Arrangements of Bacterial Cells
Courtesy of A. Swarthout

Eukaryotes

Eukaryotes are organisms that contain a nucleus surrounding their DNA. The organisms can be single-celled to multi-celled and include algae, protozoa, fungi, plant, and animal cells. Eukaryotes are much more complex than Prokaryotes and contain cell organelles. Eukaryotes use a variety of methods of asexual reproduction: binary fission, budding, fragmentation, spore-forming, and schizogony (nuclear divisions in a single cell, later undergo cytokinesis). Many eukaryotes reproduce sexually, which involves the formation of sex cells called gametes; the fusion of two gametes form a structure called a zygote.

Algae

Algae are widely distributed and are represented in many different kingdoms; diatoms (marine algae), seaweeds, spirogyra, green algae, red algae, brown algae, and kelps are included in the algae group. Algae are a very diverse group, from single-celled microorganisms to multicelled organisms. They differ widely in distribution, structure, reproduction, and biochemical traits. Algae are phototrophic organisms (they use photosynthesis) and contain cellulose in their cell walls.. Most algae reproduce sexually, because each alga cell is a gamete and fertilizes other algae cells to create a zygote.

Fig. 2.4 Fish Killed By Red Tide
Brandon Seidel/Shutterstock.com

The classification is so unclear that the following examples of toxin-producing organisms are classified by some sources as algae, and by others as a type of protozoans called dinoflagellates.

Gonyaulax: This produces the neurotoxins saxitoxin and gonyautoxin. The algae/dinoflagellate is ingested by shellfish where the toxin accumulates. People who eat the contaminated fish experience nausea, dizziness, muscle weakness, trouble breathing and sometimes death.

Karenia brevis can cause a phenomenon known as Red Tide. Red Tide occurs when the organism overgrows, creating a "bloom" (Figure 2.5). It is unclear what the cause of Red Tide is, but the formation of blooms may be due to increased temperature or increased levels of nitrates and phosphates in the water. *K. brevis* produces a potent neurotoxin, brevetoxin, which can become concentrated in the flesh of fish and shellfish and subsequently be ingested by humans. The toxin can cause tingling, diarrhea and slow heart rate. The toxin can also kill fish and other marine life, which can then wash up on shore (Figure 2.4).

Fig. 2.5 Red Tide
Kkaplin/Shutterstock.com

Protozoa

Protozoa are single-celled organisms that are large compared to other microorganisms; they include Amoeba, Paramecium, and Euglena. Most Protozoa are multi-nucleic, contain large organelles (especially contractile and food vacuoles), and lack a cell wall. Protozoa are called "animal-like", because most protozoa are motile and feed on algae and other protozoa. Protozoa require moist environments; you will find them in our lakes, rivers, and ponds. An example of marine protozoa is the *Dinoflagellates*. The method of motility was used as the basis of classification; now they are further grouped according to many other features. Most are free-living in water and are part of the food chain as they absorb or engulf (phagocytize) large organic matter, called chemoheterotrophic. Free-living aquatic and pathogenic protozoa exist as a vegetative, motile, mature stage called a trophozoite. Many protozoa have a protective, resting stage called a cyst. Many protozoa reproduce asexually via binary fission or schizogony. Some protozoa reproduce sexually via conjugation. Examples of pathogenic protozoa in the Great Lakes area are *Cryptosporidium* and *Giardia*.

Cryptosporidium forms an oocyst that is resistant to desiccation and chlorine. The organism is difficult to identify and detect during routine testing of drinking water. The primary mode of transmission for *Crypto* is by ingesting contaminated water. In healthy hosts, the microbe causes watery diarrhea, but may also cause nausea, cramps and vomiting. Some infected individuals may develop no symptoms at all. People with compromised immune systems may develop serious illness or die from *Crypto* infection.

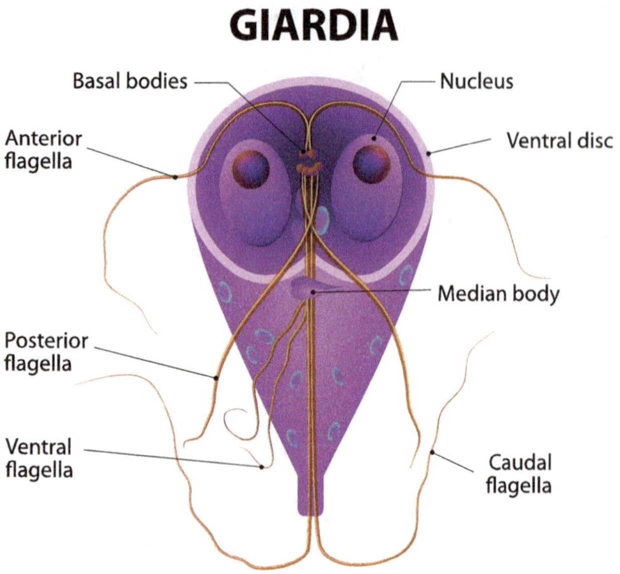

Fig. 2.6 Giardia
Designua/Shutterstock.com

Like *Cryptosporidium*, *Giardia lamblia* is a waterborne protozoan that has an outer shell that makes it resistant to chlorination. (Figure 2.6) Infection with *Giardia* causes excessive gas and diarrhea, which together may cause "explosive" diarrhea in the victim. As few as 10 *Giardia* cysts is all that is needed to infect and cause disease in a host.

Examples of other notable protozoal diseases include:

Amoebic dysentery, caused by *Entamoeba histolytica*; the parasite lives in the intestines of humans and is transmitted via the fecal-oral route. Only one in 10 who come in contact with it will acquire an infection, but people can be carriers without developing symptoms. Symptoms typically take one to four weeks to appear. This is usually found only in tropical and subtropical regions.

Primary amoebic meningoencephalopathy is cause by amoebae that live in warm, moist environments such as lakes, ponds, puddles, humidifiers and contact lens solutions. They enter through cuts/abrasions of eye/skin or via inhalation, and then migrate to the brain, causing encephalitis and destroying brain tissue. It is almost always fatal (there are only three known survivors!). There have been recent cases of this in the southern U.S.

Toxoplasmosis is caused by *Toxoplasma gondii*, an intestinal parasite of animals, particularly cats. Infection is usually harmless and asymptomatic in healthy individuals, but can cause disease in immunocompromised people. The protozoan can also cross the placenta and infect the unborn baby; this can lead to blindness and mental retardation in the baby. Doctors recommend that pregnant women do not clean litter boxes because of this parasite.

Recent research indicates that *Toxoplasma* infections can affect the brains and behaviors of their hosts. It seems that rodents and humans infected with *Toxoplasma* engage in more high-risk behavior than their uninfected counterparts. Scientists hypothesize that the increased risk behavior in rodents makes them easier prey for cats.

Trichomoniasis aka "trich" is caused by the flagellated protozoan *Trichomonas vaginalis*. This microbe is transmitted by sexual contact, but unlike most STDs, trich can be transmitted via indirect contact! The disease is often asymptomatic, and when symptoms appear they are different for men and women. The symptoms for women include discomfort with intercourse, itching of the vagina, vulva and inner thighs, swelling of the labia and a greenish-yellow frothy or foamy vaginal discharge with a strong foul odor. Infection in men can cause a burning sensation after urination or ejaculation, itching of the urethra and slight discharge from the urethra.

Malaria is caused by several species of *Plasmodium*. The parasite is carried by *Anopheles* mosquitos and is transmitted to the human when the mosquito feeds (Figure 2.7 and 2.8).

According to the Center for Disease Control and Prevention, there were 198 million cases of malaria and 500,000 deaths due to the disease in 2013. The CDC also states that 3.4 billion people live in areas that put them at risk of malaria infection. Scientists have so far been unsuccessful in developing a vaccine for malaria; however, using vector control measures can prevent transmission of the disease. These include the use of mosquito nets, reducing the amount of standing/stagnant water where mosquito larvae develop, and using insect repellent.

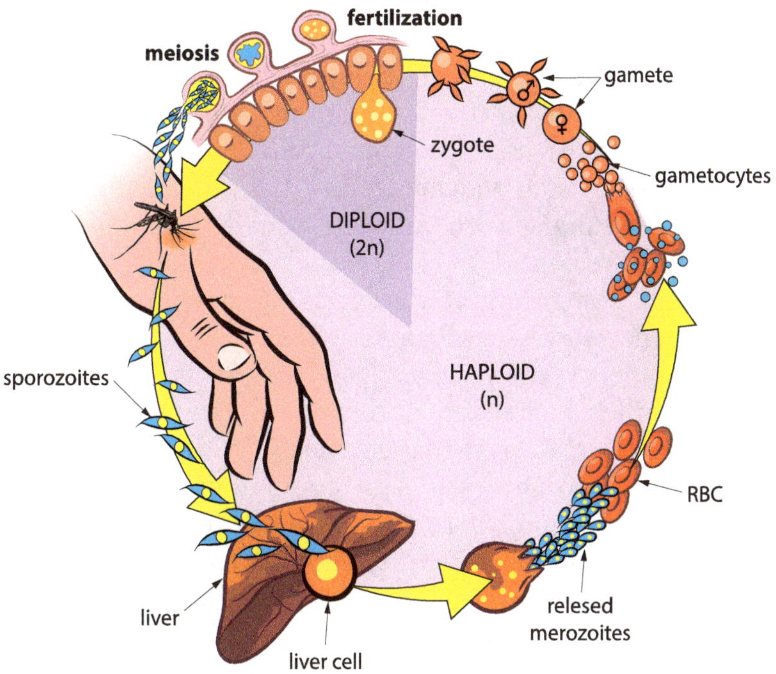

Fig. 2.7 Life Cycle of Plasmodium
LSkywalker/Shutterstock.com

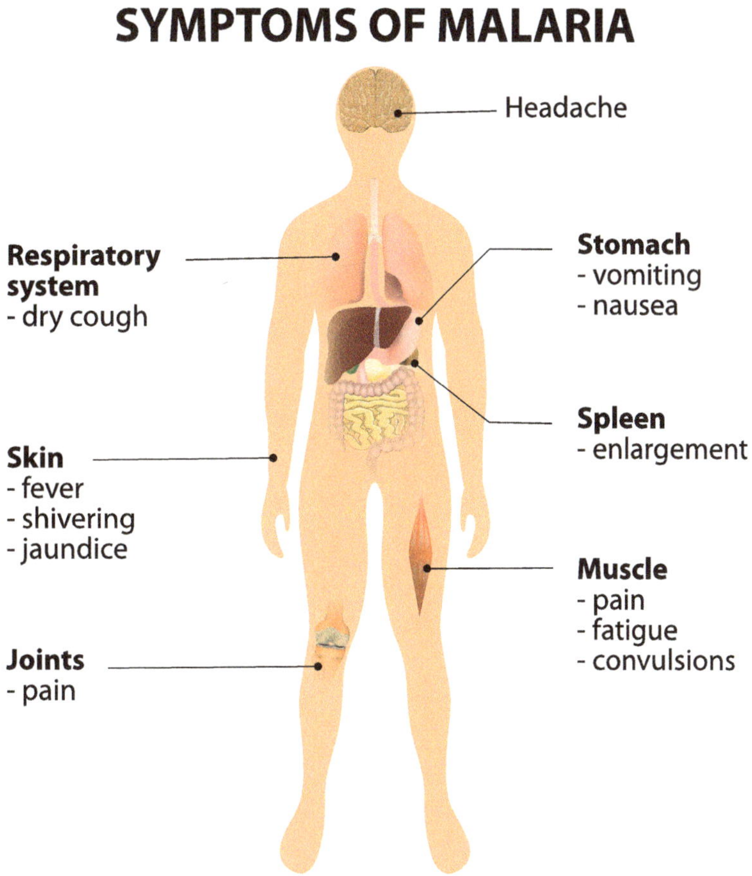

Fig. 2.8 Symptoms of Malaria
Designua/Shutterstock.com

Fungi

Fungi are a large group of organisms, which include yeast, molds, and mushrooms. Fungi are in their own kingdom and separated into four divisions based on the type of spore produced. Fungi have strong, flexible cell walls composed of chitin. Fungi acquire nutrients by absorption; they must decompose material (using enzymes to break down organic matter), and hence they are the "decomposers". When fungi absorb nutrients from dead organisms, they are called "saprobes". Because of this, fungi are considered beneficial as they degrade dead, organic matter and recycle nutrients in the soil. Some fungi derive nutrients from living plants and animals with modified hyphae, which penetrate tissue. Plants are the hardest hit by pathogenic fungi; some attack humans, and some others cause food spoilage. People at high risk for yeast or mold infections are those who have their hands frequently immersed in water. **Yeasts** are single-cellular microorganisms that are spherical or oval-shaped. Yeasts divide unevenly in shape, reproducing by binary fission or budding. *Saccharomyces* is an example of beneficial yeast, used in baking and brewing. Example of pathogenic yeast is *Cryptococcus*. Example of opportunistic yeast is *Candida albicans* (Vulvovaginal candidiasis). An example of a yeast-like pathogen is *Pneumocystis pneumonia*. **Dimorphic Fungi** are fungi that produce both yeast-like structures and mold-like structures; many pathogenic fungi are dimorphic including *Coccidioides*, *Histoplasma*, and *Blastomyes*. **Molds** are multi-cellular, filamentous organisms, with hair-like structures called hyphae. Examples of molds in our area are *Aspergillus*, *Penicillium*, and *Rhizopus*. The portion of the mold which grows is known as the vegetative hyphae. As hyphae grow, sporangia form, which hold the spore; the portion of the mold which creates spores is the reproductive hyphae. As mold grows, hyphae intertwine (like a web), forming a mycelium. Mold can reproduce sexually or asexually and most develop spores during the reproduction process. Molds need three items for growth, namely, food source, moisture, and proper living conditions.

Significant Roles of Fungi In the Environment:

- Decompose dead organisms and recycle their nutrients
- Help plants absorb water and minerals
- Are used for food and in manufacture of foods and beverages
- Produce antibiotics (penicillin is a natural product produced by the mold *Penicillium*)
- Serve as important research tools
- Cause 30% of diseases of plants, animals, and humans
- Can spoil many types of food

Mycoses

Diseases caused by fungi are called mycoses. Mycoses are classified by the location of the fungus on the body and the ability of the fungus to penetrate deeper into the body, as follows:

- Superficial mycoses grow **on** the hair, skin and nails
- Cutaneous mycoses grow **in** the skin
- Systemic mycoses occur when the fungus colonizes the body, most likely the respiratory, nervous or reproduction systems
- Wound/Subcutaneous mycoses occur when an opportunistic fungus colonizes a wound

Fungi can also cause disease by destroying the food supply, causing malnutrition and paving the way for other opportunistic microbial infections or by producing damaging toxins that are ingested by the host.

Examples of Mycoses

Jock itch, ringworm and athlete's foot are all cutaneous mycoses caused by dermatophytes. (Figure 2.9) The cell-mediated immune response to the dermatophytes causes damage to the deeper tissues of the host.

One of the most common fungal infections is the superficial mycosis, vaginal candidiasis, caused by the yeast *Candida albicans*. This condition usually results from the overgrowth of normal yeast microbiota due to antibiotic use. It affects the mucus membranes of the vagina and the skin of the labia and causes irritation and white mucus production. Many people are not aware that yeast infections are transmissible via direct contact.

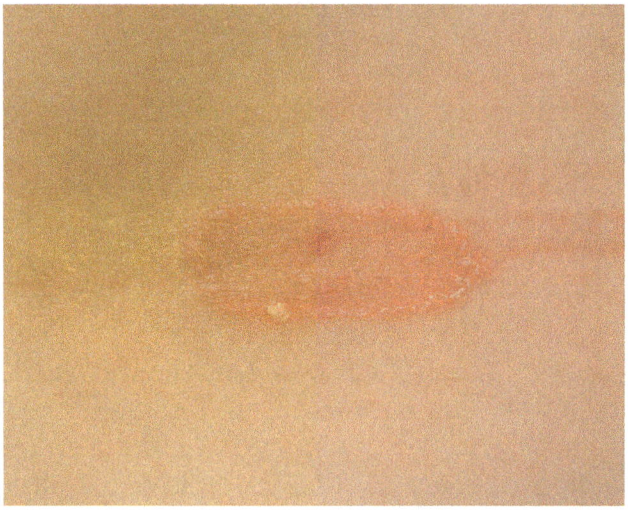

Fig. 2.9a Ringworm and Athlete's Foot
Schankz/Shutterstock.com

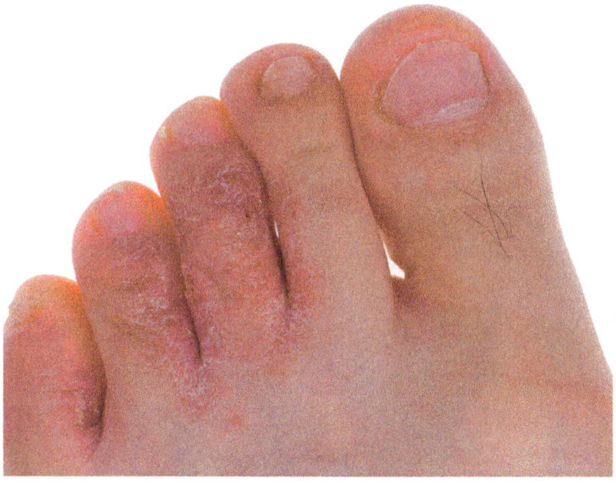

Fig. 2.9b
Carroteater/Shutterstock.com

A few systemic mycoses of note include:

Cryptococcal meningitis produces symptoms similar to bacterial meningitis. It is transmitted by inhalation of dried yeast cells and rarely affects people who have healthy immune systems, but frequently causes illness in terminal AIDS patients and in organ transplant recipients.

Coccidiomycosis occurs primarily in the southwestern U.S. and northern Mexico. The fungus lives in soil and is transmitted by inhalation of dust containing the microbe. The symptoms are similar to those of tuberculosis.

Histoplasmosis can occur when humans inhale airborne spores from the soil. This disease is more prevalent in the eastern U.S. Infections in immunocompetent individuals typically resolve without treatment, but infections in immunocompromised individuals can be serious.

***Pneumocystis* Pneumonia (PCP)** caused by *Pneumocystis jirovecii* is transmitted by inhalation of droplets containing the fungus and causes difficulty in breathing, anemia, hypoxia, and fever. This is a common disease in AIDS patients. Unfortunately it is impossible to prevent infection by *P. jirovecii.*

Helminths

Helminths are parasitic worms. There are three types of helminths: nematodes (round worms), cestodes (tapeworms) and trematodes (flukes). Although many of these worms are macroscopic they are studied in the field of microbiology since many are very small, they have microscopic stages of life, and they can impact humans in a negative way.

Tapeworms are flat-segmented worms that get their nutrients by absorbing them through their integument. Tapeworms are usually species-specific and have a limited range of hosts. For example, beef tapeworms live in cattle and are acquired by ingesting undercooked beef. There are also pork tapeworms and fish tapeworms. Pets can get tapeworms when they eat fleas that have ingested tapeworm eggs. Tapeworms attach to the intestine of the host by a sucker-like structure, the scolex (Figure 2.10). They absorb the nutrients from the host directly through their skin.

The most common worm infection in the U.S. is **pinworm** or ***Enterobiasis.*** Pinworms are transmitted via the fecal-oral route and are common in daycares and lower-elementary school classes. Both humans and animals can harbor the worms. The adult female worms are roughly the size of a staple and are nocturnal. They exit the anus of the infected individual at night and lay their eggs in the perianal region. The migration of the worm can cause discomfort and disturbed sleep. The eggs themselves itch and are transmitted to the hands when the victim scratches. The eggs are then either ingested by the original victim, or they can be transferred to an inanimate object, and then to the hands of a new victim and then ingested. The eggs can persist in the environment (bedding, pajamas) for up to two weeks.

Guinea Worm Disease, also known as dracunculiasis is notable here because it has almost been eradicated. The disease is caused by a nematode, *Dracunculus medinensis.* The larvae of *D. medinensis* are ingested by "water fleas". When humans drink water contaminated with the water fleas, they ingest both the flea and the larvae. The fleas are destroyed by the stomach acid, which releases the larvae, which then penetrate the stomach and intestinal walls of host, migrating to the abdominal cavity where they reproduce. The male worms die and female worms migrate to the subcutaneous tissue of the skin. The worm causes uncomfortable blisters. When the victim seeks to ease the pain of the blisters by soaking them in water, the worm erupts from the blister releasing larvae and re-contaminating the water, starting the cycle over again.

Fig. 2.10 Tapeworm scolex
Decade3d—anatomy online/Shutterstock.com

There is no treatment or vaccine for Guinea Worm Disease, and the host does not develop immunity to the infection. The worms can only be removed when they erupt from the blister to release their larvae. The blisters can cause painful and dangerous secondary infections. However, because the larvae are transmitted to the human inside the water fleas, very simple means of filtration can prevent the spread of Guinea Worm Disease.

Trichinosis (also called Trichinellosis) is a disease that people can get from eating contaminated meat that is raw or undercooked. Trichinosis is usually associated with contaminated pork. The disease is caused by a variety of species of *Trichinella*, a type of nematode. The worms form cysts in the muscle tissue of the host, which are then transmitted when the meat is ingested. Transmission can be controlled by freezing meat, which can kill the larvae, or cooking meat to the appropriate temperature.

Vectors

Vectors are organisms that do not cause disease themselves, but rather spread infection by carrying and transmitting microbial pathogens. Biological vectors are vectors which are necessary to the lifecycle of the pathogen. The *Anopheles* mosquito is a biological vector for *Plasmodium*, the causative agent of malaria. A mechanical vector is one that does not play a role in the life cycle of the microbe, but just carries the microbe from place to place.

Disease vectors belong to two classes of arthropods:

1. **Arachnida**: These have four pairs of legs. These include spiders, ticks and mites. Of the arachnid vectors, hard ticks are the most prominent carriers of disease. Arachnids are most notorious for carrying Lyme Disease and rickettsial diseases (i.e., Rocky Mountain Spotted Fever).
2. **Insecta:** These have three pairs of legs.

Insects and the Diseases they Transmit

- fleas—carry bubonic plague
- lice—typhus
- flies
- mosquitos—malaria, West Nile Virus, Yellow Fever
- kissing bugs—Chaga's disease

Because it is difficult to develop vaccines and treatments for protozoal and helminthic diseases, often the most effective prevention is often vector control. This includes using mosquito nets, spraying to control the reproduction of insects and arachnids, wearing preventive clothing and using insect repellents.

Microscopy

Microscopy is the field of using microscopes to observe microorganisms and other small objects that are too small to be seen by the unaided eye. A major component of this course is staining and observing various types of cells and microorganisms using a microscope. Micrometer (μm) is the unit to measure the length of cells (1,000,000 μm = 1 meter). The smallest size the unaided eye can see is 200 μm. The average bacterium is 1μm and a human red blood cell is 10 μm. Therefore, the microscope is a necessary tool in microbiology.

The type of microscope used in this class is a bright field, binocular, compound microscope. These microscopes project bright light through the specimen on the slide.The microscope is referred to as a compound because it uses a series of lenses to achieve magnifications of up to 1,000 times.

Parts of the Microscope (Figure 2.11)

Ocular lens or Eyepiece: The ocular lens has a 10X magnification. The ocular lenses can be adjusted for the distance between your eyes; this distance is called the interpupilary distance.

Objectives: Each of the four lenses has a different magnification. The revolving nosepiece is used to change objective lenses. The 4X objective or scanning-lens, which will give a total magnification of 40X when used with the eyepiece, is used for scanning a slide and is usually the first lens to be used. The 10X objective or low-power lens, which will give a total magnification of 100X when used with the eyepiece, is used to focus the object on the slide. The 40X objective or high-power lens (also known as high-dry lens), which will give a total magnification of 400X when used with the eyepiece, will enlarge the object four times more than the 10X lens. The 100X objective or high-power lens, which will give a total magnification of 1,000X when used with the eyepiece, is used to view cells or microorganisms that are smaller than human cells.

Mechanical Stage: This is where the slide is placed for viewing. Notice the stage clip that is used to secure the slide in place. The stage can be moved by using the stage motion knobs.

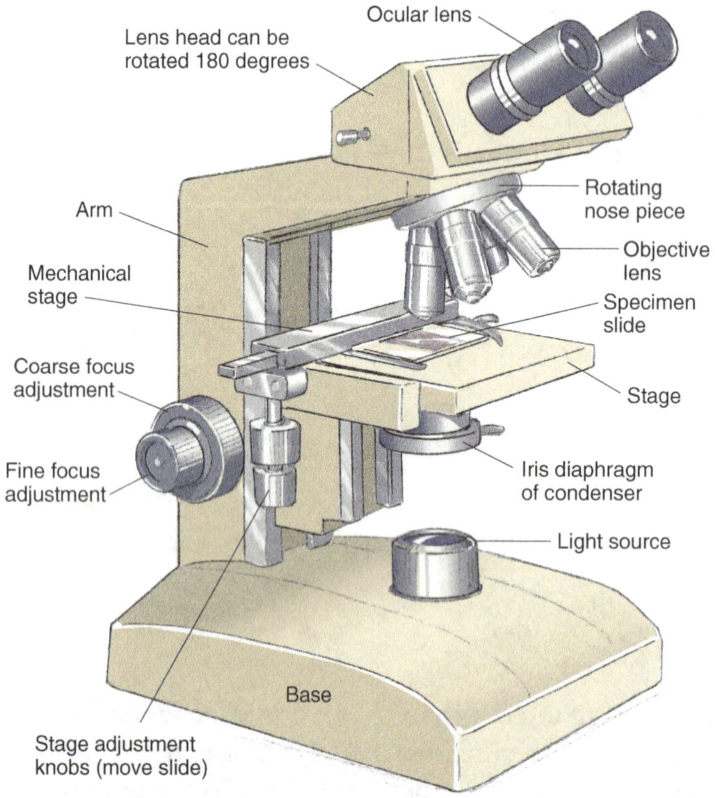

Fig. 2.11 Parts of a Compound Light Microscope

Coarse and Fine Adjustments: These knobs are located on the sides of the body. The coarse adjustment knob is the larger of the two and is used to move the objective lens up and down at a fast pace. The coarse adjustment knob can only be used with the 4X and 10X objective lenses. The fine adjustment knob is the smaller of the two and is used to move the objective lens up and down in smaller increments. The movement of the fine adjustment knob is so minuscule that you can barely see the stage move at all. The fine adjustment knob is used to focus in fine detail and is the only adjustment knob used with the 40X and 100X objective lenses.

Light source/Illumination: The light source is positioned near the base of the microscope. The most common problem that students encounter with their microscopes is improper illumination. If you are having trouble seeing the object, the light adjustment should be one of the first things you check.

Rheostat (brightness control dial): Next to the on/off switch is a knob that is used to adjust the amount of illumination passing through the specimen.

Condenser: Beneath the stage is a lens in which the light is focused through the specimen. The condenser contains two lenses that are necessary to produce clarity of the specimen. The height of the condenser can be adjusted with the condenser knob. Keep the condenser close to the stage, especially when using the oil immersion (100X) objective lens.

Iris Diaphragm: The light passing through the specimen can be regulated by adjusting this part of the microscope. When using the microscope at 4X or 10X objectives, adjust the opening of the diaphragm so that a minimum amount of light passes through the slide. As the magnification is increased, increase the light through the slide by increasing the size of the opening of the iris diaphragm. At 40X objective, the iris diaphragm is widely used to create a clearer image.

Other Aspects of Light Microscopy

Parfocal Capability

Our microscopes have parfocal capability, which means that once they have been focused, they should remain focused as you switch to other objective lenses. The reality is that you will need to adjust your focus slightly as you move to each objective.

Field of View

The field of view is what you see when you look through the microscope. It is important to understand that as you increase the magnification, the field of view actually decreases in size. This is similar to using the "zoom" function on a word document or map. As the object gets bigger, you decrease what you can see of the object. For this reason, it is important that you use the stage adjustment knobs to keep what you are looking at in the center of the field of view, or it may be lost on the periphery.

Properties of Microscopy

The usefulness of the microscope depends on three properties:

1. **Magnification**, defined as the apparent increase in the size of the object you are viewing. The total magnification of the object is calculated by multiplying the magnification of the ocular lens by the magnification of the objective lens that you are using. Luckily for us, our ocular lenses magnify 10 times. So, using the 4X objective, the total magnification =

$$10X \text{ (ocular)} \times 4X \text{ (objective)} = 40X \text{ (total magnification)}$$

2. **Resolution**, defined as the ability to resolve two objects that are very close together. **Resolving power (resolution)** is defined as the minimum distance between two objects where those objects still appear as separate objects. In other words, resolution determines how much detail can be seen. Resolution is improved by changing the wavelength of the illuminating light. Therefore, resolution is enhanced with lenses of higher magnification (100X) by the use of **oil immersion**. Light bends (refracts) as it moves from glass to air. Oil reduces light refraction by bridging the gap between the slide and the objective lens. Also, immersion oil has nearly the same refractive index as glass. Oil allows the maximum amount of light to be transmitted through the ocular lens because the light is not refracted away. The more light to reach the lens, the clearer the image you will see. Figure 2.12 shows how the oil will refract light to increase resolving power. The maximum resolving power of your laboratory microscope under optimal conditions is 0.2 mm. This means that two objects separated by at least 0.2 mm are distinguished as two objects. If they are closer than 0.2 mm, they will appear as a single object.
3. **Contrast**, which is the difference in intensity of two objects. Microbes usually appear colorless on the clear glass of the microscope slide, making them very difficult to see.

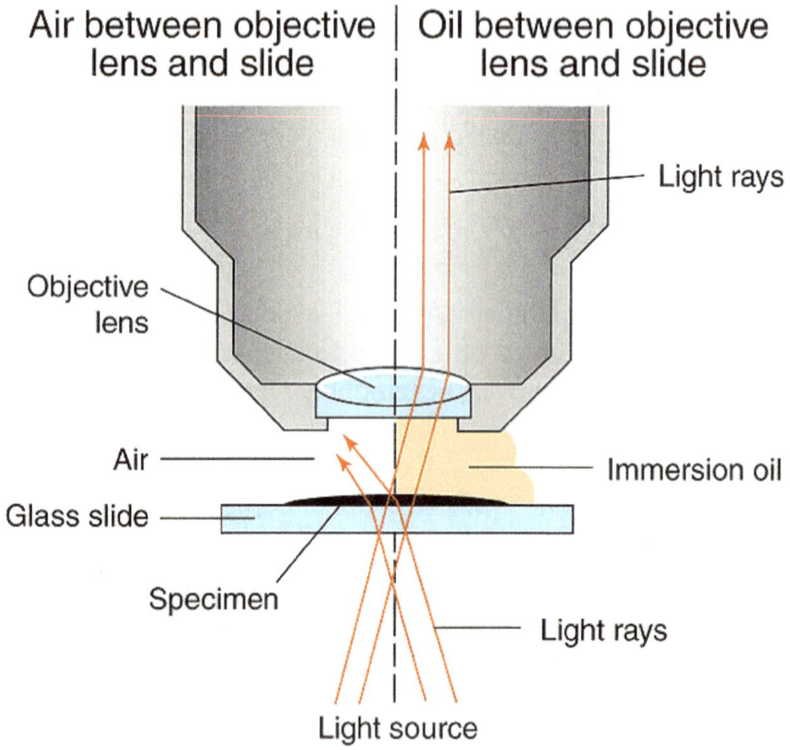

Fig. 2.12 Immersion Oil Increasing Resolving Power
©Kendall Hunt Publishing Company

Increasing Contrast by Staining

Staining improves the resolution and contrast between the cell and the background, thus improving the visibility and physical characteristics of the specimen.

A **simple stain** involves the use of only one stain or dye to give an organism contrast against the background. To understand what a dye is, we must first understand what a stain is. A **stain** is a salt, which means that it is composed of positive and negative ions. What makes it a stain is that one of the ions (the positive or the negative) is colored. The colored ion is called the **chromophore**, or the color-bearing group.

Another type of staining procedure is a **differential stain**. A differential stain uses two or more solutions including stains to differentiate different organisms from each other. Some examples of differential stains are the Gram Stain, Endospore Stain, and Acid-Fast Stain. Lab procedures describing these staining procedures are outlined in later labs in this textbook.

In **acidic dyes**, the chromophore is the negative ion. Acidic dyes include eosin, nigrosine, and India ink. In **basic dyes**, the chromophore is the positive ion. Basic dyes include methylene blue, crystal violet, safranin, malachite green, and carbol fuschin.

Simple staining uses one stain to view the specimen. Simple staining involves positive or negative staining procedures.

Positive staining procedures use a basic dye, meaning the chromophore is the positive ion. Bacteria carry a slightly negative charge and respond best to basic dyes. The positively charged chromophore of the basic dye is attracted to the negatively charged bacterial cells and colors the cells.

Negative staining procedures use an acidic dye, meaning the chromophore is the negative ion. Since bacteria also carry a negative charge, the dye and the bacteria repel one another (i.e., like charges repel). As a result, the background is stained instead of the bacteria. Another name for this procedure is background staining, since that is what is stained. The negative staining technique is used to determine the size, shape, and grouping arrangement of microorganisms. The size and shape of bacteria are not distorted, because the heat of the Bunsen burner is not used to fix the cells on the slide.

Electron Microscopes

Electron microscopes allow us to view objects much smaller than those we can see with a typical compound light microscope. Images viewed with electron microscopes can be measured in nanometers (nm) rather than micrometers (μm). Electron microscopes use concentrated beams of electrons rather than light to create an image, which is then transmitted to a computer screen for viewing. **Transmission electron microscopes (TEM)** allow the electrons to pass through the specimen and can be used to visualize internal components of cells, such as organelles. **Scanning electron microscopes (SEM)** are used to scan the surface of cells and view the external topography of the cells or viruses.

Figure 2.13 on the following page shows a virus, which would not be visible using compound light microscopy.

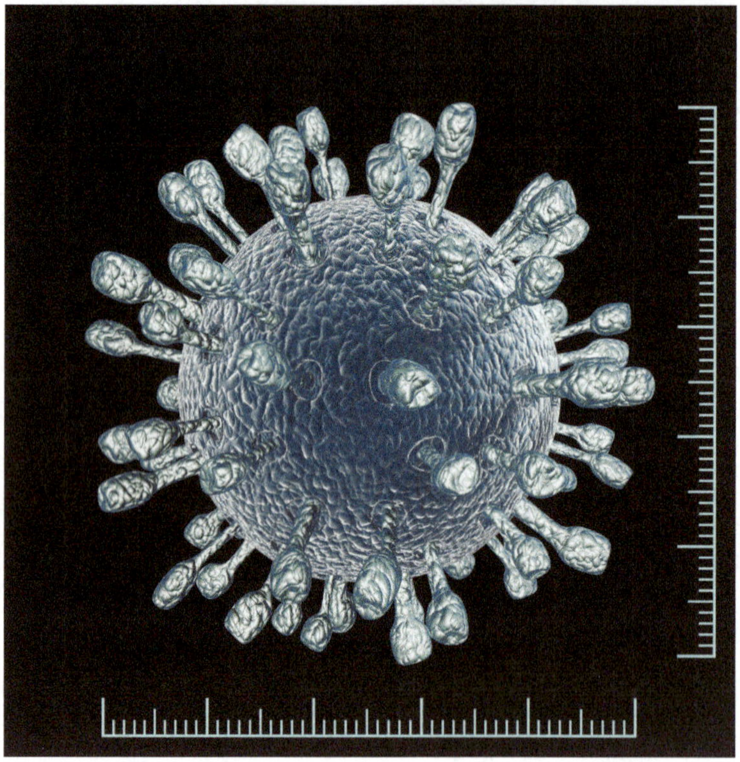

Fig. 2.13 Electron Micrograph of Virus Particle
Peter Simoncik/Shutterstock.com

Use and Clean-Up of the Microscope

Proper handling:

1. Each student is assigned a microscope number.
2. Ensure your lab table is clear and all activities are complete prior to retrieving your microscope.
3. Remove the plastic covering of the microscope and leave the cover on the microscope counter.
4. Always carry the microscope with two hands; one hand carries the base and one hand carries the arm.
5. Use lens paper and lens cleaner solution to clean your microscope before and after each use.
6. You will be held responsible for keeping your microscope in good condition. At your lab table, inspect your microscope. Report major discrepancies to your instructor.

Microscope procedures:

1. Plug in the electrical cord and turn the light on. Adjust the rheostat for brightness.
2. Adjust the iris diaphragm to allow light through the condenser.
3. Place the 4X objective (scanning lens) over the stage.
4. Adjust the eyepiece for maximum comfort.
5. Place your slide on the stage and secure the slide with the clips.
6. Use the stage control knobs to center the specimen below the objective lens.
7. Using the coarse adjustment knob, move the stage all the way up.
8. Look through the eyepiece. You will see a bright, round circle; this is the field of view.
9. Slowly rotate the coarse adjustment knob to move the stage downward while looking through the eyepiece, until the specimen is in focus. ***This may be the only time you will use the coarse adjustment knob.
10. Rotate the fine adjustment knob to improve the focus of the specimen.
11. Adjust the rheostat to improve quality of the field of view.
12. Rotate the revolving nosepiece to bring the 10X objective (low lens) into place. After increasing the power, the specimen will still be in focus; this property is called **parfocal capability**. Use the fine adjustment knob to improve the focus of the specimen.
13. Rotate the revolving nosepiece to bring the 40X objective (high lens) into place. Use the fine adjustment knob to improve the focus of the specimen.
14. Look through the eyepiece. Adjust the iris diaphragm lever of the condenser to improve the quality of the field of view. This will adjust the background light and enhance the focus of your specimen.

Oil-Immersion Objective:

1. To view prokaryotes, you may need to use the highest magnification on your microscope. You will use the 100X objective to increase the overall magnification to 1,000X.
2. *Be certain that you have a clear view of the specimen at the 40X objective lens before moving to the 100X objective lens.*
3. Rotate the revolving nosepiece to bring the 100X objective lens into place. Use the fine adjustment knob to improve the focus of the specimen. Your specimen's image will have poor quality, because of poor resolution.

4. Rotate the revolving nosepiece backwards so that it is halfway between the 40X and the 100X objective lens.
5. Add one drop of immersion oil onto the slide directly over the viewing area.
6. Rotate the revolving nosepiece so that the 100X objective lens is back in place. The tip of the 100X objective lens will enter the oil. The 100X lens is specially designed to be used with immersion oil.
7. Carefully use the fine adjustment knob to improve the focus of the specimen. Adjust the brightness control to adjust the amount of light. Use the stage motion knobs to move the slide for viewing.

Clean-up procedure and proper storage:

1. When finished with your slide, move the stage downward with the coarse adjustment knob. Remove your slide and place the used slide in the disinfectant container.
2. Using the stage motion knobs, center the stage within the scope.
3. Use the power switch to turn off the light. Unplug the power cord and wind it up on the back of the scope.
4. **DO NOT TRY AND CLEAN THE 4X or 10X OBJECTIVES**. Clean 40X objective lenses BEFORE cleaning the oil-immersion lens (100X). If you used oil, wipe the oil-immersion lens three or four times to make certain all the oil has been removed. (Do not wipe used lens paper on the other objective lenses.)
5. Use lens paper and lens cleaner to clean the eyepiece, objectives, and stage. If your scope is contaminated, use alcohol pads to disinfect your scope.
6. Place the 4X objective lens into place, over the stage.
7. Carry the scope with both hands to the assigned storage area. Place the dust cover back on the scope.

Microscope and Staining Activity

Materials

Laboratory materials: Letter 'e' slide, clean slides, sterile swab, methylene blue stain, cover slips, microscope, lens paper, lens cleaner, transfer pipettes, prepared slides of protozoa and fungi, live mixed cultures of protozoans, prepared slides of bacteria shapes, and colored pencils.

I—Prepared slide of the Letter 'e'

1. Using the proper microscope procedures, view the letter 'e' slide at the 4X, 10X, and 40X objective lenses.
2. Draw the letter 'e' at each magnification in the lab report. Draw the exact image you see in the field of view.

II—Prepared slides of Fungi

1. Using the proper microscope procedures, view prepared slides of fungi.
   - View slides of two of the following molds: Penicillium, Aspergillus, Rhizopus.
   - View a slide of *Candida* cells.
2. View the fungi slides at the 4X, 10X, and 40X objective lenses.
3. Prepared slides have cover slips. Therefore, the highest magnification normally used for viewing them is the high-dry (40X) lens. You will NOT use the oil (100X) lens.
4. Draw the prepared slides at the 10X or 40X objective lens in the lab report.
5. Place the prepared slides back in the appropriate slide tray.

III—Prepared slides of Protozoa: Amoeba, *Plasmodium*

1. Using the proper microscope procedures, view prepared slides of protozoa.
2. View the slides at the 4X, 10X, and 40X objective lenses.
3. Prepared slides have cover slips. Therefore, the highest magnification normally used for viewing them is the high-dry (40X) lens. You will NOT use the oil (100X) lens.
4. Draw the prepared slides at the 10X or 40X objective lens in the lab report.
5. Place the prepared slides back in the appropriate slide tray.

IV—Live Slides of Protozoa and Algae

1. For a wet mount, use the transfer pipette. Place a drop of the live, mixed culture of protozoans on the slide at its center. Bring the edge of a coverslip up to the edge of the liquid culture. Slowly lower the coverslip onto the culture, trying not to trap water bubbles.
2. Choose three protozoa to observe.
3. Using the proper microscope procedures, view the wet mounts or prepared slides at the 4X and 10X objective lenses.
4. Draw the wet mounts or prepared slides at 10X objective lens in the Lab Report.

V—Prepared slides of Helminths: Pinworms, *Trichinella* and *Taenia*

1. Using the proper microscope procedures, view prepared slides of helminths.
2. View the slides at the 4X, 10X, and 40X objective lenses.
3. Prepared slides have cover slips. Therefore, the highest magnification normally used for viewing them is the high-dry (40X) lens. You will NOT use the oil (100X) lens.
4. Draw the prepared slides at the 10X or 40X objective lens in the lab report.
5. Place the prepared slides back in the appropriate slide tray.

VI—Swab, stain, and observe cheek cells

1. Obtain a sterile swab. Insert the swab inside your mouth and vigorously swipe the inner cheek back and forth for 10 seconds.
2. Roll the applicator on the middle portion of a clean slide. Ensure the middle portion is completely covered with your specimen.
3. Discard the applicator in the trash.
4. Place the slide on a paper towel and allow the slide to sit for one minute.
5. Place the slide on the staining rack of the sink.
6. Add one drop of methylene blue on your specimen. Allow the slide to sit for one minute.
7. Tip the slide to allow excess fluid to drip off.
8. Place the slide back on the paper towel and dry the bottom of the slide.
9. Place a cover slip on your specimen.
10. Place the slide on the microscope. Observe your epithelial cells, as well as any prokaryotic cells. Prokaryotic cells will be very small compared to your epithelial cells. Identify the nucleus in an epithelial cell.
11. View the slide at the 4X, 10X, and 40X objective lenses. Draw your observations at each magnification in the Lab Report.

VII—Prepared slide of bacteria shapes

1. Using the proper microscope procedures, view the prepared slide of the three different bacteria shapes: cocci, bacilli, and spiral. All three shapes are on the same slide, so you must move the stage to observe all three.
2. View the slide at the 4X, 10X, and 40X objective lenses.
3. Prepared slides have cover slips. Therefore, the highest magnification normally used for viewing them is the high-dry (40X) lens. You will NOT use the oil (100X) lens.
4. Draw each bacteria shape at the 40X objective lens in the lab report.
5. Place the prepared slides back in the appropriate slide box.

Name ______________________

Exercise 2.1 Lab Report

MICROSCOPIC DRAWINGS—LETTER 'e'

4X objective lens

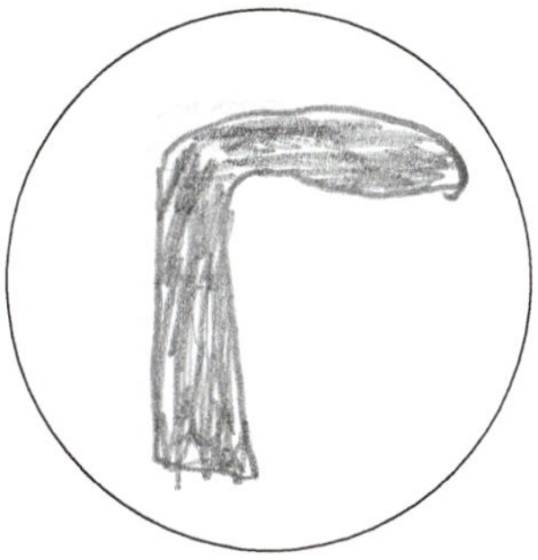

10X objective lens

40X objective lens

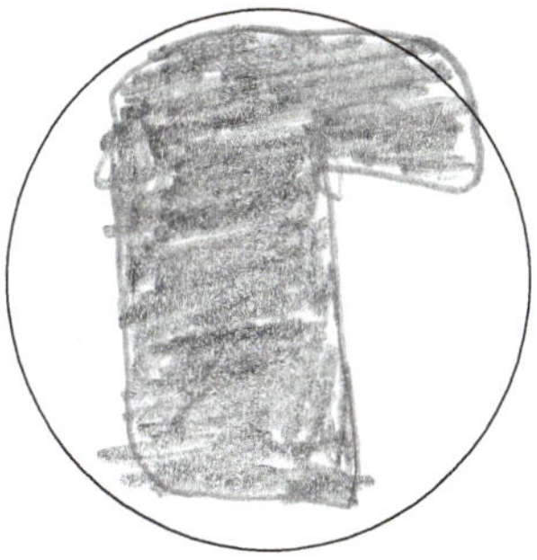

MICROSCOPE DRAWINGS—FUNGI

Mold

Name the organism:
Penicillium

Mold

Name the organism:
____

Yeast

Name the organism:
_____*Candida*_____________

Don not use the focus knob with the 40X
Fine focus – little knob.
Oil not good for 40X
Turn 100X into the oil

MICROSCOPE DRAWINGS—PROTOZOA

Living

Amoeba	Paramecium	Euglena

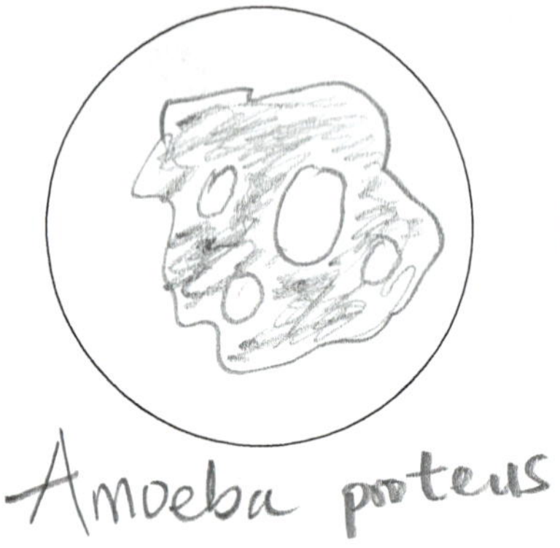

PROTOZOA

Prepared Slides

Plasmodium

MICROSCOPE DRAWINGS—POND WATER (Protozoa & Algae)

4X objective lens	10X objective lens

MICROSCOPE DRAWINGS—HELMINTHS

Pinworms | ***Trichinella*** | ***Taenia* scolex**

MICROSCOPE DRAWINGS—CHEEK SWAB

{Label the cells and nuclei}

4X objective lens | **10X objective lens** | **40X objective lens**

MICROSCOPE DRAWINGS—BACTERIA SHAPES

40X objective cocci | **40X objective bacilli** | **40X objective spiral**

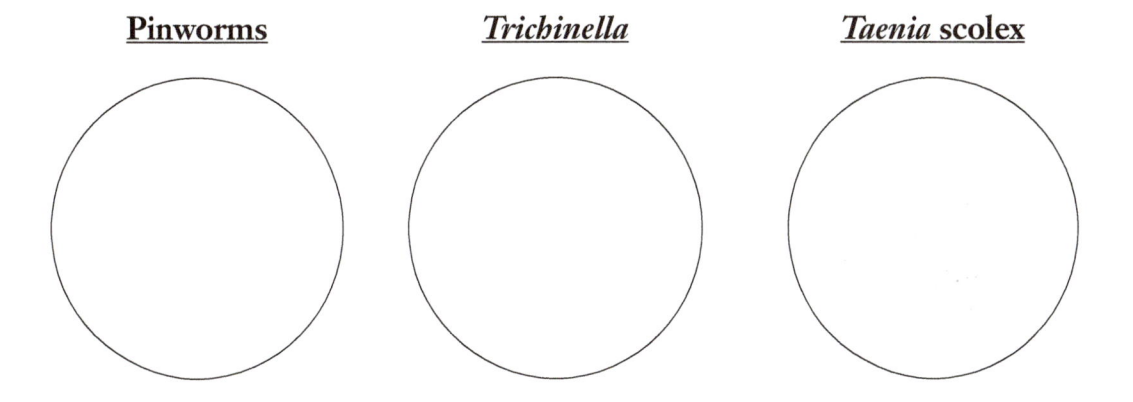

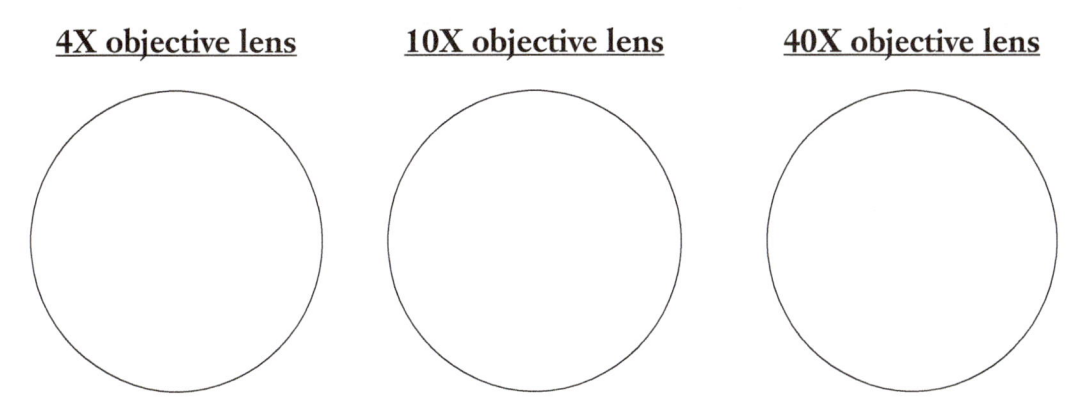

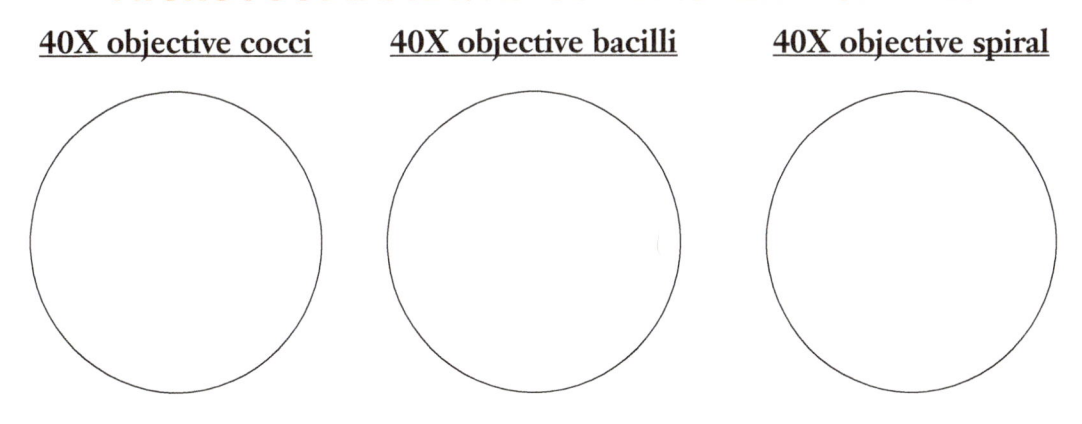

1. Describe the size of a human epithelial cell compared to a bacterial cell. Which one has a nucleus? What is the main difference between a prokaryotic cell and eukaryotic cell?

2. What is the length of a typical bacterial cell? What is the length of a human red blood cell?

3. When moving the stage left, what direction does the field of view move in the eyepiece?

4. Describe the general characteristics of a bacterial cell.

5. Name and describe the various morphologies of bacterial cells.

6. Describe the general characteristics of a eukaryotic cell. List the various methods of reproduction.

7. Describe the common characteristics of algae.

8. Describe how algae might cause disease in humans and give examples.

9. Describe the common characteristics of protozoa.

10. Describe the life cycle of *Plasmodium*.

11. Describe the transmission of *Toxoplasma* and name the most susceptible host(s).

12. Which protozoa listed in your text are transmitted by contaminated water?

13. Describe the symptoms of trichomoniasis.

14. Describe the characteristics of yeast.

15. Describe the characteristics of mold.

16. List and describe the ways in which fungi cause disease in humans.

17. Describe how antibiotic use contributes to the development of candidiasis due to *Candida albicans*.

18. List and describe the three types of helminths.

19. Describe the factors that make pinworm the most prevalent worm infection in the U.S.

Explain the Functions of the Parts of a Compound Microscope

PART	FUNCTION
OCULAR LENS	
OBJECTIVE LENSES	
STAGE	
STAGE CONTROL KNOBS	
CONDENSER CONTROL KNOB	
IRIS DIAPHRAGM	
COARSE ADJUSTMENT KNOB	
FINE ADJUSTMENT KNOB	

Explain the Properties of a Compound Microscope

PROPERTY	PURPOSE
FIELD OF VIEW	
PARFOCAL CAPABILITY	
LOW POWER	
HIGH POWER	
TOTAL MAGNIFICATION	
RESOLVING POWER	
IMMERSION OIL	

Bacterial Cell Structure and Function

Bacteria are prokaryotes as they lack nuclei. Bacterial cells are contained by a cell membrane (aka plasma membrane or cytoplasmic membrane) that is made of a **phosopholipid bilayer** (Figure 3.1 and 3.7). This makes it similar to our cell membranes, which are also made of a phospholipid bilayer. The job of the cell membrane is to "contain" the cell and to control what enters and leaves the cell. The membrane is also the site of energy generation by means of the electron transport system.

DNA

Most prokaryotes have only one **chromosome** that is made up of a circular molecule of DNA. Bacteria contain far fewer genes than humans. A typical *E. coli* bacterium contains around 3,500 genes, whereas a human contains around 30,000 genes.

Bacteria may also have **plasmids**, which are smaller circular segments of extrachromosomal DNA. Plasmids vary in size and may contain only a handful of genes or a few hundred genes. The genes found on plasmids vary widely and may be used to aid in bacterial survival by providing antibiotic resistance, the ability to produce toxins or new metabolic capabilities. While all

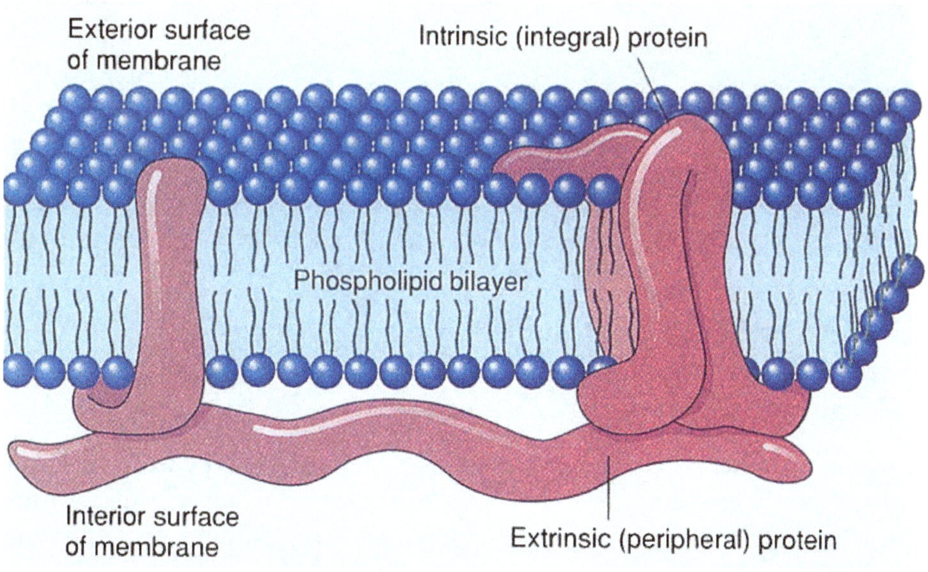

Fig. 3.1 Bacterial Cell Membrane

bacteria have chromosomes, plasmids are "optional". Some bacteria have many, some a few and some have no plasmids.

Nucleoid/Nucleoid Region

Prokaryotes do not have membrane-bound nuclei, but the area surrounding the chromosome is called the nucleoid or nucleoid region (meaning "nucleus like"). The cytoplasm of the nucleoid region tends to be thicker and more gelatinous than the rest of the cytoplasm.

Ribosomes

Ribosomes are the site of protein synthesis. Bacterial ribosomes are considered to be 70s, and consist of two parts: the small ribosomal subunit (30s) and the large ribosomal subunit (50s). The "s" indicates the density of each subunit when spun at high speeds. Prokaryotic ribosomes are smaller than eukaryotic ribosomes. Eukaryotic ribosomes are 80s consisting of a 40s subunit and a 60s subunit.

Cytoplasm and Cytosol

The cytoplasm refers to the entire content of the cell within the cell membrane, and the cytosol refers specifically to the liquid portion of the cellular contents.

Granules

Many bacteria have granules within the cytoplasm. These granules can be made of starch, lipids, or any substance that the cell is storing for later use.

External Structures

The Cell Wall

It is vital to your success in this class to get a handle on the structure of the bacterial cell wall now! Understanding the structure of the bacterial cell wall is necessary to understand the following:

- Bacterial replication
- Antibiotic susceptibility—which antibiotics work and why
- The reaction of the immune system to different types of bacteria
- How to identify unknown bacteria in the lab

The bacterial cell wall is made of a unique substance, **peptidoglycan**, which is found **only** in bacteria. Peptidoglycan is made up of alternating subunits of N-acetylglucosamine (NAG) and N-acetylmuramic acid (NAM), strung together like beads on a string. But the really important part of peptidoglycan is that several of these chains line up in rows and then are cross-linked to each other via adjacent NAM molecules. This cross-linking makes the peptidoglycan very strong, essentially forming a "chain-mail armor" or mesh around the cell. (Figure 3.2) This allows the cell to maintain its shape and to resist bursting due to osmotic pressure.

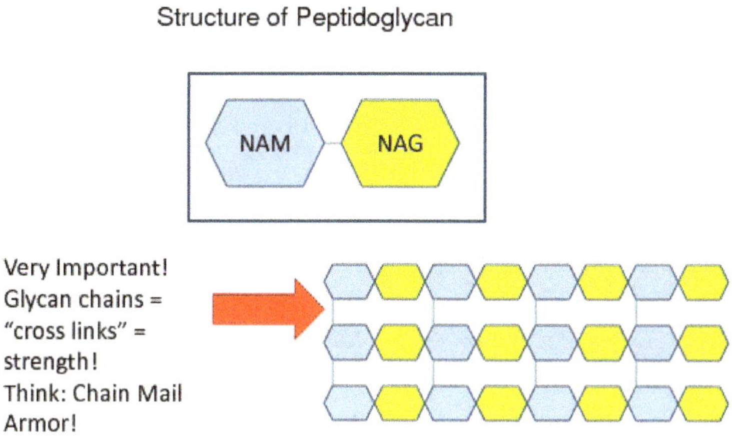
Structure of Peptidoglycan
NAM
NAG
Very Important!
Glycan chains =
"cross links" =
strength!
Think: Chain Mail
Armor!

Fig. 3.2 Structure of Peptidoglycan
Courtesy of A. Swarthout

2 Varieties cell walls: Gram-positive and Gram-negative

The **Gram-positive cell wall** is characterized by a thick layer of peptidoglycan. Up to 50% of the total cell weight is its cell wall. The Gram-positive wall also has teichoic acids and lipoteichoic acids embedded in it. The lipoteichoic acids anchor the peptidoglycan to the cell membrane, while the teichoic acids aid the passage of ions through the cell wall. Remember: the cell membrane is NOT a part of the cell wall! (Figure 3.3) Certain genera (not all!) of Gram-positive organisms, such as Mycobacterium, contain up to 60% mycolic acid (a waxy substance) in their cell walls. Cells that contain mycolic acid are referred to as Acid Fast bacteria. The waxy mycolic acid allows the bacteria to resist effects of certain disinfectants and protects the cells from dessication.

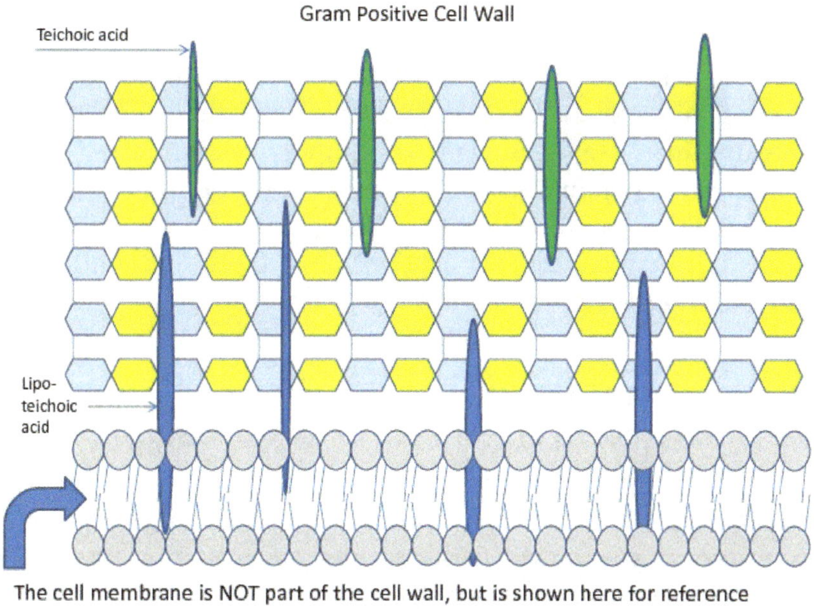
Gram Positive Cell Wall
Teichoic acid
Lipo-
teichoic
acid
The cell membrane is NOT part of the cell wall, but is shown here for reference

Fig. 3.3 Gram-positive Cell Wall
Courtesy of A. Swarthout

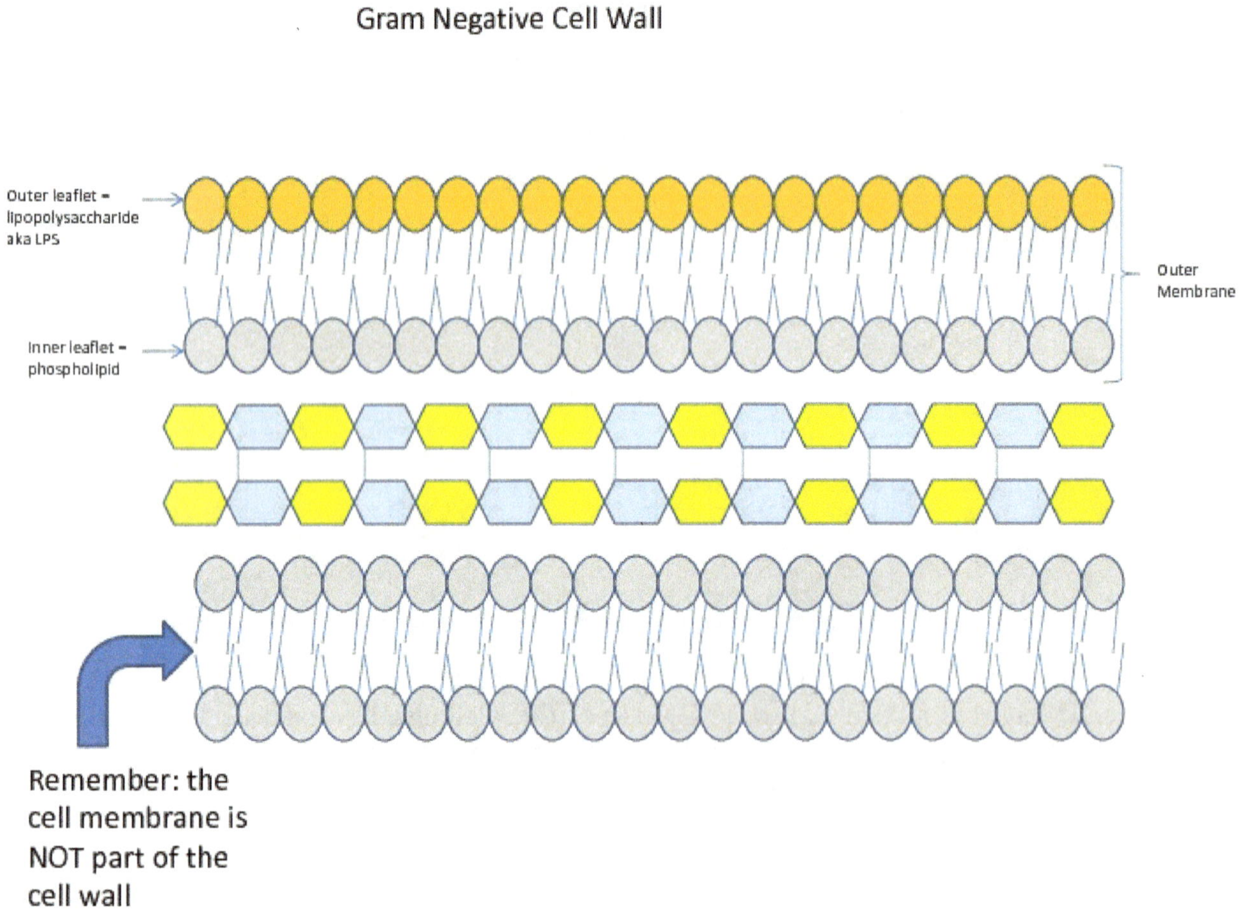

Fig. 3.4 Gram-negative Cell Wall
Courtesy of A. Swarthout

The **Gram-negative cell wall** is characterized by a thin layer of peptidoglycan, which is then surrounded by an **outer membrane**. It is important to note that the outer membrane is part of the cell wall and is not considered "another cell membrane". The outer membrane is made up of two layers or "leaflets". The inner leaflet is made of phospholipids (just like the cell membrane), but the outer leaflet is made of lipopolysaccharide (LPS). LPS is made of a polysaccharide (O antigen) plus a lipid (lipid A). Lipid A is an **endotoxin** and triggers inflammation, fever and shock in victims of Gram-negative infections. The outer membrane also contains **porins** which are highly selective transport proteins. The space surrounding the peptidoglycan is called the **periplasm.** (Figure 3.4)

Special Cell Wall Circumstances

One genus of bacteria, Mycoplasma, lacks a cell wall. Because of this, Mycoplasma are smaller than other types of bacteria and are pleomorphic (they do not have a standard shape). Their shape changes to fit the space they are in. In addition, this lack of cell wall makes them intrinsically resistant to antibiotics that work by disrupting the formation of peptidoglycan, such as penicillin, cephalosporin and vancomycin.

Another bacterium, Mycobacterium, has a special cell wall that is discussed in Lab 3.3.

Antibiotics and Peptidoglycan

Several antibiotics work by interfering with the synthesis of peptidoglycan. If the peptidoglycan is not made properly, the cell wall is weak and the cell is overcome by osmotic pressure and lyses. Because the outer membrane of Gram-negative cells surrounds and protects the peptidoglycan of those cells, antibiotics like penicillin, cephalosporin and vancomycin, which work by interfering with the synthesis of peptidoglycan, may not work on Gram-negative cells, because they won't have physical access to the peptidoglycan.

An enzyme present in human tears and sweat, **lysozyme**, can damage peptidoglycan and cause lysis of bacterial cells.

Glycocalyx

The **glycocalyx** is a slimy layer of polysaccharides secreted by some bacteria outside their cell walls. If the glycocalyx is thin, not highly organized and not firmly attached to the cell wall, it is called a **slime layer**. Slime layers allow the bacteria that have them to glide along solid surfaces as a mode of motility.

If the glycocalyx is thick, highly organized and firmly attached to the cell wall, it is called a **capsule (Figure 3.7).** Capsules help to hide bacteria from the phagocytic white blood cells of the immune system, thus aiding in the ability of the bacteria to cause disease.

Flagella

Flagella (Figure 3.7) are long filamentous structures that propel bacteria, giving them motility. (Figure 3.5) Flagella can come in a variety of arrangements, and these arrangements are characteristic of the species of bacteria that have them. (Figure 3.6)

Pili and Fimbriae

Pili and fimbriae (Figure 3.7) are protein projections from the surface of the bacterial cell. They are shorter than flagella, more rigid and not used for motility. Pili and fimbriae are used by bacterial cells to attach to surfaces in the environment and inside our bodies. Pili are generally longer than fimbriae.

A specialized type of pili, the **sex pilus**, is used to transfer DNA from one bacterial cell to another.

A Generalized Bacterial Cell

All bacterial cells have a cell membrane, a nucleoid, a chromosome, ribosomes and cytoplasm. Most bacteria have a cell wall. The most notable exception is the genus *Mycoplasma*.

Structures such as pili, fimbriae, flagella, plasmids and the glycocalyx are not found in all bacterial cells, but are specific to certain types of bacteria, and their presence or absence can aid in bacterial identification.

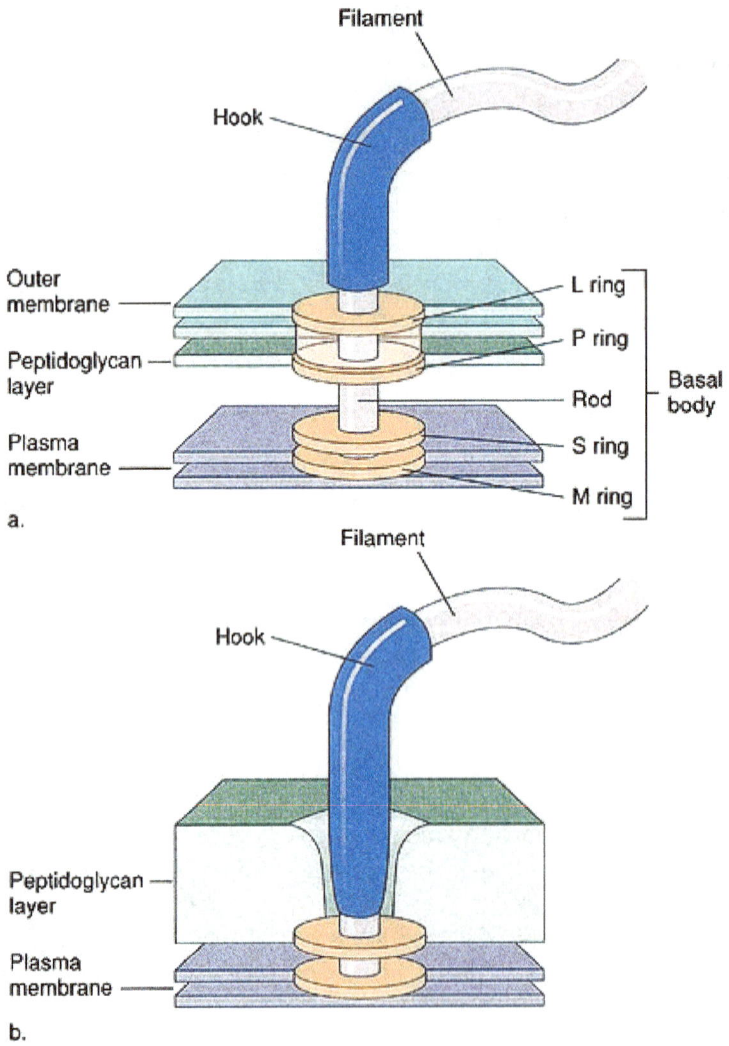

Fig. 3.5 Structure of Flagellum

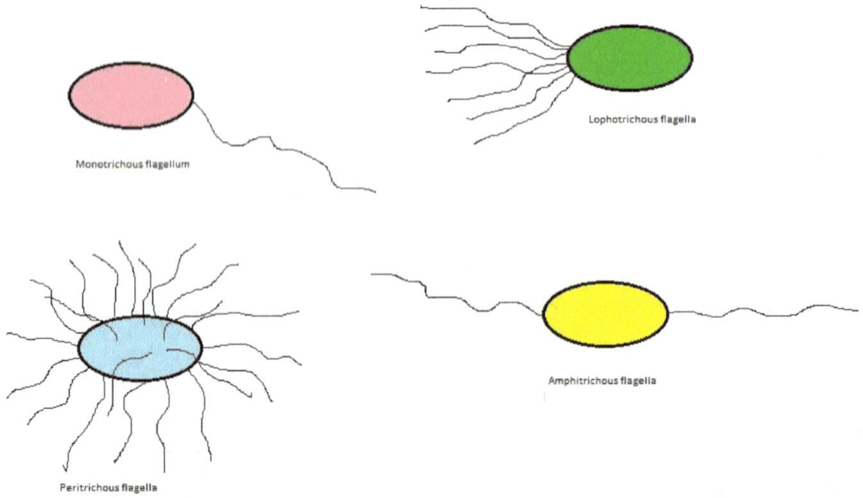

Fig. 3.6 Flagellar arrangements
Courtesy of Authors

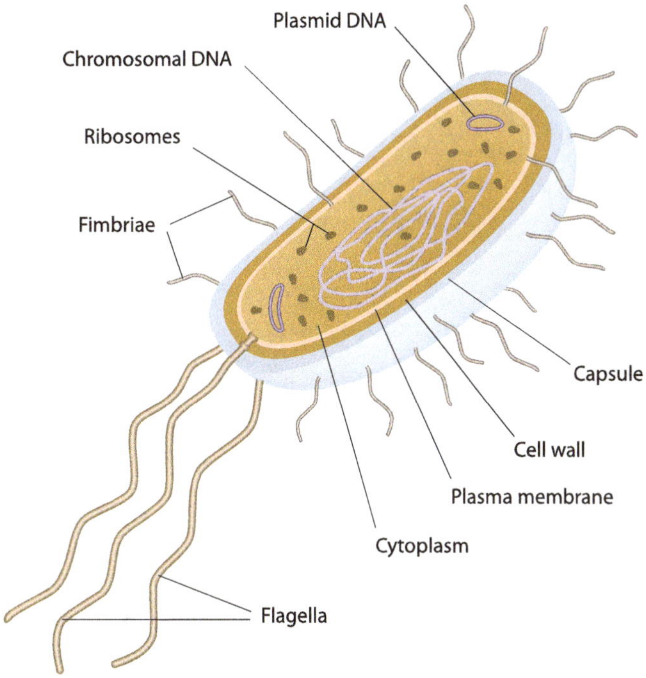

Fig. 3.7 A Generalized Bacterial Cell
Alila Medical Media/Shutterstock.com

Special Bacterial Cell Structures

Endospores

Some Gram-positive bacteria are capable of forming survival structures called **endospores.** Endospore forming bacteria that are clinically significant are formed by members of the genus *Clostridium* and the genus *Bacillus.*

Endospores provide the cells with resistance to heat, chemicals, dessication, UV light and nutrient deprivation. Cells that produce endospores exist and function as vegetative cells when conditions are favorable. A vegetative cell is an active, metabolizing, and reproducing cell (not a vegetative state!). When nutrients become limited and the population is overcrowded endospore forming bacteria begin a process called sporulation. Sporulation is about an 8 hour process and is the process of forming the actual endospore. Once the endospore is formed there is no detectable metabolism and is in a dormant state. During this dormant endospore state is when the organism is resistant to the environmental conditions listed at the beginning of the paragraph. (Figure 3.8)

An endospore returns to its vegetative state by a process called **germination.** Germination is triggered by physical or chemical damage to the endospore's coat. The endospore's enzymes then break down the extra layers surrounding the endospore, water enters, and metabolism resumes. Since one vegetative cell forms a single endospore. The fact that an endospore can undergo germination is significant to those in the healthcare field, since there are many organisms that form endospores that cause disease. The endospore can lay dormant for long periods of time and resist disinfection procedures, once exposed to suitable conditions (i.e. the human body) the endospore can germinate and cause disease. Sporulation in bacteria is not a means of reproduction, meaning that this process does not increase the number of cells. Bacterial endospores differ from spores formed by eukaryotic fungi and algae, which detach from the parent and develop into another organism and, therefore, represents reproduction. (Figure 3.9)

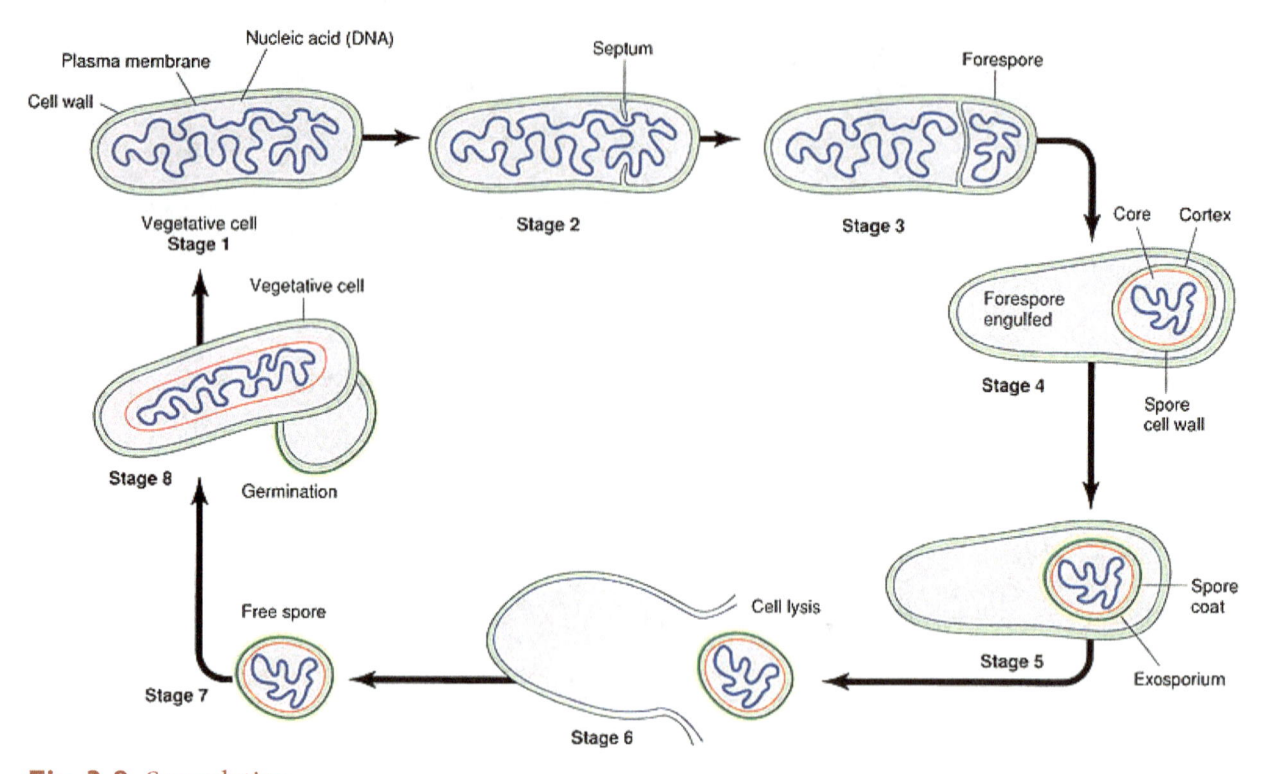

Fig. 3.8 Sporulation
© Kendall Hunt Publishing

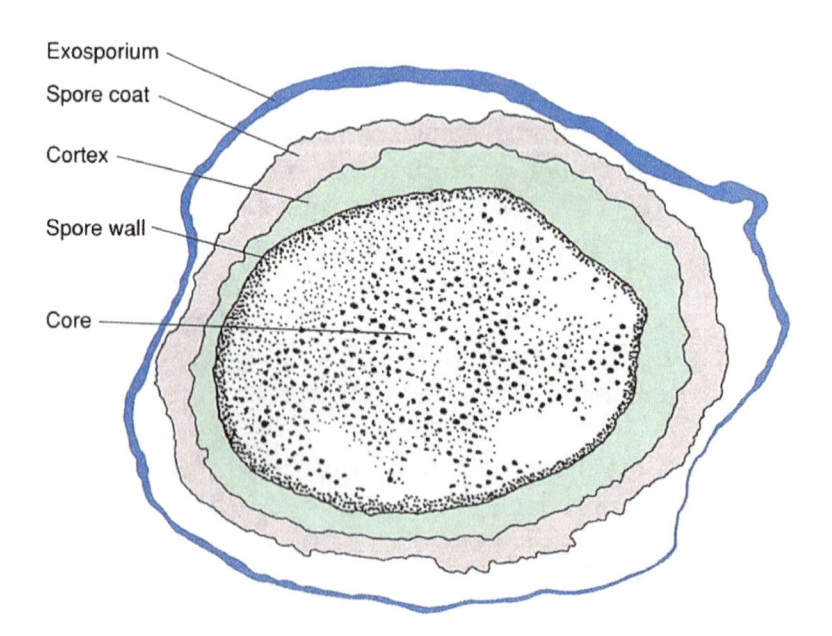

Fig. 3.9 Structure of An Endospore
© Kendall Hunt Publishing

When released into an environment, endospores can survive extreme heat, lack of water, exposure to toxic chemicals (disinfectants), and radiation. For example, a 7500-year old endospore of *Thermoactinomyces vulgaris* from the freezing muds of Elk Lake Minnesota germinated after being warmed and placed in nutrient media, and 25–40 million year old endospores found in the gut of a stingless bee entombed in amber were found in the Dominican Republic. After being placed in nutrient media these 25–40 million year old endospores reportedly germinated!

Endospores are important from a clinical viewpoint and in the food industry, because they are resistant to processes that normally kill vegetative cells. Such processes include heating, desiccation (drying), use of chemicals, and UV radiation. Most cells are killed when temperatures reach above 70° Celsius; endospores, however, can survive boiling water for several hours or more. Endospores of **thermophilic** (heat-loving) bacteria can survive in boiling temperatures for up to 19 hours. Endospore-forming bacteria are a problem in the food industry because they are likely to survive under-processing, and, if conditions for growth occur, some species produce toxins and disease. Special methods for controlling endospores will be discussed later in the semester.

Notable Endospore-Forming Bacteria

Bacillus anthracis

The endospores of *Bacillus anthracis* have a fine, powdery consistency that aerosolize easily, making *B.* anthracis a candidate for bio-weapon development. Once these airborne spores are inhaled, they germinate in the body and cause flu-like symptoms that eventually worsen into severe breathing problems and shock. Left untreated, there is an 85–90% mortality rate. With treatment, the mortality rate drops to around 45% (Center for Disease Control and Prevention). Vaccination is available for those at high risk of exposure, and the antibiotic ciprofloxacin is effective if administered promptly after exposure.

Clostridium botulinum

Clostridium botulinum is a soil-dwelling, endospore-forming bacterium that is a frequent contaminant of canned foods. If the canning process is not done correctly, endospores of *C. botulinum* will survive the process and germinate in the can or jar of food, and then begin to produce and secrete botulinum toxin into the food. This **neurotoxin** interferes with the function of the nervous system.

Nerve cells release acetylcholine into the synaptic cleft. The acetylcholine binds to receptors on muscle cells and serve as a signal for muscle contraction. Botulinum toxin prevents the release of acetylcholine from nerve cells, resulting in flaccid paralysis in the victim. Symptoms begin 12–36 hours after ingestion of the contaminated food. If left untreated, vital circulatory and respiratory muscles stop functioning, leading to death of the individual.

Treatment of botulism involves administration of preformed botulism antitoxin, which is a special form of antibody against the neurotoxin. There is no vaccine against botulism, so the best cure is prevention, which can be done by boiling foods at 100°C for 10 minutes to inactivate the neurotoxin.

Infant botulism is the most common form of botulism in the United States. Honey and corn syrup may contain bacterial endospores, which may germinate in the body. Older children and adults, with fully functioning immune systems, seem to be able to fight off the effects of the toxins released by the germinating endospores. Infants during their first year of life, and immunocompromised adults, are at risk of infection.

Botulinum toxin is used pharmaceutically as Botox®.

Clostridium tetani

A close relative of *C. botulinum* is *Clostridium tetani*. Endospores of *C. tetani* may enter a wound and then germinate in the body, where they begin to produce **tetanospasmin**, a neurotoxin that binds to muscle cells causing prolonged muscle contraction. The common name for this condition, "lock jaw", is the result of facial muscle contraction. It can also lead to impairment of the respiratory and cardiac systems.

Tetanus immunoglobulin (TIG), an antitoxin, may be administered as a preventative measure in response to acute injury. However, the DTaP (Diphtheria, Tetanus, and Pertussis) vaccine is also available. It consists of a series of four shots given at the ages of two months, four months, six months, and 12–18 months. Booster shots are given at four to six years of age, at 11–12 years, and recommended every 10 years thereafter.

Clostridium difficile

Antibiotic-associated colitis, or "C. diff," is caused by the microorganism *Clostridium difficile*. This organism is found as normal fecal microbiota in about 3% of healthy people, but is most often associated with nosocomial (hospital-acquired) infections in nursing homes and long-term care facilities. Extensive antibiotic therapy disrupts normal healthy gut microbiota, allowing *C. difficile* the opportunity to colonize the intestinal tract. Antibiotic-associated colitis results in extreme diarrhea and abdominal pain, and is difficult to treat with antibiotics. Healthcare workers need to employ both standard and contact precautions when working with *C. difficile* patients.

Clostridium perfringes

Gas gangrene may result from exposure to *Clostridium perfringes* through surgical wounds, injury, or severe burns. Individuals with poor cardiovascular or respiratory function, such as diabetic patients, are at greatest risk. If the endospore enters the person's body through a wound opening, an anaerobic or oxygen-poor environment encourages endospore germination. Within a few days, pain at the site of injury is accompanied by gaseous, discolored, and foul-smelling tissue. Treatment involves debridement of the wound, administering antibiotics, hyperbaric oxygen, and performing limb amputation if the disease has progressed. There are no vaccines or antitoxins against *C. perfringes* at this time.

Biofilms

Biofilms are complex aggregates of microbes—prokaryotes and eukaryotes—that live in polysaccharide-encased communities. These communities may be suspended in aqueous environments or attached to surfaces. Biofilms are responsible for many annoying conditions such as slime in sink drains and toilets, but they are also responsible for many types of more serious infections. Dental plaque is the result of biofilm formation. Patients who receive implants (artificial hips or heart valves) or have in-dwelling devices such as venous or urinary catheters may get infections caused by bacteria growing on/in these devices in the form of biofilms. These infections are difficult to treat because the polysaccharide matrix encasing the microbes protects them from the effects of antibiotics.

Biofilms, like microorganisms in general, may also be beneficial. They are being used in wastewater treatment and their activities are necessary for nutrient cycling.

Biofilms can be formed by bacteria, archaea, and yeast. They can be homogeneous—composed of only one type of organism, or heterogeneous—composed of a mixture of microbes.

Biofilm Formation

Biofilm formation begins when one or a few microbes attach to a surface by van der Waals forces (weak attractions between molecules). The microbes then strengthen their attachment by using fimbriae and pili. The attached cells send out molecular signals to recruit other cells to the biofilm. Finally, they begin to produce the extracellular polysaccharide matrix—a sticky, slimy substance that allows them to adhere better and allows other cells from the surrounding fluid to adhere to the surface as well. The biofilm grows as attached cells go through binary fission and other cells from the environment are "recruited" to join the developing biofilm. (Figure 3.10)

The matrix protects the cells in the biofilm community from exposure to antimicrobial drugs, antiseptics, and chemical disinfectants, making it difficult to treat infections caused by microbes living in biofilms. The polysaccharide matrix also has channels running through it that the microbes can use to share nutrients, share genetic information (like antibiotic resistance genes), and communicate.

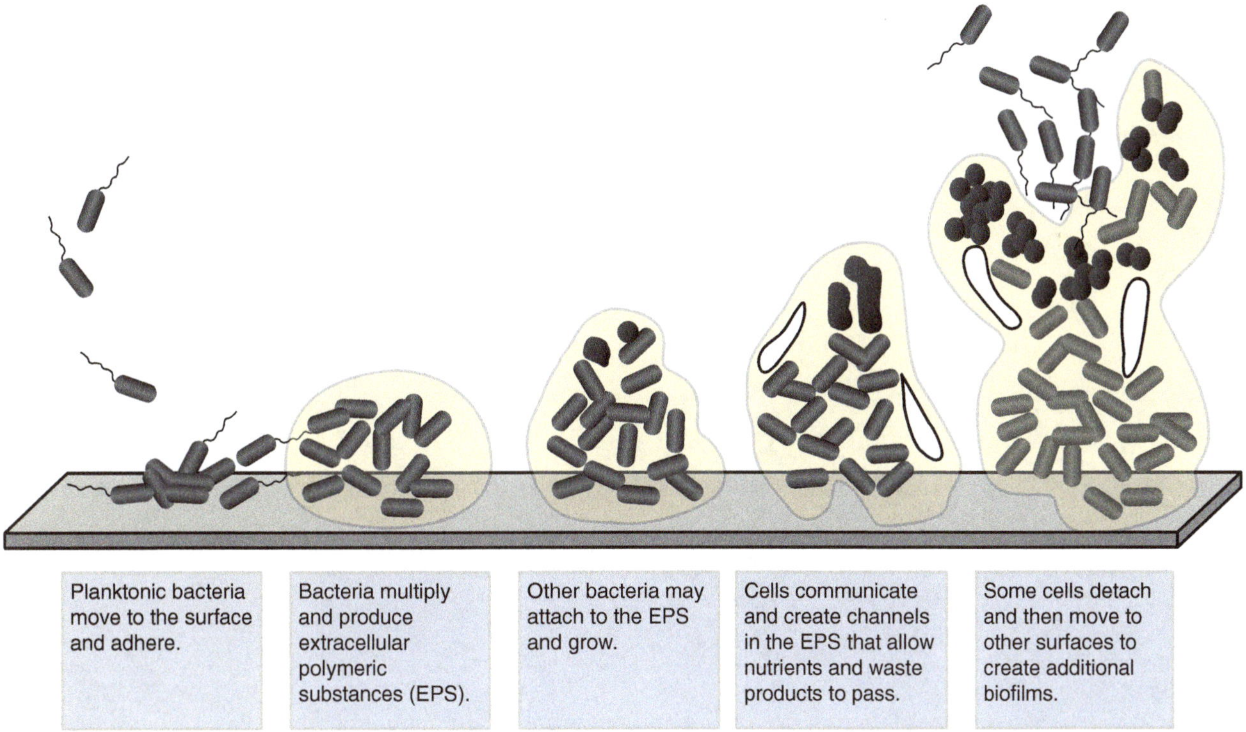
Planktonic bacteria move to the surface and adhere.
Bacteria multiply and produce extracellular polymeric substances (EPS).
Other bacteria may attach to the EPS and grow.
Cells communicate and create channels in the EPS that allow nutrients and waste products to pass.
Some cells detach and then move to other surfaces to create additional biofilms.

Fig. 3.10 Biofilm formation

Lab Exercise 3.1

Smear Preparation, Gram Staining, and Laboratory Diagnosis

Student Objectives

1. Describe how to aseptically transfer a bacterial sample and inoculate a slide with the sample.
2. Perform a bacterial smear from a liquid and a solid culture. Know the step-by-step smear preparation procedures.
3. Understand the reasons for employing stains in the laboratory.
4. Evaluate the steps in Gram stain procedure, and the advantages and disadvantages of procedure.
5. Identify the terms primary stain, mordant, decolorizer, and secondary stain.
6. Evaluate the medical importance of the Gram stain (i.e., the types of information that clinicians gain through Gram staining.)
7. Describe the structure of the cell wall and the differences between the Gram-positive and Gram-negative cell walls.
8. Explain Gram staining patterns and basic morphology for the microorganisms used in this laboratory exercise.

INTRODUCTION

The **Gram stain** is a very useful stain for identifying and classifying bacteria. The Gram stain is a differential stain that allows you to classify bacteria as either Gram-positive or Gram-negative. It works by taking advantage of the differences in cell wall structure between Gram-positive and Gram-negative bacteria. The Gram staining technique was discovered by Hans Christian Gram in 1884, when he attempted to stain cells and found that some lost their color when excess stain was washed off. The staining technique consists of the following steps (Table 3.1):

1. Apply **primary stain** (crystal violet). All bacteria are stained purple by this basic dye.
2. Apply **mordant** (Gram's iodine). The iodine combines with the crystal violet in the cell to form a crystal violet-iodine complex (CV-I).

Table 3.1 Gram Stain Procedure

Reagents	Time Applied	Reactions	Appearance
Unstained smear			Cells are colorless and difficult to see. 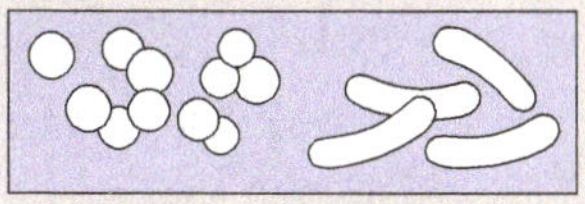
Crystal violet	1 minute, then rinse with water	Crystal violet penetrates all cells and stains them violet. This is a simple stain at this point.	Both gram-negative and gram-positive cells are deep violet. 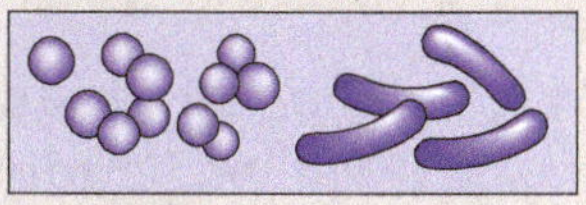
Gram's Iodine (mordant)	1 minute, then rinse with water	Iodine is a mordant and attaches to the crystal violet molecule and creates a large structure referred to as the CV-I (crystal violet-Iodine complex).	Both gram-negative and gram-positive cells remain deep violet. 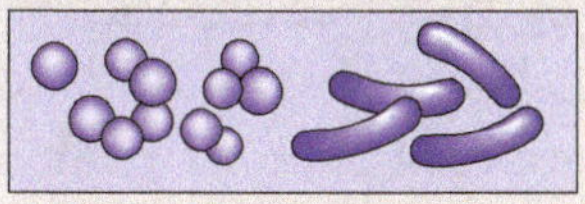
Alcohol or acetone-alcohol mix (decolorizer)	10 to 15 seconds, then rinse with water	Decolorizer leaches the crystal violet and iodine from the cells. The color diffuses out the gram-positive cells more slowly than out of gram-negative cells because of the chemical composition and thickness of the gram-positive cell walls. Gram negative cells have a thin peptidoglycan layer therefore the CV-I is rather easy to remove.	Gram-positive cells remain deep violet, but gram-negative cells become colorless and difficult to see. 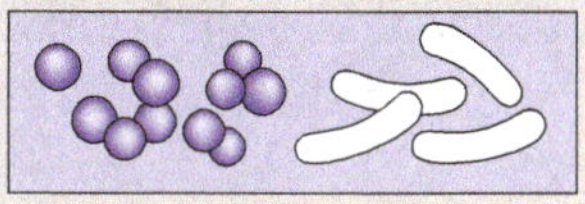
Safranin (counterstain)	1 minute; then rinse throughly, blot dry, and observe under oil immersion	Safranin is added and penetrates into the colorless Gram negative cells and become pink. Gram positive cells are stained with safranin as well, however, the crystal violet is the more dominate color.	Gram-positive cells remain deep violet, whereas gram-negative cells are stained pink or red. 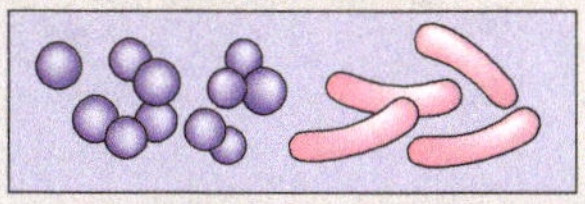

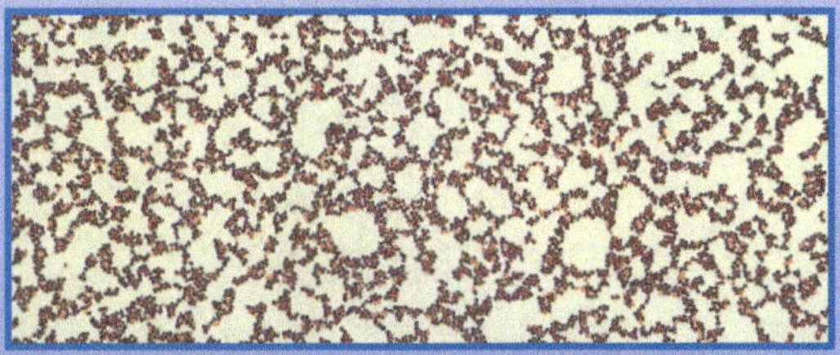

Gram stain of *Staphylococcus aureus,* a gram-positive coccus (x252). *Daniel Lim.*

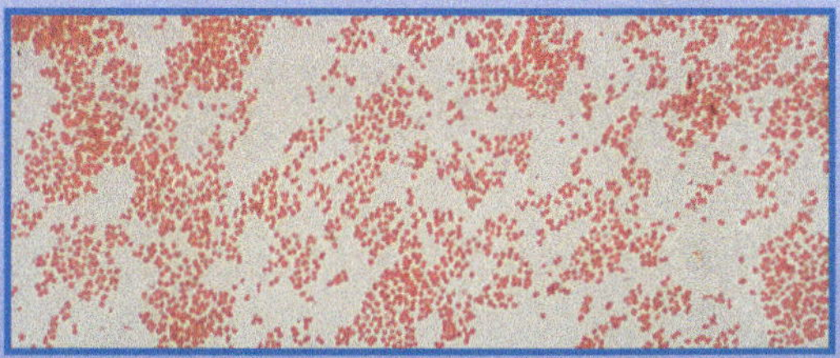

Gram stain of *Neisseria flavescens,* a gram-negative coccus (x252). *CDC.*

3. Apply **decolorizing agent** (ethyl alcohol or ethyl alcohol-acetone). The primary stain is washed out (decolorized) of the cell walls of Gram-negative bacteria.
4. Apply **secondary stain** or **counterstain** (safranin). This basic dye stains the decolorized bacteria pink.

Why It Works

The CV-I complex is larger than the crystal violet or iodine molecules that initially entered the cell and cannot pass through the thick peptidoglycan layers during decolorization. In Gram-negative cells, the alcohol dissolves the outer **lipopolysaccharide layer**, and the CV-I washes out though the thin layer of peptidoglycan. All cells will appear purple until the decolorization step which is where they differentiate. Gram-negative cells which are no longer stained after the addition of ethanol will become pink when the safranin is added. Gram-positive cells which remain purple after decolorization, will also be stained with the pink safranin stain but appear purple because the purple color dominates the pink stain.

The Gram stain is most consistent when done on young bacterial cultures (<24 hours old). As the cells age, cell wall structures deteriorate and aren't able to "hold on" to the CV-I complex as well, making the Gram stain results variable and difficult to interpret.

Carrying Out the Procedure

Since Gram staining is usually the first step in identifying bacteria, the procedure should be memorized as it will be performed many times throughout the semester.

We will practice the procedure using both Gram-positive and Gram-negative bacteria. **Gram-positive bacteria appear purple on our slides and Gram-negative bacteria appear pink.** On both of your slides (you're doing a "back-up"), you will want to be able to observe a Gram stain of each of your bacteria by itself, plus a Gram stain of the two bacteria together. Observing the bacteria separately helps you to understand the characteristics of the individual bacterium: if it's Gram-positive or Gram-negative, its shape (bacillus or coccus), and its arrangement (clusters like grapes, chains like a necklace, etc.). Observing the bacteria together helps you to recognize and differentiate between Gram-positives and Gram-negatives. Both experiences have value!

The first part of *any* staining procedure is the smear preparation (smear prep). The smear prep accomplishes two things: it adheres the sample to the slide, so that the cells are not washed away during staining, and it kills the bacteria.

APPLICATION

Uses of Gram Stains

(1) The Gram stain is the most frequently used bacterial test in the healthcare profession.

Usually when a specimen is taken from a patient, the Gram stain is part of the battery of tests that are run on that specimen. All microbiologists (clinical, academic, industrial, etc.) use the Gram stain in the identification process of a microorganism. The main resource for identification of an unknown or suspected bacterium is *Bergey's Manual of Determinative Microbiology*. You will use this same manual later in the semester to identify an unknown bacterial specimen.

(2) The second vital piece of information gained through the Gram stain deals with patient treatment regimens.

Antibiotics work upon a bacterial cell by interfering with a vital process of the cell. The drug may work by disrupting the cell membrane, inhibiting synthesis of the bacterial cell wall, inhibiting the production of a needed enzyme, or inhibiting DNA replication. Antibiotics which inhibit cell wall synthesis usually do so by inhibiting the synthesis of a specialized layer of the cell wall known as the peptidoglycan layer. The results of a bacterial Gram stain particularly help in determining the type of antimicrobial to use on the patient.

(3) The third vital piece of information gained through the Gram stain involves information about the type of infection present.

When a clinical microbiologist does a Gram stain, it is a direct Gram smear of the patient's specimen. This means it will contain more than just the microorganism causing the infection. There could be normal body cells (often epithelial), normal microbiota (bacteria not suspected of causing the infection), and cells of the immune system (most commonly neutrophils and macrophages). To the clinician, this gives additional information about the type of infection. In a smear of cerebrospinal fluid, large numbers of macrophages indicates viral or fungal meningitis, but if the smear contains large numbers of neutrophils, this indicates bacterial meningitis.

Smear Prep Activity

Materials

Laboratory materials: Clean flat microscope slides, inoculating loop, inoculating needle, sterile saline, clothes pin, Bunsen burner, igniter, and sterile swab.

Cultures: Broth cultures of *E. coli* and *Staphylococcus aureus*, mixed broth culture of *E. coli* and *Staphylococcus aureus*.

Use Aseptic Transfer and Inoculation Procedures from the Previous Lab Exercise

Prepare all four of your smears using four clean slides prior to performing the Gram stains. Doing them one at a time will not allow you to finish the lab on time.

Part 1: **Making smears from liquid media**

1. Label one slide "EC" for *E. coli*, another slide "SA" for *S. aureus*, and one "mix" for *E. coli* and *S. aureus*. Prepare a smear from each of the broth cultures using an **inoculating loop**.
2. Place the slide on the slide warmer. Remember that at this point the organism is still "alive" on the slide. Use care in transferring it back and forth from the slide warmer.
3. Use a clothespin and pass the slide through the middle of the flame five or six times. This procedure fixes the organism to the slide while killing it.

Now you are ready to stain.

Gram Stain Activity

Materials

Laboratory materials: Smears, clothes pin, crystal violet, Gram's iodine, 95% ethyl alcohol, safranin, and distilled water.

Prepare a Gram stain of each of your heat-fixed smears. Stain over your laboratory sink!

1. Add the primary stain, crystal violet, to the slide and leave for **60 seconds**.
2. Rinse off the unbound crystal violet for a maximum of five seconds. Tilt the slide and gently squirt water above the smear so that the water runs down the slide over the smear.
3. Cover the smear with Gram's iodine for **60 seconds**. This is a mordant, or an agent that fixes the crystal violet to the bacterial cell wall.
4. Rinse off the iodine by tilting the slide and gently squirting water above the smear so that the water runs over the smear.
5. Decolorize with 95% ethyl alcohol. Let the alcohol run through the smear until no more purple color drips off the slide into the sink (usually ~**10 seconds**). Decolorize each slide individually and rinse immediately with distilled water (3–5 seconds). Since ethanol is colorless it is easy to forget the wash step and over-decolorize your slide. **NOTE:** The degree of alcohol decolorizing depends on the thickness of the smear. This step is critical. Try not to over-decolorize. Experience is the only way you will be able to determine how long to decolorize.
6. Add the secondary stain, safranin, to the slide and leave for **60 seconds**.
7. Gently rinse with distilled water for a maximum of five seconds.
8. Blot the slide dry with bibulous paper. Be sure not to rub your smear!
9. Examine the stained slide microscopically using first the low, then the high-dry, and finally the oil immersion objectives. Record your observations. Do they agree with those given in your textbook? If not, try to determine why. Some of the most common sources of Gram stain errors are forgetting a step (such as heat fixing or adding iodine) or decolorizing too long. Additionally, the age of the cells and condition of the cell walls are important for consistent results.

We don't expect perfect results for your first try at Gram staining, but it is important to recognize and explain errors that may have given you unexpected results. Three of the cultures you are using in this lab are pure, so your instructor will know the expected results for each of these. **To receive the highest score for the lab you should look up the expected morphology, arrangement, and Gram reaction for all three pure cultures.** Any variation in the stains you prepared should be explained in question #2 on the back page of your lab report. Your body swab culture may have more than one bacterium and perhaps more than one Gram reaction. Record the results you see. Not all cells will stain perfectly so it is important to look around the slide to get a good overview of the results and be sure to look where cells are nicely scattered. Cells that are spread out are more likely to provide the most representative Gram reaction due to proper contact time with the stains and decolorizer, and will provide the best indication of the arrangement of the cells.

Name ______________________________

Exercise 3.1 Lab Report

1. **Gram stain reaction**: Sketch a few bacteria under 1,000X magnification. Identify whether it is G+ or G− and describe the morphology (shape) and arrangement of cells relative to one another.

Broth Cultures:

Staphylococcus aureus

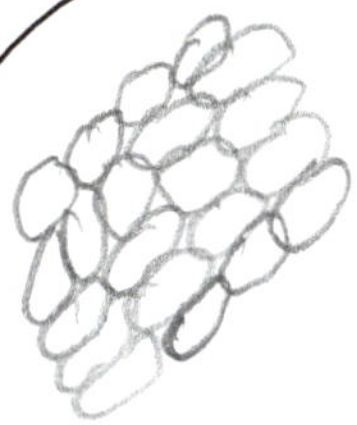

Gram reaction ________________
Morphology ________________
Arrangement ________________

Escherichia coli

Gram reaction ________________
Morphology ________________
Arrangement ________________

Mixed

bacilli + cocci

Gram reaction ________________
Morphology ________________
Arrangement ________________

100x } clean
40x

3 lens paper
dry wet dry

2. Upon investigating your mixed sample under a microscope, describe what you discovered.

____

____

3. List the reagent/stain used Gram staining procedure in order (omit washings) and fill in the ***color or appearance*** of Gram-positive cells and Gram-negative cells after each step.

Step	Chemical Used	Appearance of Gram-positive cells	Appearance of Gram-negative cells
1			
2			
3			
4			

4. Why do Gram Negative cells lose the CV-I?

____

____

5. Why do Gram positive cells retain the CV-I?

____

____

6. Explain why old Gram-positive cells might stain Gram-negative.

____

____

7. From a *pure bacterial culture* (not contaminated!), suppose you observed a field of pink and purple cocci. Cells were spread out evenly on the slide and adjacent cells were not always the same color. What ***two*** things can you conclude about this culture?

____

____

8. Suppose you are viewing a Gram stained field of pink rods and purple cocci through the microscope. What can you conclude about this culture?

9. Suppose two patients are in the hospital with bacterial infections. One has been diagnosed with a Gram-positive infection, while the other has a Gram-negative bacterial infection. Specifically, what do you know about each type of bacteria before you provide antibiotic treatment for each of these individuals? What antibiotic treatment would you use for each?

   a. Gram-positive infection:

   b. Gram-negative infection:

10. Suppose it is Friday afternoon. You're tired and it's been a long week. The doctor gives you a specimen for Gram stain identification. You decide to hold the specimen and do the stain procedure on Monday morning. What effect might this decision have upon the successful treatment of the patient?

11. What is the difference between a simple stain and a differential stain?

Lab Exercise 3.2

Endospores and Hypothesis Testing

Student Objectives

1. Understand the structure and sporulation/germination cycle of an endospore.
2. Identify common endospore-forming genera and the diseases associated with them.
3. Understand the clinical importance of endospore-forming bacteria.
4. Identify common diseases endospore-forming bacteria can cause.
5. Identify reasons why endospores are formed.
6. Construct and test a hypothesis.
7. Analyze results and accept or reject a hypothesis.

Name ____________________

Pre-Lab Exercise 3.2

1. What is a vegatative cell?

2. What is sporulation and about how long is the process?

3. Under what conditions are endospores formed?

4. What is Germination?

5. A **hypothesis** is a proposed explanation for a natural phenomenon. A good hypothesis should be written to limit variables for testing.

   Form a hypothesis for each of the two experiments:

   Heat:

   E. coli ____________________

   B. subtilis ____________________

   Freezing:

   E. coli ____________________

   B. subtilis ____________________

6. Which broad group of bacteria are capable of forming endospores?

7. Under what conditions do certain microorganisms form endospores?

____

____

8. How does the ability to form an endospore contribute to the ability of a microbe to cause disease?

____

____

9. Why is the ability of an organism to form an endospore significant for people who can fruits and vegetables at home?

____

INTRODUCTION

When essential nutrients are depleted from an environment, certain Gram-positive bacteria, such as those of the genera *Clostridium* and *Bacillus*, form specialized "dormant" cell structures called **endospores**. Some organisms from these two genera are significant in the medical field since they are known to cause disease. Species in the genus *Clostridium* are known to be the causative agent for diseases such as gangrene, tetanus, and botulism. Species in the genus *Bacillus* cause anthrax and food poisoning. Endospores are unique to bacteria and they are highly durable with thick walls and additional layers to help protect them from varying environmental factors. The endospore is formed internal to the bacterial cell membrane. A similar structure to the endospore can be formed by the Gram-negative organism *Coxiella burnetii*; it is not, however, a true endospore. Only certain Gram-positive organisms are known to form endospores.

In this lab exercise, *Bacillus subtilis* and *Escherichia coli* will be tested for the ability to resist the effects of extreme temperatures.

Testing the Effectiveness of Heat

Materials

Laboratory materials per table: Two Tryptic Soy agar plates, inoculating loop, Bunsen burner and hot plate.

Cultures: *Bacillus subtilis* liquid culture and *Escherichia coli.*

1. Construct a hypothesis for this experiment and record it in your pre-lab report.
2. A boiling water bath will be placed at your lab bench. Cultures with *B. subtilis* and *E. coli* will be provided.
3. Two TSA plates will be needed: one for *E. coli* and one for *B. subtilis*. Divide your plate into fifths, as seen below.
4. Place both culture tubes into the boiling water. Offset the time you place them in the water bath to allow time for streak plating.
5. TSA plates will be streaked at Time 0, Time 1 minute, Time 2 minutes, Time 3 minutes, and Time 5 minutes. The total amount of time the test tubes should be in the water is five minutes. When streaking in each section, do a single streak back and forth towards the middle of the plate (See image below).

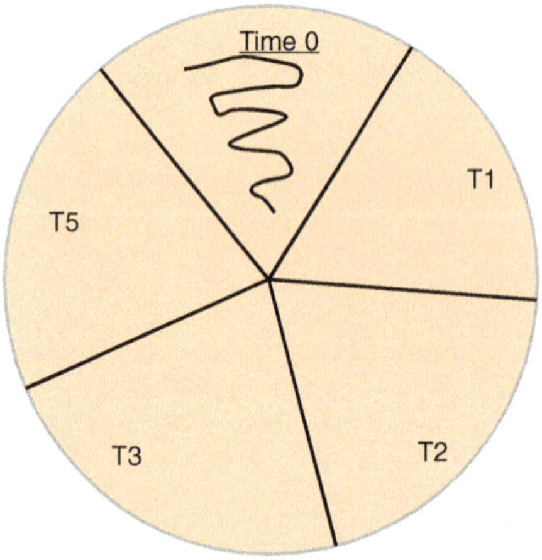

Courtesy of Authors

Testing the Effectiveness of Freezing

Materials

Laboratory materials per table: One Tryptic Soy agar plates,inoculating loop, Bunsen burner.

Frozen Cultures: *Bacillus subtilis* liquid culture and *Escherichia coli.*

1. Construct a hypothesis for this experiment and record it in your pre-lab report.
2. Cultures of *B. subtilis* and *E. coli* have been frozen in TSB. Allow the cultures to fully thaw and then streak them on separate sides of a TSA plate.

Name ______________________________

Exercise 3.2 Lab Report

Record your observations below for each organism at each time point following incubation (use + for little growth, ++ for medium growth, and +++ for heavy growth).

B. subtilis	**Amount of Growth**
Time 0	
Time 1 minute	
Time 2 minutes	
Time 3 minutes	
Time 5 minutes	
Freezing	

1. Were your predictions for *B. subtilis* correct? Explain.

____

____

E. coli	**Amount of Growth**
Time 0	
Time 1 minute	
Time 2 minutes	
Time 3 minutes	
Time 5 minutes	
Freezing	

2. Were your predictions for *E. coli* correct? Explain.

____

____

3. What can you conclude about *E. coli's* ability to form an endospore? Explain how you came to this conclusion.

____

____

4. What can you conclude about *B. subtilis'* ability to form an endospore? Explain how you came to this conclusion.

____

____

5. List the conditions endospores are resistant to:

____

____

6. Explain the endospore cycle: use the terms vegetative cell, sporulation, and germination in your answer.

____

____

____

____

7. When Bacillus is placed in the boiling water does that trigger sporulation? When do you think the endospores were formed?

____

____

____

8. When you sampled Bacillus from your boiled test tube and plated the sample on TSA, explain what is happening if you observe growth.

____

____

____

9. Did you accept or reject your hypothesis?

____

10. What are the clinically significant genera of endospore forming bacteria?

____

11. Explain the significance of the following organisms and provide details on the diseases they cause:

    a. Clostridium tetani:

    ____

    ____

    b. Clostridium difficile:

    ____

    ____

    c. Clostridium perferinges:

    ____

    ____

    d. Bacillus anthracis:

    ____

    ____

12. Why is the ability of an organism to form an endospore significant for bioterrorism?

____

____

Lab Exercise 3.3

Acid-Fast Stain

Student Objectives

1. Understand the structure of an acid-fast cell wall.
2. Identify common acid-fast genera and the diseases associated with them.
3. Explain the chemical basis for each step in an acid-fast stain.
4. Perform the acid-fast staining procedure.
5. Explain acid-fast staining patterns and differentiate between the acid-fast and non-acid-fast microorganisms used in this laboratory exercise.

INTRODUCTION

Acid-fast organisms are generally Gram-positive bacilli with an unusual cell wall structure. In addition to the traditional thick layers of peptidoglycan, their cell wall contains a waxy lipid material called mycolic acid. Mycolic acid allows the bacteria to resist various chemicals, and contributes to their pathogenicity.

The **acid-fast staining** procedure is important as a diagnostic test for patients suspected of having tuberculosis (*Mycobacterium tuberculosis*). The test can be run on sputum samples from patients with respiratory symptoms. During the staining procedure, heat is used to weaken the mycolic acid, allowing the primary stain to penetrate the cell wall. The waxy lipid is not affected by the decolorizing agent acid alcohol, so these organisms are said to be "acid-fast." All other microorganisms, which are more easily decolorized, are non-acid-fast.

Set-up and Smear Preparation

Materials

Laboratory materials: Four microscope slides, inoculating loop, Bunsen burner and sparker, two small cans for staining.

Cultures: *Mycobacterium smegmatis* solid culture, *Serratia marcescens* solid culture

1. Each table should obtain one hot plate. Fill two small cans ¾ th full with tap water, and place on the hot plate to begin warming. The water should release steam, but need not be boiling during the staining procedure.
2. Create smear preparations of *Mycobacterium smegmatis* and *Serratia marcescens*, following the "Smear Preparation from a Solid Culture" procedure described in *Laboratory Exercise* 5. Remember to make two preparations for each organism; one slide will be your backup. *Mycobacterium smegmatis* will be your acid-fast positive organism, and *Serratia marcescens* will be your non-acid-fast (control) organism.

Ziehl-Neelsen Acid-Fast Stain

Materials

Laboratory materials: Acid-fast stain reagents: carbol fuchsin, acid-alcohol, and methylene blue, bibulous paper, microscope and immersion oil

Cultures: *Mycobacterium smegmatis* and *Serratia marcescens* smear preparations

This staining technique consists of:

a. A *primary stain*, carbol fuchsin, applied over steam to weaken the cell wall and allow the stain to enter the cytoplasm; both acid-fast and non-acid-fast organisms are stained red.
b. A *decolorizing agent*, acid-alcohol, to remove the dye from non-acid-fast cells.
c. A *secondary stain*, methylene blue, to color the non-acid-fast cell walls blue.

1. Place the prepared slides over the steaming juice can.
2. Flood the slides with carbol fuchsin over steam for five minutes. Closely observe your slides during this time and add additional carbol fuchsin as necessary to keep the slides moist as the stain appears to dry.
3. Remove the slides from the steaming juice can and allow them to cool on the sink staining bar for several minutes.
4. Holding the slides at an angle directed into the sink, carefully drop acid-alcohol (not ethyl alcohol) onto the slides for 15–20 seconds.
5. Rinse the slides with distilled water for 30 seconds.
6. Flood the slides with methylene blue and allow the dye to integrate for 30 seconds.
7. Rinse the slides with distilled water for 30 seconds.
8. Blot the slides dry with bibulous paper.
9. Observe the acid-fast stain slides under the microscope, working your way up to the oil immersion lens. Note the color of the acid-fast positive as compared to the non-acid-fast cells.

APPLICATION

Sterilization of Acid-Fast Bacteria

The mycolic acid in acid-fast bacteria makes them resistant to low-level disinfectants and quaternary ammonium compounds (QUATS). However, most powerful chemical and physical treatments can effectively destroy acid-fast bacteria.

Diseases Caused by Acid-Fast Organisms

Tuberculosis (TB)

Mycobacterium tuberculosis is the causative agent of TB. The bacteria are Gram-positive, endospore-negative, nonmotile, obligate aerobes that grow well in the oxygen-rich upper lobes of the lungs. The bacilli-shaped cells often lie in close parallel rows, referred to as a *palisade arrangement*, which increases their virulence. This species grows slower than many other bacteria, with a generation time of 12–20 hours.

Tuberculosis is an airborne infection spread by coughing and sneezing. The infection is latent in most infected individuals; only about 10% of people harboring *M. tuberculosis* actually develop the acute form of the disease.

Treatment of tuberculosis involves a six-month regimen employing several antibiotics to avoid resistance. However, newer, more resistant forms of TB require a longer, more aggressive course of therapy.

The tuberculin skin test, also known as purified protein derivative (PPD), is required annually of healthcare workers. The PPD is injected into the dermis, causing a *bleb*, or small raised bump, to develop in the skin. A zero to five millimeter diameter bleb indicates a negative reaction, while a 10 millimeter or larger diameter bleb indicates TB exposure. However, a positive PPD test does not distinguish between present infection, past infection, or TB vaccination. (For this reason, the Bacille Calmette-Guerin, or BCG, vaccinating agent is not routinely used in the United States, as use of the vaccine would conflict with the PPD screening tool). A positive reaction to the PPD test is followed by a confirmatory sputum culture and chest x-ray.

MAC Disease

MAC disease, or *Mycobacterium avium* Complex Disease, is caused by an organism present in almost every environment. These bacteria are found in water, dust, soil, and bird droppings. They may enter the body through food, water, and sometimes through the lungs. Symptoms of MAC disease include weight loss, fever, chills, night sweats, swollen glands, abdominal pains, diarrhea, and overall weakness, which are more severe in immunocompromised individuals. MAC bacteria exposure can be reduced by boiling water, peeling fruits and vegetables, avoiding raw food consumption, and reducing contact with animals, especially birds and bird droppings.

Leprosy (Hansen's Disease)

Mycobacterium leprae is the species responsible for causing leprosy, also known as Hansen's disease. The disease is found primarily in tropical regions, but is increasing in prevalence in the United States. Leprosy is treatable with sulfone drugs used in combination with rifampin.

Name ______________________________

Exercise 3.3 Lab Report

1. Explain the cell wall structure of acid-fast bacteria, and explain what advantages this structure confers to the microorganism.

   ____

   ____

   ____

   ____

2. Sketch your observations of acid-fast and non-acid-fast bacteria. Pay close attention to the differences in color between the two species, and be sure to use colored pencils in your drawings.

ACID-FAST	NON-ACID-FAST
Microorganism: ________________	Microorganism: ________________
Total magnification: ____________	Total magnification: ____________

3. Describe the color of the acid-fast positive organisms. ______________________

   Is this the color of the primary or the secondary stain? ______________________

   Describe the color of the acid-fast negative organisms. ______________________

   Is this the color of the primary or the secondary stain? ______________________

4. Why must heat be applied during the acid-fast staining procedure?

5. Why is the term "acid-fast" used?

6. Name the effective control methods for destroying acid-fast organisms.

7. Describe the diseases caused by each of the following acid-fast microorganisms:

   Mycobacterium tuberculosis: ______

   Mycobacterium avium: ______

   Mycobacterium leprae: ______

Lab Exercise 3.4

Endospore Stain

Student Objectives

1. Understand the structure and sporulation/germination cycle of an endospore.
2. Identify common endospore-forming genera and the diseases associated with them.
3. Perform a bacterial smear preparation from a solid culture.
4. Explain the chemical basis for each step in an endospore stain.
5. Perform the endospore staining procedure.
6. Explain endospore staining patterns and differentiate between the endospore and vegetative cell forms used in this laboratory exercise.

INTRODUCTION

The **endospore stain** identifies bacteria that are capable of entering a dormant state in response to physiologic and environmental stressors. In this laboratory exercise you will perform an endospore stain to differentiate between the endospore and vegetative cell forms.

Set-up and Smear Preparation

Materials

Laboratory materials: Four microscope slides, inoculating loop, Bunsen burner and sparker, two small cans for staining

Cultures: *Alcaligenes faecalis* solid culture, *Bacillus subtilis* solid culture

1. Each table should obtain one hot plate. Fill two small cans 3/4th full with tap water, and place on the hot plate to begin warming. The water should release steam, but need not be boiling during the staining procedure.
2. Create smear preparation of *Bacillus subtilis* and *Alcaligenes faecalis*, following the "Smear Preparation from a Solid Culture" procedure described in *Laboratory Exercise 5*. Remember to make two preparations for each organism; one slide will be your backup. *Bacillus subtilis* will be your endospore-positive organism, and *Alcaligenes faecalis* will be your endospore-negative (control) organism.

Schaeffer-Fulton Endospore Stain

Materials

Laboratory materials: Endospore stain reagents: malachite green and safranin, bibulous paper, microscope and immersion oil.

Cultures: *Alcaligenes faecalis* and *Bacillus subtilis* smear preparations

This staining technique consists of:

a. A *primary stain*, malachite green, applied over steam to facilitate penetration of the dye through the tough endospore coat. Both endospores and vegetative cells are stained green.
b. A *decolorizing agent*, water, to wash excess dye from the slide and remove the dye from vegetative cell components.
c. A *secondary stain*, safranin, to color the vegetative cell walls pink.

1. Place the prepared slides over the steaming juice can.
2. Flood the slides with malachite green over steam for 15 minutes. Closely observe your slides during this time and add additional malachite green as necessary to keep the slides moist as the stain appears to dry.
3. Remove the slides from the steaming juice can and allow them to cool on the sink staining bar for several minutes.
4. Rinse the slides with distilled water for one minute.
5. Flood the slides with safranin and allow the dye to integrate for one minute.
6. Rinse the slides with distilled water for one minute.
7. Blot the slides dry with bibulous paper.
8. Observe the endospore stain slides under the microscope, working your way up to the oil immersion lens. Note the size, shape, and color of the endospores as compared to the vegetative cells.

APPLICATION

Sterilization of Endospores

Endospores are of particular concern in healthcare settings and the food industry because they are resistant to control by common physical and chemical methods. Endospores are not destroyed by heat (including boiling), ultraviolet radiation, antimicrobial agents, and most chemical disinfectants. Evidence exists that endospores may survive and re-enter the vegetative state after 40 million years!

The best way to achieve sterilization is through autoclaving, or the use of moist heat under pressure. Autoclaving at 121°C with 15 PSI for 15–20 minutes destroys endospores, and is the method of choice for sterilizing critical instruments.

Name ______________________________

Exercise 3.4 Lab Report

1. Define the terms "endospore" and "vegetative cell."

____

____

____

2. Explain the processes of sporulation and germination in the life cycle of an endospore-positive microorganism.

____

____

____

____

3. Sketch your observations of endospore-positive and endospore-negative bacteria. Pay close attention to differences in size, shape, and color between the endospores and the vegetative cells. Be sure to use colored pencils in your drawings.

ENDOSPORE-POSITIVE

Microorganism: ____________________

Total magnification: ________________

ENDOSPORE-NEGATIVE

Microorganism: ____________________

Total magnification: ________________

4. Describe the color and size of the endospores. ______________________________

   Is this the color of the primary or the secondary stain? ______________________

   Which structure does this dye stain? ______________________________

   Describe the color and size of the vegetative cells. ______________________________

   Is this the color of the primary or the secondary stain? ______________________

   Which structure does this dye stain? ______________________________

5. Why must heat be applied during the endospore staining procedure?

   ____

   ____

   ____

6. Name the most effective control method for destroying endospores.

   ____

7. Describe the diseases caused by each of the following endospore-forming microorganisms:

   Bacillus anthracis: ______________________________

   ____

   Clostridium botulinum: ______________________________

   ____

   Clostridium difficile: ______________________________

   ____

   Clostridium perfringes: ______________________________

   ____

   Clostridium tetani: ______________________________

   ____

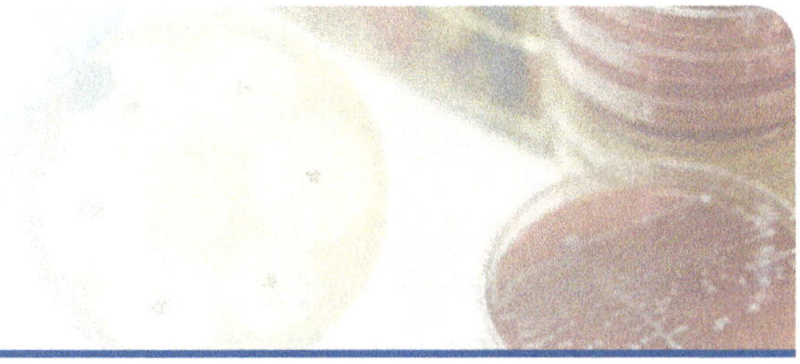

Lab Exercise 3.5

Capsule and Flagella Stains

Student Objectives

1. Understand the structure of a capsule.
2. Explain the role of the capsule in bacteria.
3. Evaluate the pathogenicity of the bacterial capsule.
4. Understand the structure of a flagellum.
5. Explain the role of flagella in bacteria.
6. Evaluate the pathogenicity of the bacterial flagellum.

INTRODUCTION

Capsules

Many bacteria secrete chemicals that adhere to their surfaces, forming a viscous coat. This structure is called a **capsule** when its shape is round or oval, and a **slime layer** when it is irregularly shaped and loosely bound to the bacterium. The ability to form capsules is genetically determined, but the size of the capsule is influenced by the medium on which the bacterium is growing. Most capsules are composed of polysaccharides, which are water-soluble and uncharged. Therefore, simple stains will not adhere to them. Most capsule-staining techniques stain the bacteria and the background, leaving the capsules unstained. Essentially, this is a "negative" capsule stain (see below).

Flagella

Many bacteria are motile, which means they have the ability to move in a directed manner. Most motile bacteria possess flagella, but other forms of motility occur. *Myxobacteria* exhibit gliding motion, and spirochetes use axial filaments to create an undulating motion.

Flagella, the most common structure used for motility, are thin proteinaceous structures that originate in the cytoplasm and project out from the cell wall. They are very fragile and are not visible with a light microscope. They can be stained after coating them with a mordant, which increases their diameter. The staining process will often cause several flagella to clump together giving the appearance of one large flagellum. The presence and location of flagella are helpful in the identification and classification of bacteria. Flagella are of two main types: **peritrichous** (all around the bacterium) and **polar** (at one or both ends of the cell).

Motility may be determined by flagella stains, observing hanging-drop preparations of unstained bacteria, or inoculation of soft (or semisolid) agar.

Capsule and Flagella Activity

Materials

Laboratory materials: Capsule stain slide and flagella stain slide.

1. Observe the capsule stain slide under the microscope, working your way up to the oil immersion lens.
2. Observe the flagella stain slide under the microscope, working your way up to the oil immersion lens.
3. Draw and explain your observations in the lab report.

Name ______________________________

Exercise 3.5 Lab Report

1. Explain the structure of a capsule, and explain what advantages this structure confers to the microorganism.

____

____

____

2. Explain the structure of a flagellum, and explain what advantages this structure confers to the microorganism.

____

____

____

3. Sketch your observations of the capsulated bacteria and bacteria with flagella.

CAPSULE

Total magnification: _______________

FLAGELLA

Total magnification: _______________

4. How might a capsule contribute to pathogenicity?

____

____

____

5. How might flagella contribute to pathogenicity?

____

____

____

Lab Exercise 3.6

Biofilm and Microbial Communities

Student Objectives

1. Evaluate the role of bacterial capsules, slime, and biofilms.
2. Recognize that biofilms grow in many common household serttings.
3. Demonstrate that when cells are removed from a biofilm and suspended in water, they can attach to a new surface to produce a new biofilm even if the new growth conditions are very different from conditions at the original biofilm site.
4. Recognize that microbes often grow in complex, diverse aggregates that contain more than one type of organism.
5. Recognize that biofilm growth is very different from the growth of pure cultures of floating cells in broth or colonies on agar plates.
6. Work cooperatively in groups to discuss their results and to design possible future experiments to investigate environmental factors that influence biofilm growth.

Name ______________________________

Pre-Lab Exercise 3.6

1. Define biofilm.

2. What are some of the advantages of living in a biofilm community?

3. What are some of the disadvantages of living in a biofilm community?

4. Describe how periodontal diseases are related to dental plaque. Include examples of bacteria that cause this disease.

5. What will you be looking for when you observe your biofilm using the microscope?

INTRODUCTION

Traditionally, microorganisms have been studied in pure cultures, just as we have been doing in the laboratory. Recently, however, scientists have begun to realize and study the impact of biofilms. **Biofilms** are complex aggregates of microbes—prokaryotes and eukaryotes—that live in polysaccharide-encased communities. These communities may be suspended in aqueous environments or attached to surfaces. Biofilms are responsible for many annoying conditions such as slime in sink drains and toilets, but they are also responsible for many types of infections. Dental plaque is the result of biofilm formation. Patients who receive implants (artificial hips or heart valves) or have in-dwelling devices such as venous or urinary catheters, may get infections caused by bacteria growing on/in these devices in the form of biofilms. These infections are difficult to treat because the polysaccharide slime encasing the microbes protects them from the effects of antibiotics. Biofilms, like microorganisms in general, may also be beneficial. They are being used in wastewater treatment and their activities are necessary for nutrient cycling.

Biofilms can be formed by bacteria, archaea, and yeast. They can be homogeneous, meaning they are composed of only one type of organism, or heterogeneous—composed of a mixture of microbes. Biofilm formation begins when one or a few microbes attach to a surface by van der Waals forces (weak attractions between molecules). The microbes strengthen their attachment by using fimbriae, and then begin to produce the extracellular polysaccharide matrix—a sticky, slimy substance that allows them to adhere better and allows other cells from the surrounding fluid to adhere to the surface as well. The biofilm grows as attached cells go through binary fission and other cells from the environment are "recruited" to join the developing biofilm.

The polysaccharide matrix of the biofilm protects microbes from exposure to antimicrobial drugs, antiseptics, and chemical disinfectants, making it difficult to treat infections caused by microbes living in biofilms. The polysaccharide matrix also has channels running through it that the microbes can use to share nutrients, share genetic information (like antibiotic resistance genes), and communicate.

In this exercise we will perform a simple stain on our collection and explore some common biofilms. You will become familiar with the collection and cultivation of biofilms and be aware of the distinct appearances of different types of biofilms.

Week One: Collection of the Sample

Materials

Per student: Zip-top bag, 1 sterile swab, and a screw-cap vial containing 5 mL of sterile water.

You will receive a zip-top bag containing a sterile swab and a collection tube. Collect a biofilm sample from your body or household **within 24 hours of your next lab session**. Good collection sites that are likely to have biofilms include teeth (dental plaque), contact lenses, sink drains, toilet bowls, flower vases (with old flower water), dog dishes, bird baths, and swimming pool filters.

Surfaces that are **not submerged** in water will be easier to sample. To collect material from a biofilm that is not submerged:

- Moisten the swab with the sterile water and then rub the biofilm surface.
- Insert the swab into the vial of sterile water and stir the swab vigorously.
- Discard the swab.
- Place the cap on the vial.
- Place the vial in the zip-top bag for storage and transport.

To collect material from a biofilm that **is submerged** in water:

- Rub the biofilm with the swab.
- Stir the swab vigorously into the sterile water collection tube and then discard the swab.
- Place the vial in the zip-top bag for storage and transport.

Week Two: Culturing the Sample

Materials

Per student: 1 sterile Petri plate with 1 sterile microscope slide, 1 tube containing 10 mL of 5% trypticase soy broth (TSB).

Warning

Unidentified infectious pathogens may be present in the biofilm samples. Be particularly aware of the aseptic technique you have learned throughout the semester! Make sure that you use your PPE. Report any accidental spills or bodily contact with the biofilm fluid to your instructor immediately.

1. Obtain a Petri plate with a sterile microscope slide inside the plate.
2. Label both the top and bottom of the Petri plate, using care not to open the plate.
3. Aseptically open the Petri plate. Pour the entire contents of a tube of dilute 5% TSB inside the Petri plate, on top of the slide.
4. Pour the entire contents of your biofilm collection tube into the Petri plate. Your plate will be filled with both nutrient broth and your biofilm collection. Close your Petri plate. Do NOT tape your plate.
5. Carefully place your plate on a tray designated for incubation at the temperature that is closest to the temperature of your biofilm collection site (room temperature or 37°C). Be careful to avoid spilling as you carry your plate.
6. Dispose of the zip-top bag in the biohazard container. Place your screw-cap vial on the autoclave cart.

Week Three: Observation of the Sample

Materials

Materials per table: 500 mL beakers containing disinfectant, Gram's crystal violet, wash bottles containing distilled water, and PPE.

Reminder: **Your biofilm may contain pathogens—be aware of your PPE and aseptic technique!**

1. Retrieve your Petri plate from the incubation tray.
2. Gently pour any contents of your plate into a beaker that contains disinfectant, keeping the slide inside the Petri plate.
3. Rinse the Petri plate gently and thoroughly with a wash bottle in the beaker of disinfectant. Be careful to avoid disturbing the biofilm that has grown on your slide; direct the stream of water to the bottom of the plate but not right onto your slide.
4. Pour off all of the water into the beaker, using care not to drop the slide out of the plate.
5. Place your Petri plate on the staining rack. Add Gram's crystal violet stain inside the plate so that the entire surface of your slide is covered with dye. Let the dye set for **5 minutes**.
6. Pour off the dye into the beaker of disinfectant and rinse as in step 1.
7. Remove your slide from the plate with sterilized forceps. Wipe the bottom of the slide on a paper towel. Discard the towel in the biohazard container.
8. Place the slide on the slide warmer to dry. Sterilize the forceps and place on your table.
9. Discard the plate, your gloves, and all paper towels in a ***biohazard waste container***.
10. Wash your hands and re-glove.
11. While your slide is drying, read the questions in the Lab Report before you view your slide so that you can start thinking about your answers as you observe your biofilm.
12. After your slide is dry, remove the slide from the slide warmer and place on the microscope stage. Remove and discard gloves. Observe your biofilm with the **microscope** at each power of magnification. At high power:
    - Rotate the **fine focus** knob back-and-forth to observe the three-dimensional properties of your biofilm. Most biofilms will show several growth patterns such as scattered microcolonies (small isolated clumps that sometimes grow to considerable thickness) as well as smaller cell clusters. Look for these patterns on your slide.
    - Observe the **prokaryotes** that made this biofilm.
    - Determine whether **eukaryotic** cells (larger cells with a nucleus) are present.
    - *Observe slides that were prepared by your classmates.*
13. Draw your observations in the Lab Report. Answer the questions on your lab report pertaining to your biofilm analysis. Share your response to the questions in the lab report with the members of your lab group.
14. Place the used slides in the sharps container.
15. Use **alcohol pads** to **disinfect** the microscope stage and the slide warmer.

APPLICATION

Dental Plaque

Even if you have never heard the term "biofilm" before, you are familiar with them. Dental plaque is caused by biofilm formation from bacteria, like *Streptococcus mutans.* Plaque is the sticky, colorless film that constantly forms on your teeth. The microorganisms attach to the surface of the teeth and secrete acids that degrade the tooth enamel, resulting in the formation of dental caries.

Periodontal (gum) diseases, including gingivitis and periodontitis, are serious infections that, left untreated, can lead to tooth loss. The word *periodontal* literally means "around the tooth." Periodontal disease is a chronic bacterial infection that affects the gums and bone supporting the teeth. Periodontal disease can affect one tooth or many teeth. It begins when the bacteria in plaque cause the gums to become inflamed.

Gingivitis is the mildest form of periodontal disease. It causes the gums to become red, swollen, and bleed easily. There is usually little or no discomfort at this stage. Gingivitis is often caused by inadequate oral hygiene. Gingivitis is reversible with professional treatment and good oral home care. Untreated gingivitis can advance to periodontitis. With time, plaque can spread and grow below the gum line. Toxins produced by the bacteria in plaque irritate the gums. The toxins stimulate a chronic inflammatory response in which the body in essence turns on itself and the tissues and bone that support the teeth are broken down and destroyed. Gums separate from the teeth, forming pockets (spaces between the teeth and gums) that become infected. As the disease progresses, the pockets deepen and more gum tissue and bone are destroyed. Often, this destructive process has very mild symptoms. Eventually, teeth can become loose and may have to be removed.

The oral cavity contains some of the most varied and vast microbiota in the entire human body. Several diseases involve these two systems and manifest in the oral cavity.

In the healthy mouth, more than 350 species of microorganisms have been found. Periodontal infections are linked to fewer than 5% of these species. Healthy and disease-causing bacteria can generally be grouped into two categories:

- The harmless or helpful bacteria are usually Gram-positive bacteria.
- In periodontal disease, the bacterial balance shifts over to Gram-negative anaerobic bacteria. Inflammatory disease and injury cannot develop without these bacteria.

Actinobacillus actinomycetemcomitans and *Porphyromonas gingivalis* are some of the bacteria most implicated in periodontal disease and bone loss. These two bacteria appear to be particularly likely to cause aggressive periodontal disease. Both *P. gingivalis* and *A. actinomycetemcomitans*, along with multiple deep pockets in the gums, are associated with resistance to standard treatments for gum disease. *P. gingivalis* may double the risk for serious gum disease. *P. gingivalis* produces enzymes, such as one called arginine-specific cysteine proteinase, which may be the specific destructive factors that disrupt the immune system and lead to subsequent periodontal connective tissue destruction.

Bacteroides forsythus is also strongly linked to periodontal disease. Other bacteria associated with periodontal disease are *Treponema denticola*, *T. socranskii*, and *P. intermedia.* These bacteria, together with *P. gingivalis*, are frequently present at the same sites, and are associated with deep periodontal pockets. Some bacteria are related to gingivitis, but not plaque development. They include various streptococcal species.

An example of bacterial infection that may occur after dental surgery is *Actinomyces israelii.* This is a Gram-positive, slow-growing, non-motile, non-sporing, filamentous, facultative anaerobic bacterium. Actinomycosis is slowly progressive. There can be painful swellings (abscesses) under the skin, in the area of the jaw or neck, that eventually open and drain. The sores may heal, reappear, and drain for weeks, creating cycles of abscess formation and scarring. Proper care must be taken following dental or oral surgery to prevent this disease.

Increasing evidence indicates that oral microbiota participate in various systemic diseases. Periodontal disease permits organisms to enter deep systemic tissues, such as the carotid atheroma. An association between periodontal pathogens, such as *Porphyromonas gingivalis,* and atherosclerosis has been suggested because of the pathogen's possible direct effect on atheroma formation. *P gingivalis* has also been found in carotid and coronary atheromas. It may also invade and proliferate within heart and coronary artery endothelial cells, and, along with *Streptococcus sanguis,* it may also induce platelet aggregation associated with thrombus formation. Oral microorganisms may also enter the deeper tissue after trauma or surgery, which contributes to the disease process.

Periodontitis is a common chronic bacterial infection of the supporting structures of the teeth. The host response to this infection is an important factor in determining the extent and severity of the disease. Systemic conditions may modify the extent of periodontitis principally through their effects on normal immune and inflammatory mechanisms.

Cystic Fibrosis

Biofilms formed by *Pseudomonas aeruginosa* cause chronic lung infections in people with cystic fibrosis (CF). Chronic and recurring *Pseudomonas* infections in CF patients gradually damage the lungs, causing respiratory failure. This is the leading cause of death in CF patients.

Earache

The number one reason parents take their children to the doctor is for otitis media, or earache. Chronic otitis media has been linked to biofilm formation by *Haemophilus influenza, Streptococcus pneumoniae,* or *Moraxella catarrhalis* in the middle ear. Children who are victims of chronic otitis media may develop permanent damage to the middle ear, resulting in hearing loss.

Medical Devices

Biofilms can also form on in-dwelling medical devices. The microbes forming these biofilms are often from the patient's own skin microbiota or from the microbiota of health care personnel. Data collected by the CDC indicates that nearly all central venous catheters are colonized by biofilms. The microbes most common to these devices include *Staphylococcus epidermis, Staphylococcus aureus, Pseudomonas aeruginosa, Klebsiella pneumoniae, Enterococcus faecalis,* and the yeast *Candida albicans.*

Prosthetic valve endocarditis is a condition that results when tissue surrounding mechanical heart valves is colonized by biofilm-forming microbes such as *S. epidermis, S. aureus,* other types of *Streptococcus,* several types of Gram-negative bacilli, enterococci, and *Candida.*

Almost all patients who undergo long-term urinary catheterization develop infections due to biofilm formation in their urinary catheters. The common culprits are: *S. epidermis*, *Enterococcus faecalis*, *E. coli*, *Proteus mirabilis*, *P. aeruginosa*, *K. pneumonia*, and other Gram-negative organisms.

Biofilms may also form on contact lenses, intrauterine devices, pacemakers, prosthetic joints, tympanostomy tubes, voice prostheses, and dental implants.

Illness

Due to the architecture of the biofilm, symptoms of infections caused by biofilm-dwelling microbes usually take longer to appear than if the infection were caused by planktonic (free-living) microbes. Treatment is difficult due to the protection given the cells by the polysaccharide matrix of the biofilm, and the fact that cells in the center-most portion of the biofilm may exist in a dormant state. It is also easier for cells to share resistance genes due to their close proximity to one another in the biofilm. Higher rates of bacterial conjugation have been found in biofilm bacteria versus planktonic bacteria. Because of this difficulty in getting rid of biofilms, infections tend to be chronic.

Researchers have had limited success in preventing biofilm formation by impregnating in-dwelling devices with antimicrobial substances. Often, the only way to be sure of getting rid of an infection caused by microbes in a biofilm is to remove the in-dwelling device, but as in the case of artificial joints and heart valves, this is not always possible.

Biofilms and Industry

Biofilm formation in pipes can lead to clogs and corrosion. Microbes living in these pipes can contaminate water flowing through them. Many water-based processes such as water treatment and distribution, paper manufacturing, and the operation of cooling towers, are made more difficult and more expensive due to biofilm contamination.

However, biofilms are also being used as "biofilters." They can be used to filter air and water. Some forms of biofilms can be used to treat sewage and clean up industrial waste in streams or to clean up contaminated ground water. Biofilms can also be used to manufacture biochemicals such as medicine and cleaning products.

Name ______________________________

Exercise 3.6 Lab Report

1. MICROSCOPIC DRAWING OF YOUR SAMPLE:

2. Where was your biofilm sample collected?

3. Describe the microscopic appearance of your biofilm.

4. Describe the appearance of the prokaryotes on your slide.

5. If you found eukaryote cells, describe the appearance of the eukaryotes on your slide.

6. Observe several of your classmates' biofilms.

   a. Why do biofilms collected from different sites have similar appearances?

b. Why do some of the biofilms appear different from one another?

7. Do you think that the biofilm you sampled is beneficial, harmful, a nuisance, or of no consequence for humans? Explain your answer.

8. Could you modify this exercise to obtain additional information about your biofilm? What would you like to know about your biofilm?

9. Devise an experiment that would provide information on how an environmental factor affects the amount of growth in a biofilm. For example, describe an experiment to determine the effect of temperature on biofilm growth.

10. How will this knowledge affect your behavior as a health care professional? Explain your answer.

Media and Lab Techniques

When identifying an unknown bacterium, the first step is to determine the composition of the cell wall/membrane structure using the Gram stain. Once you have determined whether your culture is Gram-positive or Gram-negative, you will perform a specific set of biochemical tests to identify your unknown bacterium. Each type of microorganism has its own unique metabolic pathways (hydrolysis and biooxidation) and growth characteristics. The results of the biochemical tests are then compared to charts containing the expected test results for several known bacterial species. The most comprehensive source for these results is the *Bergey's Manual of Determinative Bacteriology*, but for simplification, abbreviated charts have been provided in the lab report section of this text.

Pathogenic organisms are able to cause a variety of symptoms in humans depending on how they interact with the patient. Therefore, isolation and growth of the suspected microorganism in pure culture is essential to perform tests that will identify and control the infectious agent. The specimen must always be handled very carefully or else the test results will be inaccurate.

An optimum balance of nutrients and a suitable growth environment are needed to culture a microorganism in the laboratory. The culture medium contains all the materials necessary for the microorganism to move quickly into the log growth phase of growth. Many growth media can be prepared for the microbiologist to isolate, grow, and identify a particular species in the laboratory. Some media have ingredients that encourage the growth of a suspected pathogen, whereas others include materials that discourage the growth of unwanted contaminants. Special media have been designed to identify a suspected pathogen in preparation for biochemical testing.

The proper laboratory diagnosis of an infectious disease helps the patient successfully cope with an infection and provides the health-care worker with information that may prevent others from becoming ill. Because the cultivation of a suspected pathogen is vital to successful treatment, all those involved in the process must be aware of what happens in the lab.

To make the identification of an unknown bacterium easier, microbiologists have devised "flow charts." By performing certain tests and following the chart, it is possible to flow down the lines to the most likely bacterium that meets the biochemical criteria. Results are confirmed by using *Bergey's Manual of Determinative Bacteriology*. *Bergey's Manual of Determinative Bacteriology* is a published text that provides a brief description of each organism's Family and Genus, the shape and size of the organism, Gram stain reaction, motility information, and other cell structure descriptions, biochemical characteristics, as well as habitat information. This manual allows us to identify and classify organisms we discover in the lab.

It is important to realize that the identification of an unknown bacterium is based on percentages. For example, if 87.6% of the time a bacterium acts like *Staphylococcus aureus*, we infer that it is probably *Staphylococcus aureus*. If it only acts like *Staphylococcus aureus* 50% of the time, it is probably not *Staphylococcus aureus*. Therefore, **do not expect all of your test results to match the expected results recorded in *Bergey's Manual*.**

In this chapter, we will be discussing important laboratory techniques and the media used to aid in the identification of unknown Gram-positive and Gram-negative bacteria cultures.

Pure Culture Techniques

The **streak plating procedure** is used to separate microbes from one another on the surface of a Petri plate. We try to physically separate the microbes from one another so that individual cells can grow into pure colonies that are visible on the surface of the plate.

Developing a good **streak plate technique** is important when obtaining pure cultures from clinical samples. We will use this technique throughout the semester to grow and isolate our organisms. Today we will use this technique to obtain a **pure culture** from our sample. A pure culture is a culture of microorganisms containing only **one** species of bacteria. Obtaining a pure culture is important since the body contains a great diversity of microorganisms which are either beneficial or cause no harm. When an individual has an illness caused by a specific bacterium it needs to be isolated and cultured separately from the normal microbiota.

In the health-care field you will be presented with patients with a variety of illnesses. You may be required to obtain samples from your patients for isolation and identification of the pathogenic (disease-causing) bacteria. For example, suppose a patient walks into your clinic complaining of a sore throat, painful swallowing, red throat with white patches, fever followed by chills, and headache. Where would you obtain a sample from this patient? You would take a throat swab and culture the organisms growing in the patient's throat. In that sample you would likely have a mixed culture of bacteria. How do you decide which organism is the likely culprit causing the disease? You might assume the patient has strep throat, which is caused by the bacterium *Streptococcus pyogenes.* We know *S. pyogenes* is beta hemolytic when grown on blood agar (a differential medium), so we would choose a blood agar plate as one of the media on which to grow our sample. Organisms that grow and express beta hemolysis would be selected for further analysis. Choosing the correct type of media to culture clinical specimens is very important when determining what organism is responsible for causing disease. Based on growth characteristics on the media, microbiologists are able to study such characteristics as culture morphology, metabolism, and antibiotic sensitivity.

Since one of the goals of this chapter is to study the characteristics of representative microbes, our first concern is to maintain pure cultures of the bacteria we wish to study. The methods used for preventing outside contamination are known as **aseptic techniques**. The most important aspects of aseptic technique are:

1. The inoculating needle or loop must be flame-sterilized and the lip of both the broth culture tube and the fresh medium tubes are flamed (to remove dust and microbes) before performing the transfer, to prevent contamination.
2. The inoculating utensil must be flame-sterilized after use to prevent cross contamination.
3. The tube or flask closures should never be laid down on the table top.

It is important to develop a good streaking technique. Gouging the surface of the agar will cause the organisms to be buried within the media, rather than growing as discrete colonies on the surface of the agar. If you carry too much inoculum from the loop onto the agar surface, there will be too much growth to achieve isolation. The **streak plate method** may be used to culture bacteria on the surface of the agar medium. An inoculum obtained with the wire loop

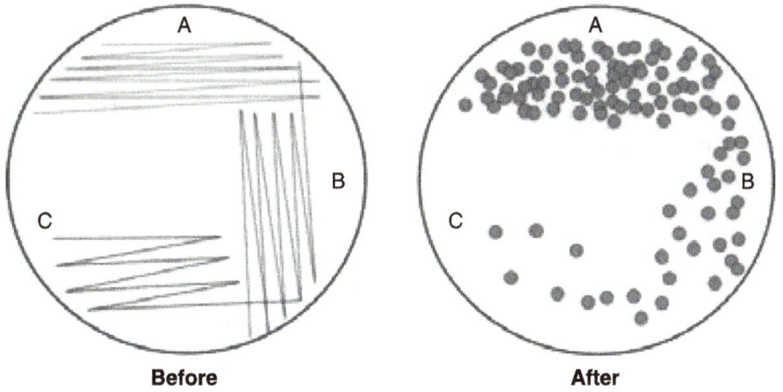

Fig. 4.1 Streak Plate Method and the Desired Outcome

is spread across the surface of the medium using one of several techniques. The most common streak plate method is the **quadrant technique (Figure 4.1)**. The purpose of this technique is to isolate bacterial cells so that they will grow and produce well-separated or isolated **colonies** (descendants of the cell that was inoculated prior to incubation). You will use the streak plate method to obtain **isolated colonies** on nutrient agar (general purpose medium).

In addition to cultivating bacteria on solid media, one may also inoculate agar in its molten (melted) state. This type of inoculation is called a **pour plate culture**. Bacteria are inoculated into a melted, tempered (45°C) nutrient agar deep, which is then poured into a sterile Petri plate. Like the streak plate method, the purpose of this method is to obtain isolated colonies from a mixture of cells so that a pure culture may be obtained. However, since the agar is inoculated while it is still liquid, cells are better distributed and colonies develop throughout the medium instead of just on top of it. Often **dilutions** are employed using known volumes of the cultures to reduce the growth enough to allow isolated colonies to develop. Since the colonies are evenly distributed and the volumes are known, pour plates provide the opportunity to estimate the number of cells in the original culture by counting the colonies and calculating the dilutions used in the process.

Transfer and Inoculation Routines

Transfer and inoculation routines are used in the context of a larger procedure, such as preparing a smear for a staining procedure or running a laboratory test. The transfers require aseptic techniques. You always sterilize your transfer instrument before doing the transfer. The purpose is to make certain that only the organism you want (and no incidental organisms) get transferred. We use instruments called inoculating loops and inoculating needles to do transfers. These are heated in the flame of a Bunsen burner prior to transfer to ensure their sterility. **The type of culture you are using (a broth culture, an agar slant, or the flat agar surface of a Petri plate) determines your transfer and inoculation routine.** The procedure varies the most between liquid (broth) and solid (slant or plate) cultures.

Transfer and Inoculation from a Broth Culture

With a broth culture, the transfer is made using an inoculating loop. If you are inoculating a slide to make a smear to observe under the microscope, then two or three loopfuls of organism will be transferred. If you are inoculating another medium to run a lab test, **usually only one loopful of the organism is transferred**.

Transfer and Inoculation from a Solid Culture

With a solid (slant or plate) culture, the transfer involves the use of both an inoculating loop and an inoculating needle. If you are inoculating a slide to make a smear, you must first place sterile distilled water on the slide. This is done using a tube of sterile water and a sterile inoculating loop. Next, you will transfer your organism from the solid medium. The transfer is made using a sterile inoculating needle. The needle allows better control in getting a small amount of inoculate. When you are taking a sample from a solid, it is much more dense (millions of organisms in close proximity), and you want to **transfer a small amount**.

Pipette Usage

The pipette is used for accurate measurements of samples.

Learning the proper technique for pipetting is an important part of this lab exercise. The following are proper pipette techniques:

1. The pipettes you will use in the laboratory are sterile; they will come in sealed packages. **You must only handle the end of the sterile pipette which will be inserted into the mechanical pipetter**. Do not touch the body or the tip of the pipette, as this would contaminate the pipette. The pipette must not come into contact with your fingers, the table, or any other object before it is inserted into your dilution.
2. Leave the pipette in its sterile package. Attach the pipette carefully into the mechanical pipetter. Your instructor will demonstrate how to use the mechanical pipetter.
3. Your instructor will demonstrate how to collect and transfer a sample with a sterile pipette. Samples are accurate when the bottom of the meniscus is on the line.
4. Whenever you have a sample within the pipette, the tip of the pipette must be facing downward; otherwise, the sample will flow upward into the mechanical pipetter. This would contaminate the pipetter. **"Elbows up"** is the rule! If you keep your elbow up, you will hopefully keep the tip of the pipette down!
5. When you have finished collecting and transferring the sample, **place the pipette back into its package and dispose of it in the Nalgene containers**.
6. **Do NOT dispose the pipetter in the Nalgene container; put this back in the pipetter storage bin**.
7. "Mouth pipetting" is strictly forbidden in the laboratory. With such a process, there is too great a risk of a portion of the specimen sample entering the mouth.

Selective and Differential Media

Choosing the correct type of media to culture clinical specimens is very important when determining what organism is responsible for causing disease. The main types of media we will employ in this lab throughout the semester are complex media, selective media, and differential media. **Complex media** are usually general purpose media in which the exact composition of each ingredient is not known. Complex media include basic media such as Tryptic Soy Agar (TSA), which can be used to grow a variety of organisms. **Selective media** have one or more ingredients (dyes, salts, chemicals, or antibiotics) which inhibit the growth of certain organisms. This allows the clinician to selectively grow only the organisms which are suspected to cause the infection for further study. **Differential media** contain indicators or ingredients that allow different organisms to display a different, characteristic pattern of growth. Two or more microorganisms grow on the same medium, but have very different, easily observable growth patterns.

Media can be further categorized as selective or differential. **Mannitol Salt Agar (MSA)** and **Eosin Methylene Blue (EMB) agar** are examples of selective and differential media frequently used in the clinical setting when determining the causative agent for a disease.

MSA: Mannitol salt agar is a **selective** medium as it contains 7.5% sodium chloride, which is a high salt content. Some organisms cannot tolerate a high osmotic pressure. Media containing higher than normal salt concentrations may inhibit the growth of these non-tolerant organisms. However, different species of *Staphylococcus* can readily grow in such a high salt environment. Mannitol salt agar is a **differential** medium as it contains the sugar mannitol. Some organisms can utilize mannitol as a food source and will produce acid end-products from this metabolism. Since this process is invisible, an indicator is added to the media to detect changes in pH. Phenol red is the indicator used in mannitol salt agar. It is red at a neutral pH but turns yellow if conditions in the media become acidic.

EMB: This is both a **selective and differential** medium. The media is selective for Gram Negative Coliforms (Fecal Bacteria). The combination of the two dyes Eosin and Methylene Blue inhibit most Gram Positive Bacteria. The media is differential for *Lactose fermenting bacteria, based on the rate of Lactose fermentation the colonies appear different in color.* Colonies of *E. coli* will be smaller and have a green metallic sheen. Colonies of *E. aerogenes* will be larger and have a rose-red color.

Application of Biochemical Tests General Tests
(Used for Gram-positive and Gram-negative bacteria)

Sugar Fermentation

Sugar fermentation tubes are used to identify bacteria that can ferment specific carbohydrates. They contain the carbohydrate source, peptone, and phenol red broth base as the pH indicator. Phenol red in the presence of acid will turn from red to yellow in color. The tubes also contain an inverted Durham tube to catch any carbon dioxide gas produced. Since the waste products of fermentation are acid and CO_2 gas, a yellow broth with a CO_2 bubble in the Durham tube indicates rapid sugar fermentation by the bacteria. The presence of a yellow broth without a CO_2 bubble indicates that sugar fermentation occurred at a slower rate. A negative test will remain unchanged. A positive test for peptone oxidation will result in turbid growth with a change to fuschia (bright pink) color. Phenol red turns fuschia (bright pink) if a sufficient alkaline reaction is produced by the oxidation of peptone, which creates amine by-products such as ammonia. The pH must rise above 8.4 for the color change to occur (Figure 4.2).

Fig. 4.2 Sugar Fermentation results: From left to right: + for acid and gas, + for acid only, + for oxidation of peptone
Courtesy of A. Swarthout

MOTILITY TEST MEDIUM

Bacteria may have a single strand of protein for a flagellum, which is the only organelle for motility in prokaryotic cells. Bacterial flagella may be stained using a special flagella stain to demonstrate their presence. This procedure is somewhat tedious. An alternate way to show that bacteria have flagella is to demonstrate their ability to move by using this motility plate. The media used is a semisolid agar. This consistency will allow bacteria to swim through the agar from the initial inoculation point if they contain flagella. Cells will be distributed along the migration route and will cause the media to become cloudy. A tetrazolium salt (TTC) is added to the medium to make the interpretation easier. TTC is colorless and soluble in the oxidized form but becomes insoluble and turns red when reduced. Since any metabolic process involves the oxidation of molecules to produce energy, the dye is readily reduced by microbial growth and other microbial activities, such as motility. When TTC is present, the trail produced by

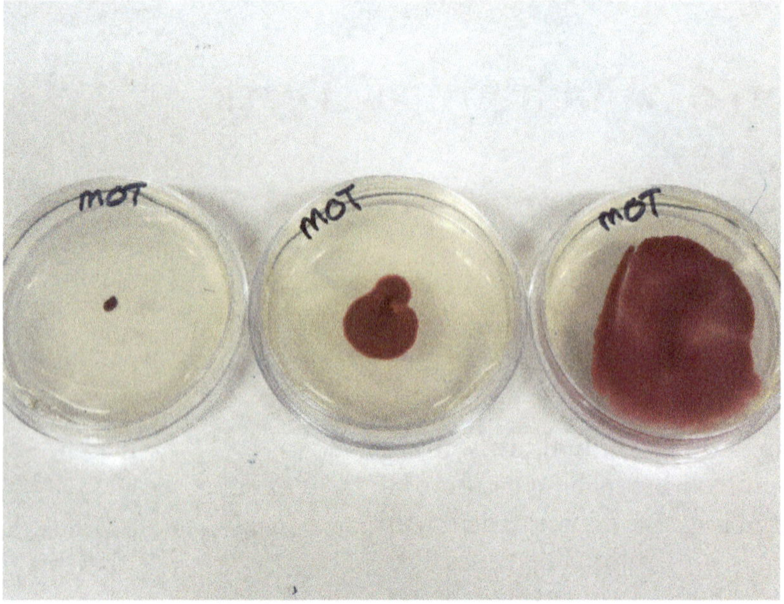

Fig. 4.3 Motility Test Results: The plate on the left is non-motile (–), the center and righ plates exhibit differing degrees of motility (+). **Note**: You should also look at a side view of the plate; a motile organism will grow throughout the thickness of the plate
Courtesy of A. Swarthout

bacteria moving through the media turns red. Another clue to motility is to look at the media from the side, the bacteria should be observed spreading away from the area of the stab line and growing through out the media. Non-motile bacteria will only produce a red color at the inoculation site (Figure 4.3).

FLUID THIOGLYCOLLATE (FT) TUBE

Fluid thioglycollate medium is used to determine the oxygen needs of a bacterium. One of the ingredients in the medium is **resazurin**, which is a redox indicator. In the presence of oxygen, it is pink; in the absence of oxygen, it is colorless. During autoclaving, oxygen is removed from the medium; however, oxygen begins to diffuse back into the medium at room temperature. Thus, the FT tube contains an oxygen gradient with the top of the tube being oxygen-rich and the bottom of the tube being oxygen-poor. Obligate aerobes will grow only on the top of the tube. Facultative anaerobes will grow throughout the tube. Obligate anaerobes will grow only at the bottom (Figure 4.4).

Fig. 4.4 Oxygen requirement test results, FT tube: Left: facultative anaerobe, Right: obligate aerobe
Courtesy of A. Swarthout

OXIDASE TEST

The oxidase test can be used to distinguish non-fermenters (oxidase positive) from fermenters (oxidase negative). The oxidase test checks for the presence of the enzyme indophenol oxidase. Tetramethyl-para-phenylenediamine (oxidase reagent) will be oxidized in the presence of atmospheric oxygen by indophenol oxidase, causing the formation of a dark purple compound known as indophenol. A bacterium that produces oxidase will react with the oxidase reagent and the **colony** will turn pink/maroon within two minutes of the addition of the reagent (Figure 4.5). Reading the test after two minutes will result in a false positive outcome since the oxidase reagent itself will react with the oxygen in the air.

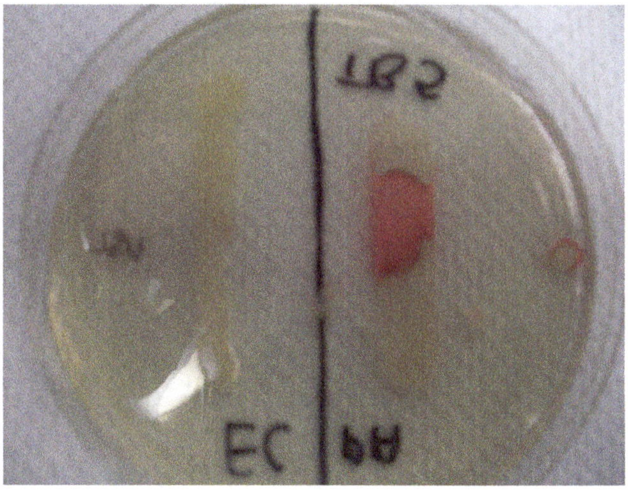

Fig. 4.5 Oxidase Test Results: Left oxidase negative, right oxidase positive

Note: The colony must turn color, not the media
Courtesy of A. Swarthout

NITRATE BROTH

The identification of some bacteria is aided by determining if the organism can reduce nitrate (NO_3^-) to nitrite (NO_2^-) or other nitrogenous compounds such as ammonia (NH_3) or nitrogen gas (N_2) via the action of the enzyme **nitratase** (also called nitrate reductase). This reaction is expressed as: $NO_3^- \rightarrow NO_2^- \rightarrow NH_3$ or N_2

In order to determine if a bacterium can reduce nitrate, the organism is inoculated into a nitrate tube containing large amounts of nitrate.

After incubation, the addition of one drop of nitrite solution A (sulfanilic acid) and 1 drop of nitrite solution B (dimethyl-alpha-napthalamine) to the test tube will cause these two compounds to react with nitrite and turn red, indicating a positive nitrate reduction test: nitrate (NO_3^-) reduced to nitrite (NO_2^-). If there is no color change, nitrite is absent. This would be a negative nitrate reduction result.

However, it is possible that the nitrate was reduced to nitrite and then was further reduced to ammonia (NH_3) or nitrogen gas (N_2). To determine if this has occurred, a small amount of zinc dust must be added. Zinc reduces nitrate to nitrite. If the organism did not reduce the nitrate to nitrite, the zinc will change the nitrate to nitrite. The tube will turn red because alpha-napthylamine and sulfanilic acid are already present in the tube. A red color after zinc is added indicates that the bacterium was unable to reduce nitrate. This is recorded as a negative nitrate reduction test.

If the tube does not change color upon the addition of zinc, then the zinc did not find any nitrate in the tube. That means the organism converted the nitrate to nitrite and then converted the nitrite to ammonia and/or nitrogen gas. Thus no color change upon the addition of zinc is recorded as a positive nitrate reduction test (Figure 4.6).

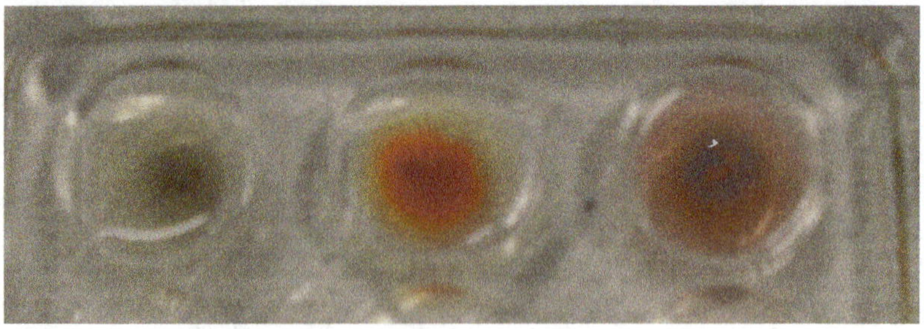

Fig. 4.6 Nitrate Reduction Test Results:

Left: Clear after of addition of reagents AND zinc = no nitrate/nitrite for reagents/zinc to react with = nitrate reduced to gas

Center: Red after addition of reagents = bacterial reduction of nitrate to nitrite

Right: Pink/red after addition of zinc = no bacterial reduction of nitrate

Courtesy of A. Swarthout

Tests Specific to Gram-Positive Bacteria

Catalase Test

This test demonstrates the ability of a bacterium to produce the enzyme catalase, capable of converting hydrogen peroxide (produced as part of oxygen usage) to water and oxygen. A bacterium which produces catalase will react with hydrogen peroxide to produce bubbles of oxygen gas (Figure 4.7). This test is one of the key tests used to differentiate between staphylococci and streptococci. All staphylococci are catalase-positive, while all streptococci are catalase-negative.

$$2\ H_2O_2 + \text{catalase} \rightarrow 2\ H_2O + O_2$$

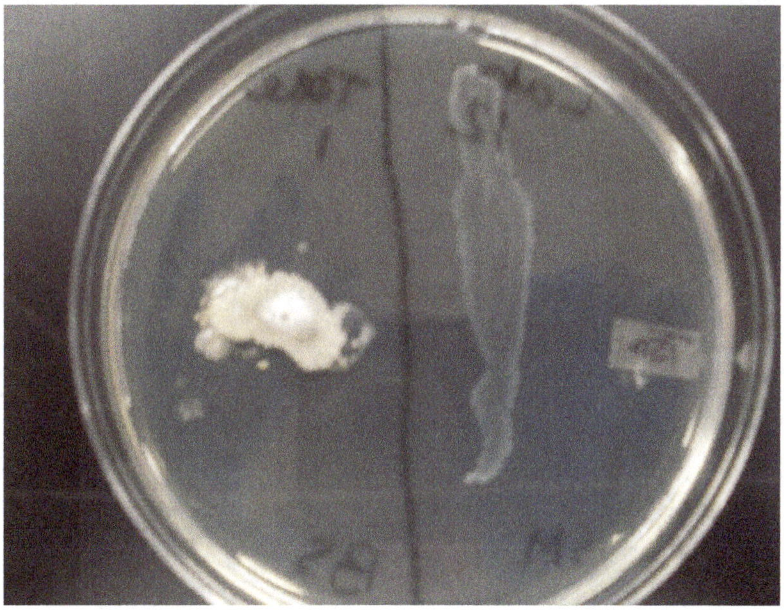

Fig. 4.7 Catalase test results: Left Catalase-positive, Right Catalase-negative

Courtesy of A. Swarthout

MANNITOL SALT AGAR (MSA)

Some organisms cannot tolerate a high osmotic pressure. Media containing higher than normal salt concentrations may inhibit the growth of these non-tolerant organisms. Mannitol salt agar contains a high salt concentration, and so only salt tolerant organisms will grow on it. Additionally, mannitol salt agar contains the sugar mannitol. Some organisms can utilize mannitol as a food source and will produce acid end-products from this metabolism. Since this process is invisible an indicator is added to the media to detect changes in pH. Phenol red is the indicator used in mannitol salt agar. It is red at a neutral pH but turns yellow if conditions in the media become acidic (Figure 4.8).

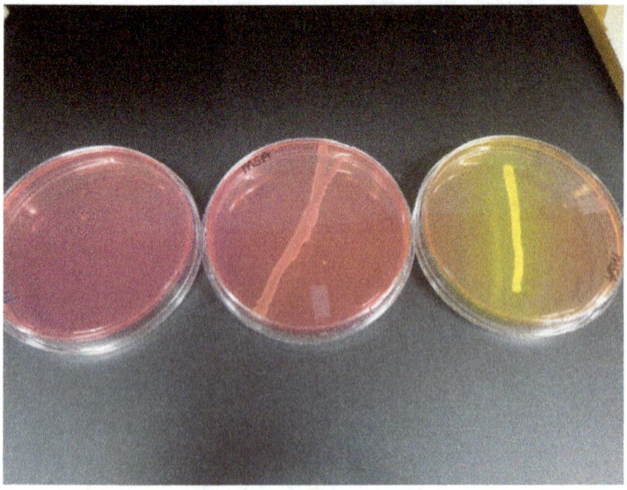

Fig. 4.8 Mannitol Salt Agar Results:

Left: Unable to grow in high salt environment, unable to determine mannitol fermentation.

Center: Tolerates high salt concentration, but unable to ferment mannitol.

Right: Tolerates salt and ferments mannitol.

Courtesy of A. Swarthout

The MSA plate is a selective medium used to detect the presence of staphylococci. It is also used to differentiate between various species of staphylococci by the ability to ferment mannitol.

MSA contains 7.5% sodium chloride, a high salt content in which most bacteria cannot grow. However, the different species of *Staphylococcus* can readily grow in such a high salt environment. *S. aureus* is also able to ferment the mannitol contained in the agar while *S. epidermidis* cannot.

BLOOD AGAR PLATE (TSA-B)

This media is used to demonstrate the ability of a bacterium to produce **hemolysin**, the exotoxin capable of attacking the hemoglobin of red blood cells. The blood agar plate contains 5% sheep blood added to TSA. It is used to determine the type of hemolysis displayed by the bacterium.

Classification Based on Hemolytic Activity

When grown on sheep blood agar, bacteria display one of three types of hemolysis of the red blood cells in the agar.

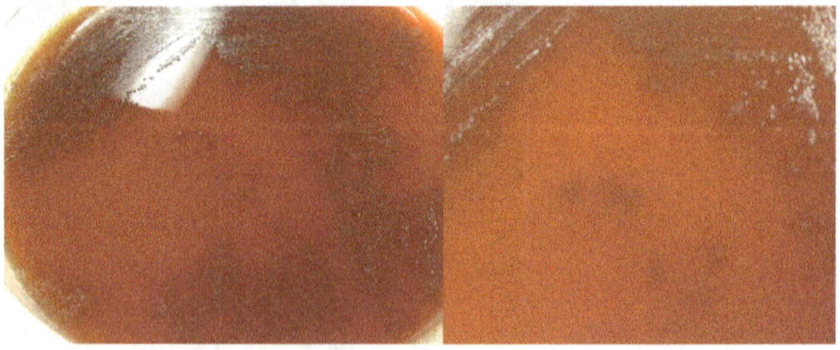

Fig. 4.9 Alpha hemolysis: The red blood cells in the media are partially digested producing a green/brown coloration of the agar directly around the margin of the colony.

Note the dark coloration in the first quadrant. To really see the hemolytic activity you must look closely at isolated colonies (Figure 4.9; image to the right); you will notice the green/brown color directly around the colony.

Courtesy of A. Swarthout

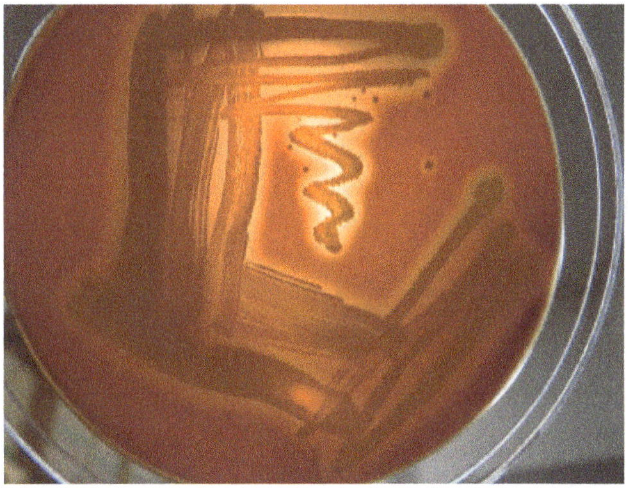

Fig. 4.10 Beta hemolysis: The red blood cells in the media are completely digested producing a clearing of the agar directly around the margin of the colony

Note that if held up to the light you can mostly see through the plate (Figure 4.10)

Courtesy of A. Swarthout

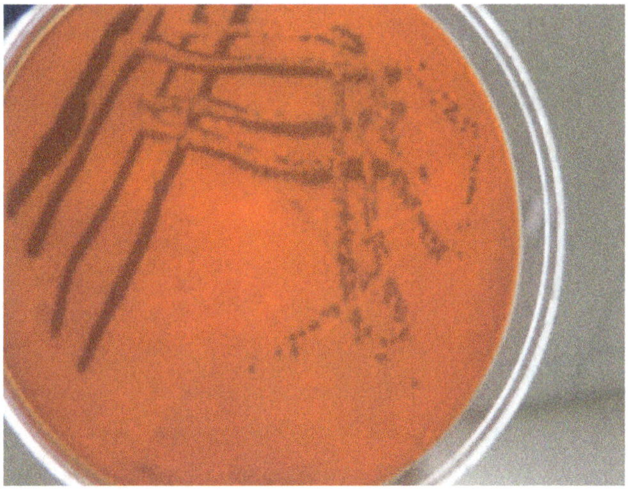

Fig. 4.11 Gamma hemolysis: No change is noted in the agar (Figure 4.11). The organism does not affect the red blood cells

Note that the media may appear dark in the first quadrant, but you must look at the isolated colonies to determine this

Courtesy of A. Swarthout

Tests Specific to Gram-Negative Bacteria

EOSIN METHYLENE BLUE AGAR (EMB)

Eosin Methylene Blue is a selective and differential media. The media is selective for Gram Negative Coliforms (Fecal Bacteria). The combination of the two dyes Eosin and Methylene Blue inhibit most Gram Positive Bacteria. The media is differential for *E. coli* as it will appear metallic green due to a rapid fermentation of the Lactose and the production of strong acids in the media. Other Gram negatives that ferment the Lactose produce acid, which can turn the colonies purple or black in color. The differences in the colony color relate to the speed at which the Lactose is fermented. Organisms from the genus Enterobacter usually appear a rosy/pink color. They do ferment lactose; however, the acids are much weaker than produced by *E. coli.* Colonies that appear colorless are unable to ferment Lactose (if you have a colorless colony, look at your sugar fermentation tubes; I bet the Lactose was negative!).

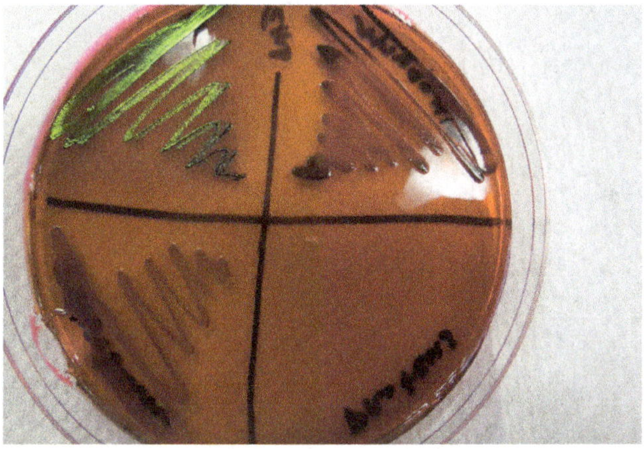

Fig. 4.12 EMB results will vary based on the organism: Clockwise from upper left corner:
Quadrant 1: *E. coli*
Quadrant 2: *Enterobacter sp.*
Quadrant 3: Unknown organism showing clear/colorless colonies
Quadrant 4: Unknown organism showing pale pink colonies
Courtesy of A. Swarthout

THE IMViC SERIES OF TESTS

IMViC is a mnemonic to remember the four biochemical tests being used: **I**ndole, **M**ethyl red, **V**oges-Proskauer, and **C**itrate. These four tests are used to differentiate between the various Gram-negative bacteria, especially the **Enterobacteriaceae family**.

INDOLE TEST (TRYPTONE BROTH)

The indole test is used to detect the production of indole by bacteria. Organisms that posses the enzyme tryptophanase can break down the amino acid tryptophan to indole. When indole reacts with an indole test reagent (Kovac's reagent), a red-colored complex is produced (Figure 4.13).

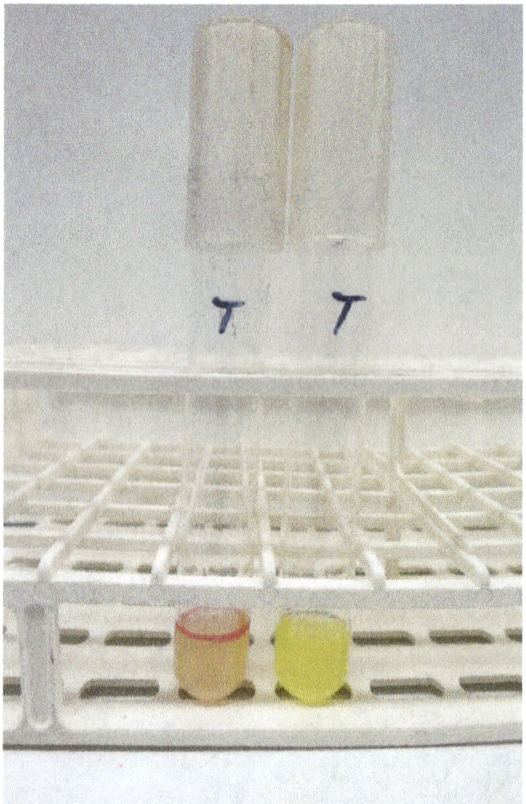

Fig. 4.13 Results of the Indole test:

Left: Positive indole test with red layer; **Right:** Negative test with no layer.

Courtesy of A. Swarthout

THE METHYL RED TEST

The methyl red test is performed to detect mixed-acid fermentation by the bacterial culture. Some organisms produce acid from the metabolism of glucose in a sufficient quantity to produce a pH of 4.4 in the media. These mixed acids are not further metabolized and are said to be stable acids. At a pH of 4.4 or less, the pH indicator methyl red is a bright cherry red. Therefore, upon addition of the methyl red reagent, the red color will appear if mixed-acid fermentation has occurred. Examples of mixed acids that are produced include: lactic, acetic, succinic, and formic, plus ethanol, CO_2 and H_2 (Figure 4.14).

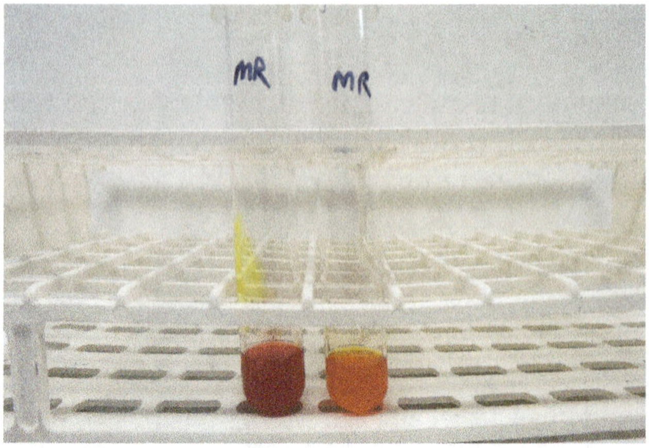

Fig. 4.14 Results of the Methyl Red Test:

Left: Positive test indicated by dark red color

Right: Negative test indicated by orange color

Courtesy of A. Swarthout

VOGES-PROSKAUER TEST

The Voges-Proskauer test is used to detect the production of acetoin by bacteria. Many bacteria can ferment glucose to pyruvate. Some of those bacteria can then quickly convert the pyruvate to acetoin. Of those, some bacteria can further convert the acetoin to 2,3-butanediol. The bacteria that cannot convert acetoin to 2,3-butanediol are the ones we are testing for in the VP test. When we add in the VP reagents, the bacteria convert the acetoin to another end product, which we are able to see by the production of the red color in the tube. The red color in the VP is considered a positive VP result. To the VP tube, add the VP-A and VP-B reagents. Carefully mix and let stand for 30 minutes. If the broth has changed to the brick-red appearance, your culture has produced acetoin or 2,3-butenediol (Figure 4.15).

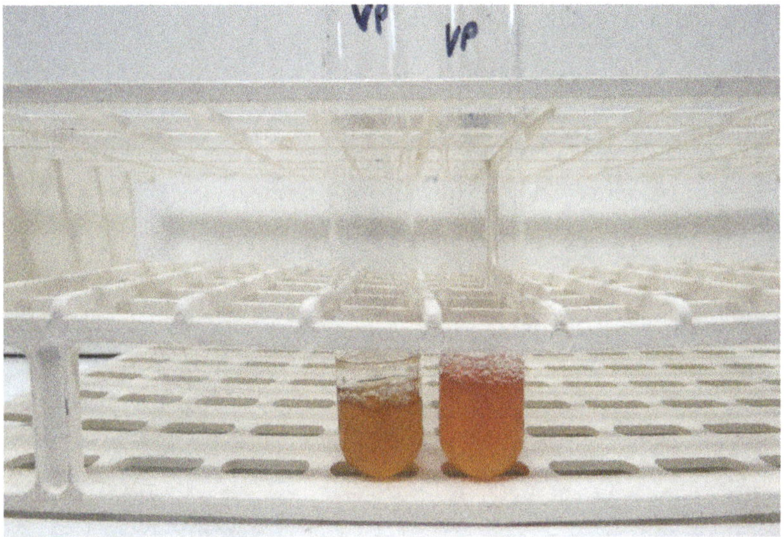

Fig. 4.15 Results of the VP test:

Left: Negative test indicated by brown color
Right: Positive test indicated by red color
Courtesy of A. Swarthout

SIMMONS CITRATE (CITRATE TEST)

Simmons citrate is used to detect citrate utilization. Citrate contains carbon. If an organism can use citrate as its only source of carbon the citrate in the media will be metabolized. Bromothymol blue is incorporated into the media as an indicator. Under alkaline conditions this indicator turns from green to blue. The utilization of citrate in the media releases alkaline bicarbonate ions that cause the media pH to increase above 7.4, which turns the agar to this blue color (Figure 4.16).

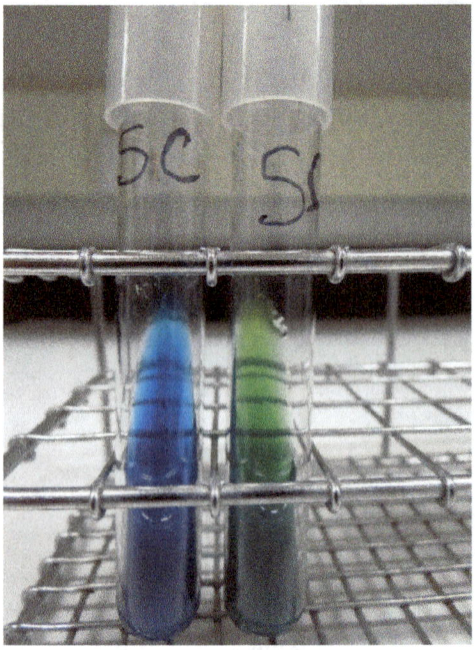

Fig. 4.16 Results of the Simmons Citrate test:
Left: Positive
Right: Negative
Courtesy of A. Swarthout

UREA BROTH (UREA HYDROYLSIS)

The urea broth is used to detect urea hydrolysis. If a bacterium produces the enzyme urease, it will break down urea to ammonia and carbon dioxide, and the broth will appear bright pink. Ammonia will increase the pH of the media to 8.0 or higher. The media contains phenol red as a pH indicator. At pH 8.0 or higher, the indicator is a bright pink color (4.17).

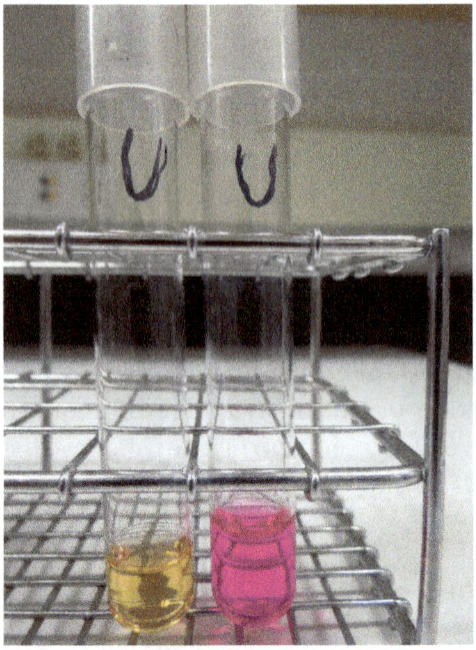

Fig. 4.17 Results of the Urea Test:
Left: Negative
Right: Positive
Courtesy of A. Swarthout

Lab Exercise 4.1

Streak Plate Technique, Pure Culture Techniques, Colony Characteristics and Selective and Differential Media

Student Objectives

1. Aseptically transfer a bacterial sample and inoculate a Petri plate with the sample.
2. Apply standard precautions in the transfer of a culture.
3. Understand the reasons for pure culture technique and streaking for isolation.
4. Perform step-by-step procedures for a common streak method called the quadrant streak.
5. Successfully isolate bacteria in a Petri plate.
6. Identify and evaluate the various colony characteristics of a bacterial species.
7. Explain the pour plate culture as a pure culture technique.
8. Describe the terms enrichment media, selective media, and differential media.
9. Evaluate the MSA and EMB media in their role as selective and differential media.

Pure Culture Technique Activity

Week One Materials

Per student: Tryptic Soy Agar (TSA) plate, inoculating loop, Bunsen burner, and igniter.

Per table: Mixed broth cultures of *Staphylococcus aureus* and *Escherichia coli*, (2) Blood agar plates, (2) MSA plates, and (2) EMB plates.

Streak Plating Procedure: Week One

The streak plating procedure is used to separate microbes from one another on the surface of a Petri plate. We try to physically separate the microbes from one another so that individual cells can grow into pure colonies that are visible on the surface of the plate.

1. Gather all the necessary materials. Obtain an inoculum of the mixed culture (*Staphylococcus aureus* and *Escherichia coli*).
2. Obtain one nutrient agar plate (TSA plate). Label the bottom (where the agar is located) with the culture name, date, and your seat number.
3. Carefully light the Bunsen burner.
4. Carefully mix the culture to re-suspend the microorganisms.
5. Hold the inoculating loop in the flame so that both the loop and the wire are being heated. Heat until they glow red.
6. Remove the cap with the little finger of the same hand in which you are holding the inoculating loop. **Do not set the cap down.**
7. Flame the lip of the test tube by passing it back and forth through the flame.
8. Insert the inoculating loop into the tube; do not touch the sides of the tube. Obtain a sample of culture on the loop.
9. Flame the lip of the tube, return the cap to the tube, and place the tube in the test tube rack.
10. Now you are ready to transfer the loopful of organism to the Petri plate.
11. Use the quadrant technique, as demonstrated by the instructor, and carefully streak this mixed culture on one of the nutrient agar plates (use **Figure 4.1** as a guide).
12. Be careful not to leave the lid open or gouge the agar surface. Practice plates are available.
13. Place the plate bottom side up into the container provided for incubation.
14. Repeat this procedure for both the EMB plate and the MSA plate. Divide the EMB and MSA plates among the lab table partners.

NOTE: Whenever you are asked to streak a plate for isolation throughout the remainder of the semester, you will use this streak plate method. In so doing, you will find that you get better and better at isolating colonies.

Streak Plating Procedure: Week Two

1. Observe the nutrient agar plate that you streaked in the last lab period. Do you have isolated colonies?
2. Draw your streak plate in the lab report.
3. Examine your streak plates. Use the Quebec Colony Counter. Note the growth of microorganisms on the plates. **You are looking for single, isolated colonies of each organism.**
4. Analyze an isolated colony and identify the colony characteristics of each species. Record your observations in the lab report.
   a. Size of the isolated colonies (small, medium, or large when compared to isolated colonies of the other microbes)
   b. Color of the isolated colonies (clear, white, tan, bright yellow, pink/red, etc)
   c. Opacity of the isolated colonies (transparent/clear, translucent/partially clear, or opaque/not clear)
   d. Configuration of the isolated colonies (see diagram on next page)
   e. Margins of the isolated colonies (see diagram on next page)
   f. Elevations of the isolated colonies (flat, raised, rounded/convex, drop-like/like a drop of water, bumpy/hilly, or crater-like with a sunken center)
5. Observe the MSA plate and the EMB plate that you streaked in the last lab period. Record your observations and answer the related questions in your lab report.

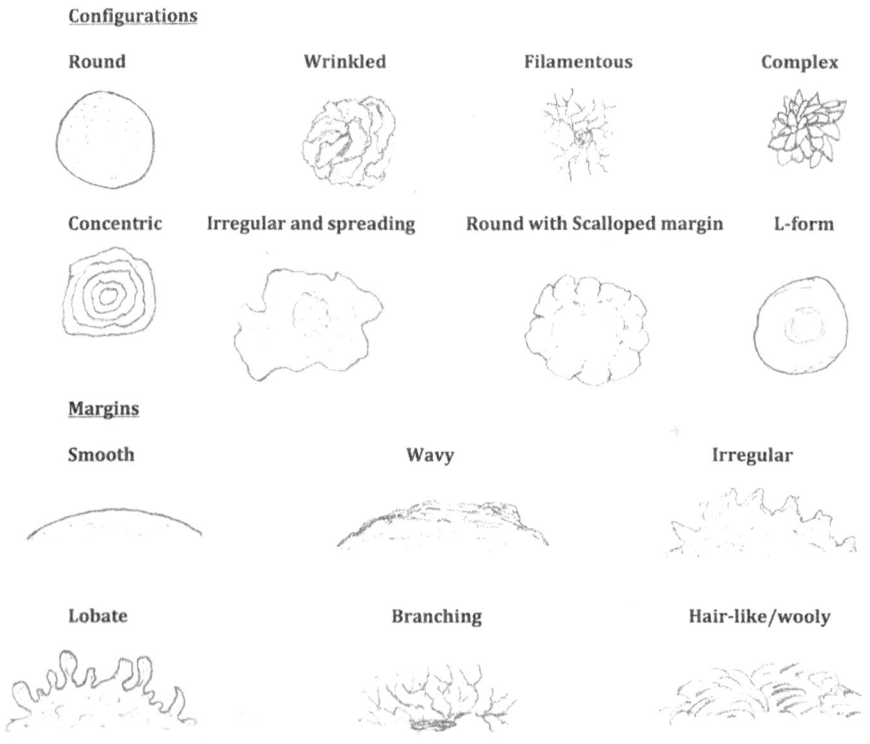

Courtesy of Authors

Name ____________________

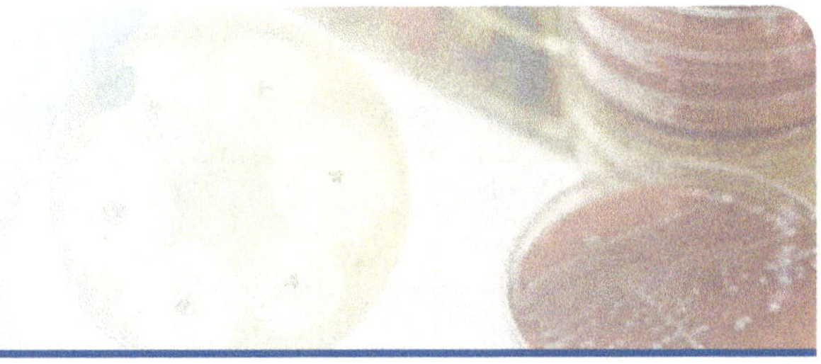

Exercise 4.1 Lab Report

Colony Morphology

1. Using the space below, **draw** the growth of your Quadrant Streak Plate from your TSA plate (use colored pencils) and answer the related questions.
   a. How many colony morphologies do you see?

   ____

   b. Did you obtain isolated colonies?

   ____

   c. How many loopfuls of bacteria did you use to inoculate the plate with?

   ____

2. Give several reasons why isolation may **not** have occurred on one or more of your plates. Explain what might have gone wrong during the streaking procedure.

   ____

   ____

3. Fill out the following table of colony morphology for the two microorganisms. (Use the TSA plate).

Name of Micro-Organism	Gram (+) Or (−)	Size	Color	Opacity	Configuration	Margin	Elevation

4. Describe the appearance of your MSA plate. How many colony morphologies do you see? Which organism is growing on this media?

____

____

____

5. In order to grow an organism on MSA the organism needs to overcome the selective property of the plate. What makes MSA a selective media?

____

____

____

6. What is in the media that causes the plate to change color?

____

____

____

7. Why is your MSA plate the color it is?

____

____

____

8. Describe the appearance of your EMB plate. Describe the colony morphology. Which organism is growing on this media?

____

____

____

9. In order to grow an organism on EMB the organism needs to overcome the selective property of the plate. What makes EMB a selective media?

10. What is in the media that causes the plate to change color?

11. Why is your EMB plate the color it is?

12. Draw your blood agar plate here and label the different colony morphologies along with the type of hemolysis they are showing.

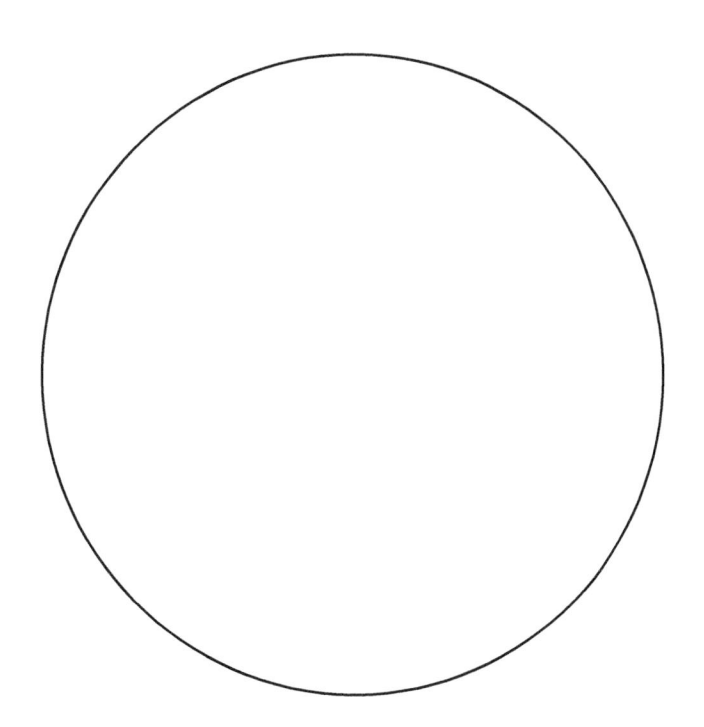

13. Describe the appearance of your blood agar plate. How many colony morphologies do you see? Describe the types of hemolysis that you observe. Do all of the colonies share the same type of hemolysis?

____

____

____

14. If a patient has a skin wound infection, which media (MSA or EMB) would preferentially be used to help identify the causative agent? Why?

____

____

____

15. Which media would be used (MSA or EMB) to help diagnose or rule out an organism that is causing bloody diarrhea? Why?

____

____

____

Lab Exercise 4.2

Gram-Positive Bacteria Biochemical Testing

Student Objectives

1. Perform proper aseptic techniques while inoculating various types of media.
2. Know how to use the "Flow Chart for the Identification of an Unknown Bacterium" to determine the sequence of tests required to identify an unknown bacterium.
3. Make the initial Gram stain determination for an unknown bacterium.
4. Make the initial morphological determination for an unknown bacterium.
5. Determine the oxygen requirements of an unknown bacterium.
6. Determine the growth temperature requirements of an unknown bacterium.
7. Use differential media to determine:
   a. Motility
   b. Sugar fermentation properties
   c. Nitrate reduction
   d. Oxygen usage
   e. Catalase production
   f. Hemolytic patterns
   g. Halotolerance
   h. Indophenol oxidase production
8. Identify one of the following unknown Gram-positive cultures: *Staphylococcus aureus*, *Staphylococcus epidermidis*, *Streptococcus mutans*, *Micrococcus luteus*, *Bacillus subtilis*, *Mycobacterium smegmatis*, *Corynebacterium xerosis*, and *Lactobacillus casei*.

Name ______________________________

Pre-Lab Exercise 4.2

1. What organelles must bacterial cells have to be motile?

2. Describe the growth of an obligate aerobe in a fluid thioglycollate (FT) tube. Which bacterium would show this?

3. Sugar Fermentation: Explain why you cannot get a carbon dioxide gas bubble in the inverted Durham tube without acid production.

4. Describe beta-hemolysis on a blood agar plate. Refer to the identification chart at the end of this lab: Which organism/s should show this characteristic?

5. What 2 groups of organisms will the oxidase test distinguish?

6. What enzyme is present in a positive catalase test? Which genera of bacteria can the catalase test be used to differentiate?

7. Fill in the following table:

ITEM	ROLE OR ACTIVITY IN TESTS
7.5% SALT	
MANNITOL	
PHENOL RED	
CATALASE	
HEMOLYSIN	
NITRATASE	

8. When inoculating your media, how might you contaminate your media, and how would those errors affect the results?

____

____

____

In this lab, you will be performing tests that will aid in the identification of unknown Gram-positive bacteria cultures.

Week One—Gram-Positive Bacteria Unknown Procedure

Materials

Materials per student; 1 of each of the following: Sugar fermentation tubes (glucose, sucrose, lactose, and mannitol), motility agar plate, fluid thioglycollate tube (FT), nitrate broth tube, blood agar plate (TSA-B), and mannitol salt agar plate (MSA).

Two of the following: TSA plates

Cultures per student: One of the following unknown Gram-positive cultures: *Staphylococcus aureus*, *Staphylococcus epidermidis*, *Streptococcus mutans*, *Micrococcus luteus*, *Bacillus subtilis*, *Mycobacterium smegmatis*, *Corynebacterium xerosis*, and *Lactobacillus casei.*

1. The first data you record is on the characteristics of your unknown as it grows in broth. This includes: (a) degree of turbidity, (b) color, (c) surface growth features, and (d) sedimentation.
   a. Degree of turbidity: View your broth culture and compare it with others in the class. Determine the relative amount of growth in your tube. This is a subjective determination and should be recorded as *"slight," "moderate,"* or *"abundant."*
   b. Color: If your cultures displays a color (i.e., red, yellow), record that information in the data sheet. You might want to hold your tube against a white paper background to help make this determination.
   c. Surface growth features: Certain bacteria form a characteristic growth pattern on the surface of liquid media. This may take the form of a: (a) "ring" around the inside of the tube; (b) "pellicle" or thick, rubbery-looking layer; (c) "membranous" or thin, filmy-looking layer; (d) "lattice" or net-like layer; or (e) no observable layer.
   d. Sedimentation: Many bacteria grow so rapidly and are so dense that they settle to the bottom of the tube (i.e., even gravity affects bacteria). This is a subjective determination and should be recorded as *"slight," "moderate,"* or *"abundant."*
2. Gather the materials and inoculate according to the *Inoculation and Interpretation Chart* below. Follow the directions in the section named *Inoculation Procedure.* Use your broth culture for all inoculations.
3. Use the broth culture to perform the Gram stain on your unknown to confirm you have a Gram-positive organism and determine the cell morphology and arrangement. Record the results in the data sheet. If you are not successful with your Gram stain, repeat the procedure. You cannot progress in the identification of your unknown without obtaining valid results with the Gram reaction. Have your instructor observe the slide under oil immersion.

INOCULATION AND INTERPRETATION CHART FOR UNKNOWNS

Media	Inoculation Procedure	Interpretation of Reactions
1. Sugar Fermentation + Peptone Oxidation a. Glucose (G) b. Sucrose (S) c. Lactose (L) d. Mannitol (M)	**Used to show ability of bacteria to ferment various sugars or peptone oxidation.** Aseptically inoculate each of these sugar tubes with one loopful of your unknown. Incubate for 24 hours at 37°C.	**Negative**: No growth or no change in the red color. **Positive for sugar fermentation**: Turbid growth with change to yellow color. Phenol red is added as the pH indicator, which turns yellow if sufficient acid is produced by the fermentation of the sugar. The pH must drop below 6.8 for the color change to occur. The inverted Durham tube is used to collect gas from the fermentation process. • ***Positive for acid production*** indicated by color change from red to yellow. • ***Positive for gas production*** indicated by at least 10% of the culture in the Durham tube being replaced by gas. • NOTE: A tube may be positive for acid without being positive for gas. However, the opposite is not true. There cannot be gas if there was not acid production due to fermentation. **Positive for aerobic oxidation of peptone (non-sacrolytic bacteria):** Turbid growth with a change to fuschia (bright pink) color. Phenol red is added as the pH indicator, which turns fuschia if a sufficient alkaline reaction is produced by the oxidation of peptone which creates amine by-products such as ammonia. The pH must rise above 8.4 for the color change to occur.

INOCULATION AND INTERPRETATION CHART FOR UNKNOWNS

Media	Inoculation Procedure	Interpretation of Reactions
2. Motility agar plate (MT)	**Used to identify ability of bacteria to move (i.e., flagellated cells).** Stab the **MT agar plate** once in the center with an inoculating **needle**. **Incubate for 5 days. If your optimal temperature is 37°C, be sure to place the MT plate in the special incubation rack labeled "INCUBATE FOR 5 DAYS."**	If there is no growth, the organism was not able to grow in the media and you cannot interpret the results. List as "no growth" and "no determination" with regard to motility. **Negative:** Growth with color change only along the line of the stab, with no pattern spreading out. **Positive:** Growth with media color change to violet/red. Spreading pattern into surrounding media; may appear as snow flakes.
3. Fluid thioglycollate tube (FT)	**Used to show oxygen usage of bacteria.** Inoculate the tube using a sterile loop. Place one loopful of the organism *just below the top surface* of the agar. This area usually has a faint pink color due to the presence of the resazurin. Incubate for 24 hours at 37°C.	The upper portion of the agar medium contains the greatest amount of free oxygen because the gas may easily diffuse into the medium from the atmosphere. The large center zone contains a decreasing amount of oxygen because the gas has more difficulty reaching that portion of the medium. Those bacteria growing ONLY in the upper portion of the tube culture would be **aerobic**. Those growing both in the upper portion and in the center zone of the tube culture would be **facultative anaerobes**.

(continued)

INOCULATION AND INTERPRETATION CHART FOR UNKNOWNS

Media	Inoculation Procedure	Interpretation of Reactions
4. TSA plate for **oxidase** test.	**Used to demonstrate the ability of a bacterium to produce the enzyme oxidase, capable of reducing oxygen during the electron transport chain of cell respiration.** Streak the plate for isolation using the quadrant method with a loopful of your unknown. Perform the quadrant method. Incubate for 24 hours at 37°C.	No growth: list as "no growth" and "no determination for oxidase". Turbid with growth: Carefully add 2 or 3 drops of the oxidase reagent onto the isolated colonies to be tested. **Be certain to have on gloves and safety glasses, as the reagent is carcinogenic.** **Positive test:** The colony will change color to pink/maroon, within 2 minutes. **Watch the time closely!** **Negative test:** Remains colorless, after the 2 minutes time interval.
5. Nitrate broth tube (N)	**Used to demonstrate the ability of a bacterium to produce the enzyme nitratase, capable of converting nitrate to nitrite.** Aseptically inoculate the tube with one loopful of your unknown. Incubate for 24 hours at 37°C.	Transfer 1 ml of the inoculated nitrate reduction media to 1 well of a spot plate. Add 2–3 drops of nitrate A and 2–3 drops of nitrate B. Appearance of a red color within 30 seconds is a positive test for nitrate reduction. Repeat for remaining organisms. If there is no red color, scoop a tiny amount of zinc dust with a toothpick and add it to the wells with no red color and mix. If the broth turns red after the zinc, the test is negative for nitrate reduction. If the broth remains clear after the addition of zinc, this indicates that nitrate was reduced all the way to **gas.** **Be certain to have on gloves and safety glasses, as the reagent is carcinogenic.**

INOCULATION AND INTERPRETATION CHART FOR UNKNOWNS

Media	Inoculation Procedure	Interpretation of Reactions
6. TSA plate for **catalase** test	**Used to demonstrate the ability of a bacterium to produce the enzyme catalase, capable of converting hydrogen peroxide (produced as part of oxygen usage) to water and oxygen.** This test is one of the simplest tests to quickly differentiate between *Streptococcus sp.* and *Staphylococcus sp.* Streak the plate for isolation using the quadrant method with a loopful of your unknown. Perform the quadrant method. Incubate for 24 hours at 37°C.	Add several drops of 3% hydrogen peroxide drop-by-drop directly onto an isolated colony **Positive test:** Bubbling occurs along the streak. The bacterium is aerobic or a facultative anaerobe. As part of its oxygen usage, it has produced H_2O_2, hydrogen peroxide. It then produced catalase enzyme to break down the H_2O_2 to water and oxygen. **Negative test:** No bubbling occurs.
7. Blood agar plate (TSA-B)	**Used to demonstrate the ability of a bacterium to produce hemolysin, the exotoxin capable of attacking the hemoglobin of red blood cells.** Streak the plate for isolation using the quadrant method with a loopful of your unknown. Perform the quadrant method. Incubate for 24 hours at 37°C.	No growth = list as "no growth" and "no determination". Turbid with growth: ***Beta hemolysis*** displays itself as a **totally clear zone** around the colony, indicating complete red blood cell destruction. ***Alpha hemolysis*** displays itself as a **"partial clearing"** of RBCs around the colony. This results in the blood in the agar becoming discolored and will show as a **greenish** color instead of the red color typical of RBCs. If **no hemolysis** takes place, the underlying agar remains bright red due to the presence of whole red blood cells. Ironically, this lack of any hemolysis is known as ***gamma hemolysis***.

(*continued*)

INOCULATION AND INTERPRETATION CHART FOR UNKNOWNS

Media	Inoculation Procedure	Interpretation of Reactions
8. Mannitol salt agar plate (MSA)	**(1) Used to demonstrate the ability of a bacterium to grow in a 7.5% salt environment. (2) Used to demonstrate the ability of a bacterium to ferment the sugar mannitol.** Streak the plate for isolation using the quadrant method with a loopful of your unknown. Perform the quadrant method. Incubate for 24 hours at 37°C.	MSA is a differential growth medium. **"MSA"** is an acronym for mannitol salt agar. **Phenol red** is the pH indicator added to the medium to help us observe if acids have been released as the result of fermentation. You must first observe the plate for the ability of the organism to grow in a salt environment. The 7.5% salt content of MSA limits the growth of most organisms. If the organism grew, it was capable of surviving in a 7.5% salt environment. If the organism was incapable of growing in the salty environment, you may **not** make any decision with regard to mannitol fermentation. If there was growth, you may now observe the plate for the ability to ferment mannitol. Mannitol is an alcohol-sugar capable of being fermented by only a few organisms. If fermentation has occurred, the medium surrounding the colonies will appear yellow. Obviously, if the organism could not grow in the salt, it could not ferment the mannitol.

Week Two—Gram-Positive Bacteria Unknown Procedure

Week Two Materials

Gram staining reagents, spore staining reagents, acid fast staining reagents, hydrogen peroxide in small dropper bottle, nitrite solution A (sulfanilic acid), nitrite solution B (dimethyl-alpha-naphthyl amine), zinc dust, and metal spatulas (to dispense zinc dust).

1. Retrieve your plates and tubes containing your "unknown" which you inoculated last week.
2. Turn to the *Inoculation and Interpretation Chart for Unknowns*. Follow along the section named *Interpretation of Reactions*. This contains directions on how to run and interpret all your tests.
3. Turn to the Data Sheet: Unknown Exercise to record the results of your work. This sheet will contain all the data collected from tests you perform to identify your unknown.
4. REMINDER: Persons who discovered they have a Gram-positive rod must do an endospore stain. It is recommended to have the instructor observe the slide under oil immersion.
5. REMINDER: Persons who did the endospore stain and discovered they have a Gram-positive non-sporing rod must do an acid-fast stain. It is recommended to have the instructor observe the slide under oil immersion.

Test Results for the Identification of an Unknown Gram-Positive Bacterium

When looking up a specific bacterium in ***Bergey's Manual***, you will find that the *Manual* only records the results of the tests they feel are the most important tests for that bacterium. Therefore, you **may not find results recorded in *Bergey's Manual* for all the tests you ran.** Not every test is important for your particular organism; but it may be critical for someone else in the lab.

Test	*Staphylococcus aureus*	*Staphylococcus epidermidis*	*Streptococcus mutans*	*Micrococcus luteus*
Glucose broth	+	+	+	–
Sucrose broth	+	+	–	–
Lactose broth	+	Weak +	–	–
Mannitol broth	+	–	+	–
Motility	non	non	non	non
FT	Fac anaerobe	Fac anaerobe	Fac anaerobe	Ob aerobe
Oxidase	–	–	–	Weak +
Catalase	+	+	–	+
Nitrate broth	+	+	–	–
TSA-B (blood)	β-hemolysis	α/γ-hemolysis	γ-hemolysis	γ-hemolysis
MSA	Salt +, mann +	Salt +, mann –	No growth	Salt +, mann –
Special notes			Cells may be slightly oval	Yellow colony color

Test	*Bacillus subtilis*	*Lactobacillus caseii*	*Corynebacterium xerosis*	*Mycobacterium smegmatis*
Glucose broth	+	−	+	−
Sucrose broth	+	−	+	−
Lactose broth	−	−	−	−
Mannitol broth	Weak +	−	−	−
Motility	motile	non	non	non
FT	Obligate aerobe	Fac anaerobe usually w/ little or no aerobic growth	Fac anaerobe	Obligate aerobe
Oxidase	+	−	−	+
Catalase	+	−	+	+
Nitrate broth	+	−	+	Weak +
TSA-B (blood)	β	γ-hemolysis	γ-hemolysis	γ-hemolysis
MSA	no growth	No growth	Salt +, mann −	No growth
Endospore/ Acid fast	Endospore +	−/−	−/−	Acid fast +
Special notes	Cells form chains	Very small colonies	Often pleomorphic	Weak + gram stain

Name ______________________________

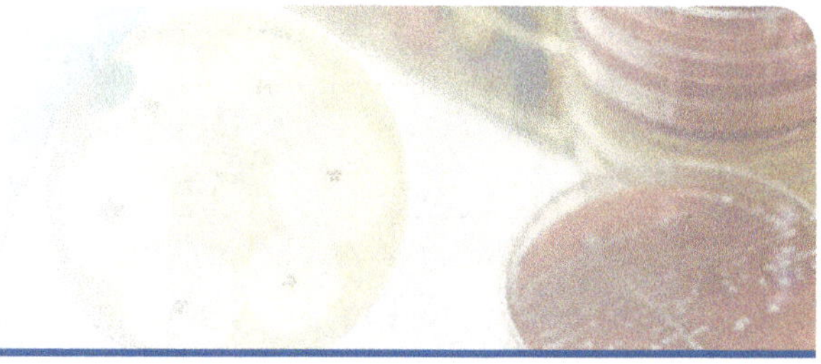

Exercise 4.2 Lab Report

UNKNOWN NUMBER:

NAME OF ORGANISM:

DATA SHEET: UNKNOWN EXERCISE

Test	Results: Circle choice or Record Results Observed	Comments
1. Gram reaction (Give type, shape and arrangement)	Gram-positive Rods Cocci Arrangement:	
2. Broth tube: a. Degree of turbidity	Slight Moderate Abundant None	
b. Pigmentation	Color:____________________	
c. Surface growth features	Ring Pellicle Membrane Lattice None	
d. Sedimentation	Slight Moderate Abundant None	
3. Sugar fermentation		
a. Glucose (G)	Acid (+) or (−); Gas (+) or (−)	
b. Sucrose (S)	Acid (+) or (−)	
c. Lactose (L)	Acid (+) or (−)	
d. Mannitol (M)	Acid (+) or (−)	
4. Motility test (MT)	Motile Nonmotile	

(continued)

DATA SHEET: UNKNOWN EXERCISE

Test	Results: Circle choice or Record Results Observed	Comments
5. Fluid thioglycollate (FT)	Aerobe Facultative anaerobe	
6. Oxidase test	Positive for oxidase enzyme Negative for oxidase enzyme	
7. Nitrate broth tube (N)	Positive for nitrate reduction Negative for nitrate reduction	
8. Catalase test	Positive for catalase enzyme Negative for catalase enzyme	
9. Blood agar plate (TSA-B)	Alpha-hemolysis Beta-hemolysis Gamma-hemolysis	
10. Mannitol salt agar plate (MSA)	Positive for growth in 7.5% salt Negative for growth in 7.5% salt Positive for mannitol fermentation Negative for mannitol fermentation	
11. Endospore stain results (if G+ rod)	Endospore-former Non-endospore-former	

1. Why must you add zinc powder if the nitrate broth remains colorless after the addition of the nitrate A and nitrate B reagents?

____

____

2. Explain why the results are inaccurate for the oxidase test if you wait too long to read the results.

____

____

3. An unknown organism is found growing throughout the media in a FT tube (top, bottom, middle). What conclusions can you draw from the results of this growth pattern?

____

4. An unknown organism is growing in small colonies on an MSA plate. The media is red following the incubation period. What can you conclude from the results of this growth pattern?

____

____

5. An unknown organism isolated from your nose is growing in small colonies on an MSA plate.

   The media is yellow.

   a. What can you conclude about this organism?

   ____

   b. You perform a Gram stain and find Gram-positive cocci in clumps. You perform a catalase test and find it to be catalase-positive. What organism do you most likely have?

   ____

6. List some notable diseases your organism may cause. If your organism is not known to cause disease list some diseases caused by a relative.

____

____

Lab Exercise 4.3

Gram-Negative Bacteria Biochemical Testing

Student Objectives

1. Perform proper aseptic techniques while inoculating various types of media.
2. Know how to use the "Flow Chart for the Identification of an Unknown Bacterium" to determine the sequence of tests required to identify an unknown bacterium.
3. Make the initial Gram stain determination for an unknown bacterium.
4. Make the initial morphological determination for an unknown bacterium.
5. Determine the oxygen requirements of an unknown bacterium.
6. Determine the growth temperature requirements of an unknown bacterium.
7. Use differential media to determine:
   a. Motility
   b. Sugar fermentation properties
   c. Nitrate reduction
   d. Oxygen usage
   e. Oxidase production
   f. EMB growth patterns
   g. IMViC test results
   h. Urease production
8. Identify one of the following unknown Gram-negative cultures: *Alcaligenes faecalis*, *Citrobacter freundii*, *Enterobacter aerogenes*, *Escherichia coli*, *Klebsiella pneumoniae*, *Pseudomonas aeruginosa*, *Serratia marcescens*, and *Proteus vulgaris*.

Name ______________________________

Pre-Lab Exercise 4.3

1. Answer the following questions *assuming* you inoculated the sugar fermentation tubes with an unknown gram-negative bacterium and you observed the following results after 24 hours of incubation:

Glucose	Sucrose	Lactose	Mannitol
(+) with gas	(+)	(+)	(+)

   a. What color did these tubes likely change to?

   ____

   b. List the sugar(s) that are fermented.

   ____

   c. Refer to the identification chart at the end of this lab, and list the organisms you would further investigate if your unknown organism produced these results?

   ____

   d. What other information can be gained from the sugar fermentation tests?

   ____

   e. Assume you put the glucose tube back into the incubator overnight, and when you returned the next day the tube was a bright fuchsia color. **Explain** what may have happened to cause the color change.

   ____

   ____

2. What enzyme is detected by the Indole test?

   ____

3. What enzyme is present in a positive oxidase test?

   ____

4. Describe a simmon's citrate slant that is positive for citrate utilization.

   ____

   ____

5. What product is detected after the addition of the VP-A and VP-B reagents in the Voges-Proskauer test?

____

6. What does the production of acid without a carbon dioxide gas bubble in the inverted Durham tube mean in a sugar fermentation tube?

____

____

7. Describe the growth of *E. coli* on an EMB Plate.

____

____

8. For what family of bacteria is the IMViC series of tests most commonly used?

____

9. When inoculating your media, how might you contaminate your media, and how would those errors affect the results?

____

____

____

In this lab, we will be performing tests that will aid in the identification of unknown Gram-negative bacteria cultures.

Week One—Gram-Negative Bacteria Unknown Procedure

Materials

Materials per student; 1 of each of the following: Sugar fermentation tubes (glucose, sucrose, lactose, and mannitol), TSA plate, motility agar plate, fluid thioglycollate tube (FT), nitrate broth tube, EMB plate, Tryptone broth tube, 2 MR-VP tubes, Simmons Citrate slant, and Urea tube.

Culture per student: One of the following unknown Gram-negative cultures: *Alcaligenes faecalis*, *Citrobacter freundii*, *Enterobacter aerogenes*, *Escherichia coli*, *Klebsiella pneumoniae*, *Pseudomonas aeruginosa*, *Serratia marcescens*, and *Proteus vulgaris*.

1. The first data you record is on the characteristics of your unknown as it grows in broth. This includes: (a) degree of turbidity, (b) color, (c) surface growth features, and (d) sedimentation.
   a. Degree of turbidity: View your broth culture and compare it with others in the class. Determine the relative amount of growth in your tube. This is a subjective determination and should be recorded as "*slight*," "*moderate*," or "*abundant*."
   b. Color: If your cultures displays a color (i.e., red, yellow) record that information in the data sheet. You might want to hold your tube against a white paper background to help make this determination.
   c. Surface growth features: Certain bacteria form a characteristic growth pattern on the surface of liquid media. This may take the form of a: (a) "ring" around the inside of the tube; (b) "pellicle" or thick, rubbery-looking layer; (c) "membranous" or thin, filmy-looking layer; (d) "lattice" or net-like layer; or (e) no observable layer.
   d. Sedimentation: Many bacteria grow so rapidly and are so dense that they settle to the bottom of the tube (i.e., even gravity affects bacteria). This is a subjective determination and should be recorded as "*slight*," "*moderate*," or "*abundant*."
2. Gather the materials and inoculate according to the *Inoculation and Interpretation Chart* below. Follow the directions in the section named *Inoculation Procedure*. Usc your broth culture for all inoculations.
3. Use the broth culture to perform the Gram stain on your unknown to confirm that you have a Gram-negative organism and determine the cell morphology and arrangement. Record the results in the data sheet. If you are not successful with your Gram staining, repeat the procedure. You cannot progress in the identification of your unknown without obtaining valid results with the Gram reaction. It is recommended to have the instructor observe the slide under oil immersion.

INOCULATION AND INTERPRETATION CHART FOR UNKNOWNS

Media	Inoculation Procedure	Interpretation of Reactions
1. Sugar Fermentation + Peptone Oxidation a. Glucose (G) b. Sucrose (S) c. Lactose (L) d. Mannitol(M)	**Used to show ability of bacteria to ferment various sugars or peptone oxidation.** Aseptically inoculate each of these sugar tubes with one loopful of your unknown. Incubate for 24 hours at 37°C.	**Negative**: No growth or no change in the red color. **Positive for fermentation**: Turbid growth with change to yellow color. Phenol red is added as the pH indicator, which turns yellow if sufficient acid is produced by the fermentation of the sugar. The pH must drop below 6.8 for the color change to occur. The inverted Durham tube is used to collect gas from the fermentation process. • ***Positive for acid production*** indicated by color change from red to yellow. • ***Positive for gas production*** indicated by at least 10% of the culture in the Durham tube being replaced by gas. • NOTE: A tube may be positive for acid without being positive for gas. However, the opposite is not true. There cannot be gas if there was not acid production due to fermentation. **Positive for aerobic oxidation of peptone (non-sacrolytic bacteria):** Turbid growth with a change to fuschia (bright pink) color. Phenol red is added as the pH indicator, which turns fuschia if a sufficient alkaline reaction is produced by the oxidation of peptone which creates amine by-products such as ammonia. The pH must rise above 8.4 for the color change to occur.

INOCULATION AND INTERPRETATION CHART FOR UNKNOWNS

Media	Inoculation Procedure	Interpretation of Reactions
2. Motility agar plate (MT)	**Used to identify ability of bacteria to move (i.e., flagellated cells).** Stab the **MT agar plate** once in the center with an inoculating **needle**. **Incubate for 5 days at 37°C, be sure to place the MT plate in the special incubation tray labeled "INCUBATE FOR 5 DAYS."**	If there is no growth, the organism was not able to grow in the media and you cannot interpret the results. List as "no growth" and "no determination" with regard to motility. **Negative:** Growth with color change only along the line of the stab, with no pattern spreading out. **Positive:** Growth with media color change to violet/red. Spreading pattern into surrounding media; may appear as snowflakes.
3. Fluid thioglycollate tube (FT)	**Used to show oxygen usage of bacteria.** Inoculate the tube using a sterile loop. Place one loopful of the organism *just below the top surface* of the agar. This area usually has a faint pink color due to the presence of the resazurin. Incubate for 24 hours at 37°C.	The upper portion of the agar medium contains the greatest amount of free oxygen because the gas may easily diffuse into the medium from the atmosphere. The large center zone contains a decreasing amount of oxygen because the gas has more difficulty reaching that portion of the medium. Those bacteria growing ONLY in the upper portion of the tube culture would be **aerobic**. Those growing both in the upper portion and in the center zone of the tube culture would be **facultative anaerobes**.

(*continued*)

INOCULATION AND INTERPRETATION CHART FOR UNKNOWNS

Media	Inoculation Procedure	Interpretation of Reactions
4. TSA plate for **oxidase** test.	**Used to demonstrate the ability of a bacterium to produce the enzyme oxidase, capable of reducing oxygen during the electron transport chain of cell respiration.** Streak the plate for isolation using the quadrant method with a loopful of your unknown. Perform the quadrant method. Incubate for 24 hours at 37°C.	No growth: List as "no growth" and "no determination for oxidase". Turbid with growth: Carefully add 2 or 3 drops of the oxidase reagent onto an isolated colony. **Be certain to have on gloves and safety glasses, as the reagent is carcinogenic.** **Positive test:** The colony will change to pink/maroon, within 2 minutes. **Negative test:** Remains colorless, after the 2 minute time interval.
5. Nitrate broth tube (N)	**Used to demonstrate the ability of a bacterium to produce the enzyme nitratase, capable of converting nitrate to nitrite.** Aseptically inoculate the tube with one loopful of your unknown. Incubate for 24 hours at 37°C.	Transfer 1 ml of the inoculated nitrate reduction media to 1 well of a spot plate. Add 2–3 drops of nitrate A and 2–3 drops of nitrate B. Appearance of a red color within 30 seconds is a positive test for nitrate reduction. Repeat for remaining organisms. If there is no red color, scoop a tiny amount of zinc dust with a toothpick and add it to the wells with no red color and mix. If the broth turns red after the zinc, the test is negative for nitrate reduction. If broth remains clear after the addition of zinc, this indicates nitrate was reduced all the way to **gas.** **Be certain to have on gloves and safety glasses, as the reagent is carcin**ogenic.

INOCULATION AND INTERPRETATION CHART FOR UNKNOWNS

Media	Inoculation Procedure	Interpretation of Reactions
6. Eosin methylene blue (EMB) plate	**Used to distinguish between members of the *Enterobacteriaceae* family.** Streak the plate for isolation using the quadrant method with a loopful of your unknown. Perform the quadrant method. Incubate for 24 hours at 37°C.	No growth = list as "no growth" and "no determination". Turbid with growth: Colonies may be smaller and have a green metallic sheen, larger and have a rosy-red color, or may be colorless. Lactose fermenters metabolize the lactose in the media and produce acid by-products, causing a color change. Strong acid production by organisms such as *E. coli* results in a green metallic sheen. Weaker fermentation of lactose results in colonies with a rosy-red color. Non-lactose fermenters remain colorless (or at least are no darker than the color of the media).
7. Indole test (I).	**Used to demonstrate the ability of a bacterium to produce the enzyme tryptophanase. This enzyme acts on the amino acid tryptophan to produce the end product indole.** **This is the "I" portion of the IMViC tests.** Inoculate a tube of tryptone (I) broth with one loopful of your unknown. Incubate for 24 hours at 37°C.	Pipette approximately 1 ml of the tryptone broth into a spot plate. Dip on Kovac's Reagent Strip into the broth. A positive test is indicated by a color change on the strip to a pink/red within 20 seconds. A negative test will have no color change to the strip. Alternatively you can simply touch the strip to an isolated colony on a TSA plate. Do not touch a portion of the plate where other reagents have been added to the plate. A positive test is a color change on the strip to a pink/red.

(*continued*)

INOCULATION AND INTERPRETATION CHART FOR UNKNOWNS

Media	Inoculation Procedure	Interpretation of Reactions
8. **MR**-VP – Tube 1: Methyl red test (MR)	**MR (methyl red) used to show mixed acid fermentation ability of bacteria.** **This is the "M" portion of the IMViC tests.** Inoculate this tube with one loop of your unknown. Label the tube "MR," to help you to remember to run the MR test with this tube next week. Incubate for 24 hours at 37°C.	No growth = List as "no growth" and "no determination". NOTE: Be sure you're using the MR tube which was incubated on the 24-hour rack! Turbid with growth: Add 1 drop of methyl red pH indicator to this tube. Methyl red is yellow at a pH greater than 6.4. It turns bright red in a high acid solution with a pH below 4.4. ****Positive test**: If the medium changes to a red color, the bacteria have performed a mixed acid fermentation. This indicates that the bacteria are functioning anaerobically, fermenting the glucose in the media and releasing a variety of organic acids including lactic, acetic, succinic, and formic, plus ethanol, CO_2 and H_2. **A color change to orange is NOT a positive result. **Negative test:** If the medium does not change to a red color, the bacteria have not performed a mixed acid fermentation.

INOCULATION AND INTERPRETATION CHART FOR UNKNOWNS

Media	Inoculation Procedure	Interpretation of Reactions
9. MR-**VP** – Tube 2: Voges – Proskauer test (VP)	**VP (Voges-Proskauer) is used show bacterial production of acetoin, also known as 2,3-butanediol.** **This is "V" the portion of the IMViC tests.** Inoculate this tube with one loop of your unknown. Label the tube "VP," to help you to remember to run the VP test with this tube next w**eek.** **Incubate for 5** days.	No growth = List as "no growth" and "no determination". **NOTE**: Be sure you're using the VP tube, which was incubated on the 5-day rack. Turbid with growth: Add 15 drops of VP-A (Barritt's Solution A) followed by 5 drops of VP-B (Barritt's solution B). Mix in vortex mixer on-and-off for about one minute. Allow to sit in the test tube rack at your table **for a half hour.** **Positive test**: Color change to brick red. This indicates that the bacteria have the genetic ability to produce the compound 2,3 butanediol or acetoin. **Negative test:** If the medium does not change color to brick red. Barritt's solution A (VP-A) is naphthol and Barritt's solution B (VP-B) is potassium hydroxide (KOH).
10. Simmon's Citrate (SC) agar slant	**Used to demonstrate the genetic ability of a bacterium to utilize sodium citrate as its sole source of carbon.** **This is the "C" portion of the IMViC tests.** Use an inoculating loop to streak one straight line on the surface of the agar. Use an inoculating needle to stab down into the media. Incubate for 24 hours at 37°C.	The medium is made so that bromothymol blue starts out as green. If the bacteria can use citrate as their sole carbon source, a reaction goes on that leads to the formation of Na_2CO_3 (sodium carbonate), a base. When the pH rises above 7.6, the medium becomes bright blue. **Positive test:** Growth on the surface and a color change from green to a bright blue. **Negative test:** Growth with no color change. In this media, sodium citrate is the only carbon present and ammonium is the only nitrogen present. Bromothymol blue is the pH indicator. It is yellow below 6.0, green in its pH range of 6.0–7.6, then turns a bright Prussian blue color at a pH greater than 7.6.

(*continued*)

INOCULATION AND INTERPRETATION CHART FOR UNKNOWNS

Media	Inoculation Procedure	Interpretation of Reactions
11. Urea (U) broth tube	**Used to demonstrate the ability of a bacterium to produce the enzyme urease, capable of hydrolyzing urea.** Inoculate the "U" tube with one loopful of unknown. **Incubate for 5 days. If your optimal temperature is 37°C, place the U tube in the incubation rack labeled "INCUBATE FOR 5 DAYS."**	**Positive test:** Growth in the medium and a color change from amber to bright fuchsia. Phenol red is used as the pH indicator. If the organism produces the enzyme urease, allowing it to hydrolyze urea as a food source, ammonia is formed as a by-product of the reaction. The pH in the tube rises as ammonia accumulates. In a pH higher than 8.4, phenol red changes to a fuchsia (bright pink) color. **Negative test:** Growth in the medium with no color change.

Week Two—Gram-Negative Bacteria Unknown Procedure

Week Two Materials

Gram staining reagents, nitrite solution A (sulfanilic acid), nitrite solution B (dimethyl-alpha-naphthyl amine), zinc dust, metal spatulas (to dispense zinc dust), indole (Kovac's) reagent, methyl red reagent, clean empty test tube, one sterile 1-mL pipette, mechanical pipetter, VP-A (Barritt's-A) and VP-B (Barritt's-B) reagents, and oxidase reagent.

1. Retrieve your plates and tubes containing your "unknown" which you inoculated last week.
2. Turn to the *Inoculation and Interpretation Chart for Unknowns*. Follow along the section named *Interpretation of Reactions*. This contains directions on how to run and interpret all your tests.
3. Turn to the *Data Sheet: Unknown Exercise* to record the results of your work. This sheet will contain all the data collected from tests you perform to identify your unknown.

Test Results for the Identification of an Unknown Gram-Negative Bacterium

When looking up a specific bacterium in *Bergey's Manual*, you will find that the *Manual* only records the results of the tests they feel are the most important tests for that bacterium. Therefore, you **may not find results recorded in *Bergey's Manual* for all the tests you ran.** Not every test is important for your particular organism, but it may be critical for someone else in the lab.

Test	*Alcaligenes faecalis*	*Citrobacter freundii*	*Enterobacter aerogenes*	*E. coli*	*Klebsiella pneumoniae*	*Proteus vulgaris*	*Pseudomonas aeruginosa*	*Serratia marcescens*
Glucose broth	−	+, gas	+, gas	+, gas	+, gas	+	+	+
Sucrose broth	−	+	+	−	+	+	−	+
Lactose broth	−	+	+	+	+	−	−	−
Mannitol broth	−	+	+	+	+	−	−	+
Motility	motile	motile	motile	motile	Non*	motile	motile	motile
FT	Obligate aerobe	Fac anaerobe	Fac anaerobe	Fac anaerobe	Fac anaerobe	Fac anaerobe	Obligate aerobe	Fac anaerobe
Oxidase	+	−	−	+ or −	−	−	+ or −	−
Nitrate broth	−	+	+	+	+	+	+	+
EMB	clear	Pink-red sometimes w/ black center	Purple/red sometimes w/ black center	Dark w metallic green sheen	Purple/red sometimes w/ black/green metallic center	clear	clear	Pink-red

(*continued*)

Test	*Alcaligenes faecalis*	*Citrobacter freundii*	*Enterobacter aerogenes*	*E. coli*	*Klebsiella pneumoniae*	*Proteus vulgaris*	*Pseudomonas aeruginosa*	*Serratia marcescens*
Urea	–	Weak +*	–	–	+	+	–	–*
Indole	–	–	–	+	–	+	–	–*
MR	–	+	–	+	+	+	–	–*
VP	–	–	+	–	-	–	–	+
Citrate	–	+	+	–	+	–	+	+
Special notes		*often appears negative			*appears motile w/slimy white center	Often darkens agar on TSA plate	Obligate aerobe except in the presence of nitrate	*red pigment makes some tests difficult

Name ______________________________

Exercise 4.3 Lab Report

UNKNOWN NUMBER:

NAME OF ORGANISM:

DATA SHEET: UNKNOWN EXERCISE

Test	Results: Circle Choice or Record Results Observed	Comments
1. Gram reaction (Give type, shape & arrangement)	Gram-negative Rods Cocci Arrangement:	
2. Broth tube: a. Degree of turbidity	Slight Moderate Abundant None	
b. Pigmentation	Color:__________________	
c. Surface growth features	Ring Pellicle Membrane Lattice None	
d. Sedimentation	Slight Moderate Abundant None	
3. Sugar fermentation		
a. Glucose (G)	Acid (+) or (−); Gas (+) or (−)	
b. Sucrose (S)	Acid (+) or (−)	
c. Lactose (L)	Acid (+) or (−)	
d. Mannitol (M)	Acid (+) or (−)	
4. Motility Test (MT)	Motile Nonmotile	
5. Fluid thioglycollate (FT)	Aerobe Facultative anaerobe	

(*continued*)

DATA SHEET: UNKNOWN EXERCISE

Test	Results: Circle Choice or Record Results Observed	Comments
6. Oxidase test	Positive for oxidase enzyme Negative for oxidase enzyme	
7. Nitrate broth tube (N)	Positive for nitrate reduction Negative for nitrate reduction	
8. Eosin methylene blue (EMB) plate	Green metallic sheen colonies Pink colonies with black center Purple/ Red colonies Clear/ Colorless colonies Other: Describe in comments	
9. Tryptone (I) (indole test)	Positive for indole test Negative for indole test	
10. **MR**-VP – 1: **MR** Test	Positive for mixed acids Negative for mixed acids	
11. MR-**VP** – 2: **VP** Test	Positive for butanediol Negative for butanediol	
12. Simmon's citrate (SC) test	Positive for citrate utilization Negative for citrate utilization	
13. Urea (U) tube	Positive for urease Negative for urease	

1. A fellow student inoculates an MR-VP tube to perform a methyl red test. He/she immediately adds 1 drop of methyl red to the tube and records the result as "negative". What did the student do incorrectly?

____

____

2. A student retrieves a sugar fermentation tube from a 24-hour incubation rack. He observes the tube to be yellow. He adds 4–5 drops of Kovac's reagent to the tube and records the test result as "negative". What did the student do incorrectly?

____

3. Another student in the lab determines that he/she has a Gram-negative rod. She performs the endospore stain and records it as "negative." She then performs the acid-fast stain and records that as "negative" also. What did the student do incorrectly?

____

____

4. List some notable diseases your unknown organism causes.

____

____

5. What are the general steps required to identify unknown bacteria?

____

____

6. Following incubation of an EMB plate you observe clear colonies. What organisms could this unknown organism be?

____

____

7. Who am I? I am a gram negative rod and I do not reduce nitrate.

____

Microbial Growth and Nutrition

Microorganisms can be found in the harshest of environments including deep sea thermal vents, polar regions, and hot springs. These organisms survive in what we consider extreme environments, but the microbes have evolved mechanisms to withstand the environmental conditions they live in. These organisms have limited environmental ranges in which the species can grow; for example, every species has a specific temperature range in which the organism can survive. Organisms found in deep sea thermal vents would be extremely difficult to culture at sea level due to the pressure and specific nutrients available.

Microbial growth refers to the increase in population of microbes. Therefore, the process of reproduction results in growth. The process by which bacteria reproduce is called **binary fission** and results in two daughter cells that are fully mature adult cells capable of the same processes as the parent cell. There is no relationship to microbe size and growth as there is when we consider human growth or plant growth.

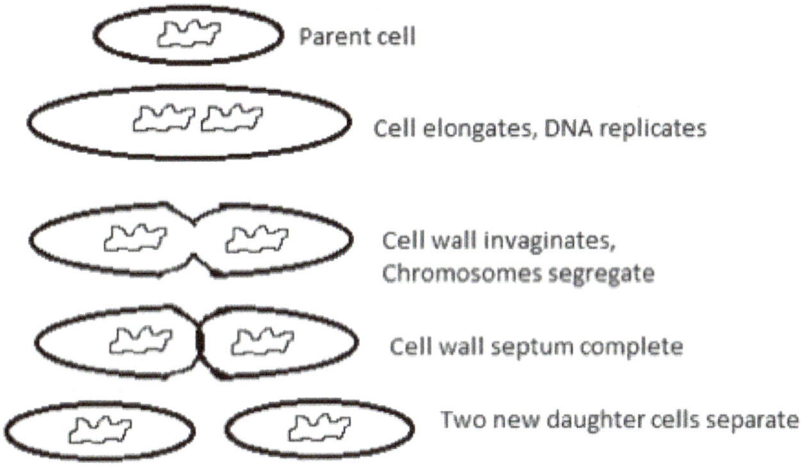

Fig. 5.1 Binary Fission
Courtesy of Authors

Microbiologists refer to the amount of time an organism takes to reproduce as the **generation time**. The generation time varies based on chemical (ex. food availability) and physical (ex. temperature) conditions. *Escherichia coli*, for example, can reproduce in as little as 20 minutes under **optimal conditions**, which are the physical and chemical factors that are the most favorable for that specific organism. *Mycobacterium tuberculosis* has a generation time of 15–20 hours under ideal conditions, and more than 48 hours in less than optimal conditions. The human body's optimal temperature is 37°C; therefore, most human pathogens have optimal temperatures for growth within a few degrees of 37°C. The **growth rate** is the number of generations

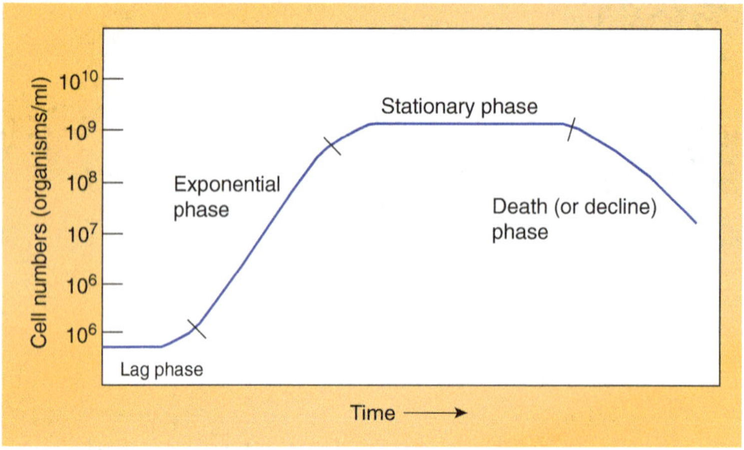

Fig. 5.2 Cell Population Growth Curve Generated Using Spectrophotometry

per hour. An important note is that growth rate and generation time are inversely related. For example, if the generation time of E. coli is 20 minutes, then the growth rate is 3 generations per hour.

Because bacteria grow this way, their population size grows very rapidly. This type of growth, where the population doubles with each generation is called logarithmic or exponential growth. Allowed to continue unchecked, exponential growth can quickly lead to a lot of cells. For example:

- Start with 1 *E. coli* with a generation time of 20 minutes.
- In 7 hours there would be roughly 1,000,000 cells.
- In 10 hours there would be around 1,000,000,000 cells.
- In 24 hours there would be about 1, 000,000,000,000,000,000,000 cells!

You can see that bacteria growing exponentially expand their populations very rapidly. Luckily, even with logarithmic growth, we are not completely taken over by bacteria. Their growth is limited by environmental factors.

Microbial growth is predictable in liquid media given the organism has access to appropriate nutrients. Once an organism is inoculated into a growth media, the organism will experience a **Lag phase**, while the organism is adapting to the environment. Following the Lag phase, the organism will begin growing at an exponential rate. This stage is referred to as the **Log phase**, during which the organism's population is doubling during each generation time. Eventually the nutrients begin to become limited and growth slows. This phase is called the **Stationary phase**, during which the number of cells reproducing equals the number of cells that are dying. Finally the **Death phase** occurs, when the number of dying cells is greater than the number of cells reproducing (Figure 5.2).

Measuring Microbial Growth

Microbiologists are able to measure and monitor cell growth by using spectrophotometry. **Spectrophotometry** is considered an indirect method to estimate a population size or density. A spectrophotometer is used to measure the turbidity of a solution, also called the optical density. A light beam is directed through a liquid sample, and a sensor records how much light passed through on the other side. The higher the cell density (correlates to a greater bacterial

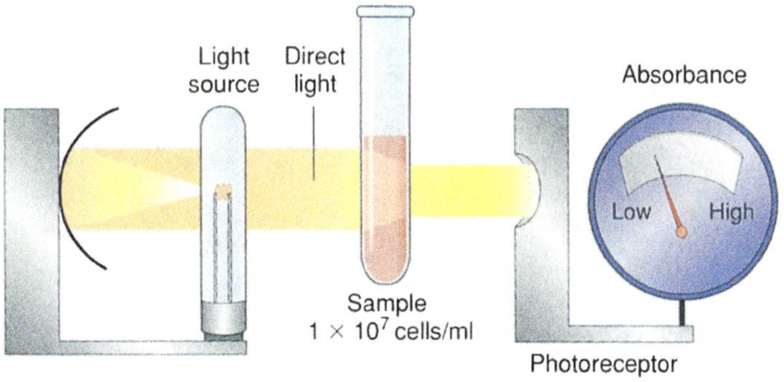

Fig. 5.3A Demonstrates a Spectrophotometer Measuring a Cell Sample with 10,000,000 Cells

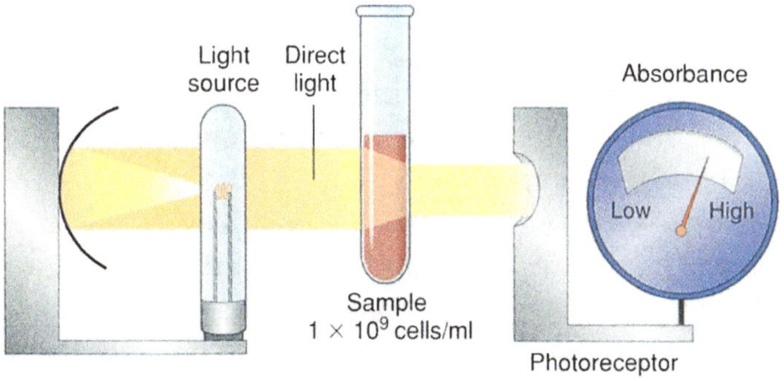

Fig. 5.3B Demonstrates a Spectrophotometer Measuring a Cell Sample with 1,000,000,000 Cells. A Sample with more Growth (Higher Cell Density) will Give a Reading on the Spectrophotometer as a Higher Absorbance

population) in a liquid culture, the more turbid the solution will be and lesser light will travel through the sample. A computer within the spectrophotometer gives you either an absorbance output reading, in which a higher absorbance reading indicates a higher cell density (Figure 5.3) or a transmittance value where a higher transmittance indicates a lower cell density. **It is possible to calculate between percent transmission (%T) and absorbance by using the following formula:**

$$\textbf{Absorbance} = -\log\ (\%T/100)$$

An important thing to note is that this technique is only useful when the cell density exceeds 1 million cells per milliliter; densities below this number do not produce enough turbidity for an accurate reading. Another important point to note is that spectrophotometry only correlates to a total number of cells and does not distinguish between living and dead cells.

Nutrients for Growth

Microorganisms are capable of using a wide variety of nutrients for their energy needs, in order to build organic molecules for cellular structures, which are required for growth. The most

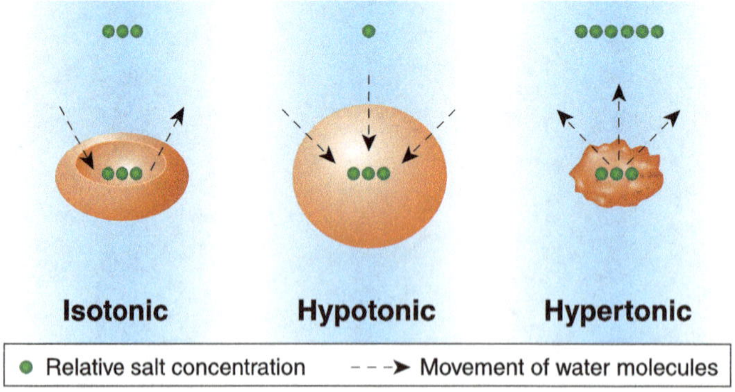

Fig. 5.4 Example of How Water Diffuses When a Cell is Placed into Varying Salt Concentrations
Blamb/Shutterstock.com

common nutrients must contain **required elements** such as carbon, oxygen, nitrogen, and hydrogen. Other elements needed for growth include phosphorus, magnesium, potassium, iodine, and sulfur. Without required elements, the organism would not be able to function or reproduce. Elements that are required in very small amounts are referred to as **trace elements** and include iron, iodine and zinc. When an organism is growing, the cells are transforming chemical elements into carbohydrates, proteins, lipids, and nucleic acids, the four macromolecules of life also considered **growth factors**. **Fastidious** organisms have very specific and complex nutritional requirements and when grown in the lab require the use of special media.

Osmotic Pressure

Because most microbes get their nutrients from surrounding water, osmotic pressure is really important. Cells will experience the following conditions depending on their environment (Figure 5.4):

Hypertonic: The concentration of solute outside of the cell is greater than the concentration of solute inside the cell. The solute is the dissolved substance such as sugar or salt.

Hypotonic: The concentration of solute outside the cell is less than the concentration of solute inside the cell.

Isotonic: The concentration of the solute outside the cell is the same as the concentration of solute inside the cell.

Certain compounds act as growth inhibitors, which affect the osmotic pressure of a cell and reduce the growth rate or prevent a microorganism from growing altogether. A notable inhibitor is salt. In a hypertonic environment, salt inhibits microbial growth by disrupting the osmotic pressure of the cell. Since water is needed for life, salt reduces the availability of water to a cell, since water will diffuse from a cell's cytoplasm when in a high salt environment.

Some genera of microorganisms can live and even thrive in the presence of salt while others quickly die due to **crenation**, or shriveling of the cell's cytoplasm. This results when water diffuses out of the cell due to a high solute concentration outside the cell membrane. Salt-tolerant organisms are considered **halophiles**. **Obligate halophiles** are adapted to grow under high osmotic pressure and have an absolute requirement for a high salt concentration. Obligate halophiles would be found in environments such as the Great Salt Lake in Utah where the salinity varies greatly from 5% to near 30%; microbes living in the Great Salt Lake would burst

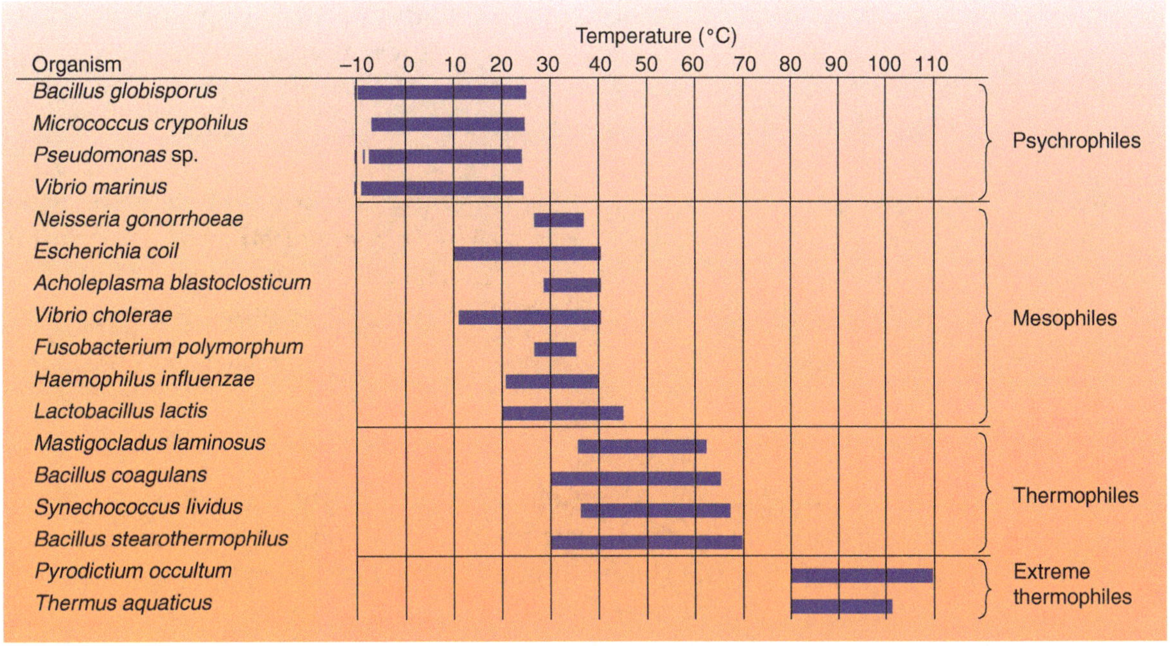

Fig. 5.5 Four Categories of Classification Based on Temperature Ranges for Growth, with Example Microorganisms

if placed into freshwater. **Facultative halophiles** do not require high salt concentrations; however, they can tolerate high salt conditions. A high salt condition would be considered 5% and above. *Staphylococcus aureus* can tolerate salt up to 20%, which allows this organism to colonize the surface of the skin. The role that salt plays on the osmotic pressure of cells accounts for the preserving action in jerky and salted fish.

Temperature

When cultivating microorganisms, especially novel organisms, microbiologists perform a battery of tests which are used to characterize and classify the microorganism. Organisms can be classified into one of six categories based on their temperature requirements for growth.

- **Psychrophiles** are cold-loving organisms that prefer temperatures between −5 and 20°C.
- **Mesophiles** are moderate temperature-loving organisms and are capable of surviving in temperatures between 15 and 45°C.
- **Thermophiles** are heat-loving microbes that are capable of growing in temperatures between 45 and 80°C.
- **Hyperthermophiles** are extremely heat-tolerant and can survive boiling temperatures between 70 and 105°C.
- **Psychroduric** organisms are bacteria that prefer warmer temperatures, but can survive in colder temperatures as well.
- **Thermoduric** organisms are mesophiles that can survive short periods at higher temperatures. These bacteria may contaminate food and survive canning and pasteurization processes.

An important point to note when classifying organisms by temperature requirements is that the optimal temperature for growth is usually in the middle of the range discussed above; for example, a psychrophile would have an optimal growth rate at around 10°C (Figure 5.5). Bacteria can generally survive if they are moved to a lower temperature range than optimum, as this will most likely slow their metabolic processes. However, moving bacteria to a higher temperature range usually results in death, as proteins and enzymes are denatured and nucleic acids are damaged.

Even within the given temperature ranges, there is room for variation. For example, while technically a mesophile, the microbe that causes leprosy, *Mycobacterium leprae* has an optimum temperature of 33°C, lower than the 37°C that is the human core temperature. This lower temperature accounts for the fact that *M. leprae* typically affects the extremities, the cheeks and noses of its victims.

Oxygen

As mentioned earlier, oxygen is often an important element for growth. You will learn later in the semester that oxygen is used as an electron acceptor during aerobic respiration. For now, we can learn how to classify microorganisms based on their oxygen requirements. Microbes can be classified as obligate aerobes, facultative anaerobes, obligate anaerobes, microaerophiles, or aerotolerant. **Obligate aerobes** have an absolute requirement for oxygen, meaning that they cannot grow without oxygen. **Facultative anaerobes** tend to grow better with oxygen; however, they can grow without oxygen as well. Facultative anaerobes can grow without oxygen since they either have the ability to switch their metabolism. **Obligate anaerobes** cannot grow in the presence of oxygen. Oxygen is actually toxic to these types of organisms. **Microaerophiles** have a very specific requirement for oxygen; high concentrations of oxygen can inhibit the growth of microaerophiles. **Aerotolerant** organisms, also considered **fermenters**, are organisms that are indifferent to oxygen; they can grow in the presence of oxygen, yet they do not use oxygen to transform energy. In lab we will use a special media called fluid thioglycolate (FT) media to determine the oxygen requirements of microbes (Figure 5.6).

Organisms capable of surviving in the presence of oxygen will experience damaging effects of oxygen. Toxic superoxides (O_2-) are formed and can oxidize important cell chemicals within a cell. Superoxides are neutralized by an enzyme called **superoxide dismutase**, which forms hydrogen peroxide in the cell (H_2O_2), which is also toxic to the cell. **Catalase**, an enzyme, breaks down hydrogen peroxide into water and oxygen, neutralizing the peroxide. Obligate anaerobes lack these enzymes, which is why they cannot survive in the presence of oxygen. Aerotolerant/ fermenters such as organisms from the genus *Streptococcus* are catalase-negative, yet survive just fine in the presence of oxygen. *Streptococcus* species produce a different enzyme called peroxidase which neutralizes hydrogen peroxide.

The last condition for growth we will discuss is how pH affects growth. Organisms can be classified based on their ability to tolerate different salt concentrations and acid/base levels. These concepts are important for understanding how humans preserve food. Many bacteria, including human pathogens, are considered **neutrophiles** and grow in a neutral pH around 6.6–7.9. Organisms growing in a pH above 8.0 would be considered **alkalophiles**. **Acidophiles** can tolerate acidic conditions of a pH of 6.5 and below (Table 5.1). Few organisms can grow below a pH of 4.0, which is important in preserving foods. Sauerkraut, pickles, cheeses, canned tomatoes, etc., are all preserved from spoilage by acids produced during a fermentation process, or by natural low acidity.

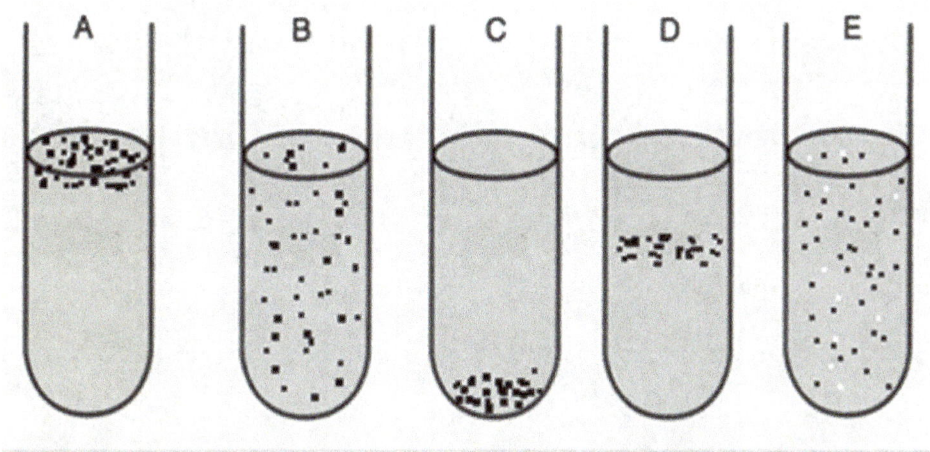

Fig. 5.6 **A)** Growth of an Obligate Aerobe in FT Media. **B)** Growth of a Facultative Anaerobe in FT Media. **C)** Growth of an Obligate Anaerobe in FT Media. **D)** Growth of a Microaerophile in FT Media. **E)** Growth of an Aerotolerant Organism/Fermenter in FT Media

Note: You cannot tell the difference between a facultative anaerobe and an aerotolerant organism; other biochemical tests would need to be performed to determine metabolic characteristics.

Courtesy of Authors

Table 5.1 Examples of pH Tolerance and Optimum Ranges for Certain Types of Prokaryotes

Organism	pH Ranges for Growth	pH Optimum for Growth
Thiobacillus Thiooxidans	1.0–6.0	2.0–2.8
Lactobacillus Acidophilus	4.0–6.8	5.8–6.6
Proteus Vulgaris	4.4–8.4	6.0–7.0
Escherichia Coli	4.4–9.0	6.0–7.0
Enterobacter Aerogenes	4.4–9.0	6.0–7.0
Clostridium Sporogenes	5.0–9.0	6.0–7.6
Nitrobacter sp.	6.6–10.0	7.6–8.6
Nitrosomonas sp.	7.0–9.4	8.0–8.8

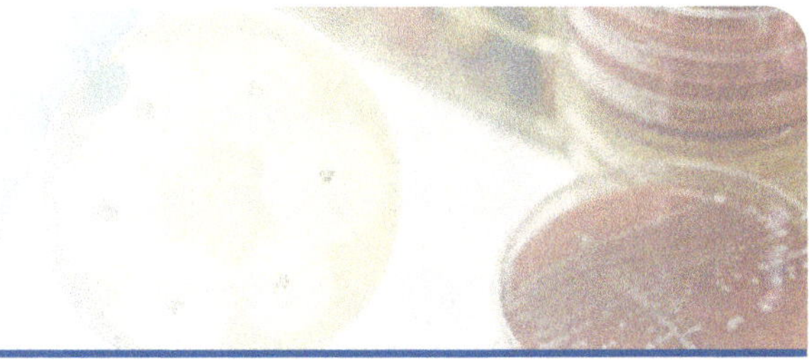

Lab Exercise 5.1

Measuring Microbial Growth, Nutritional Requirements for Growth, and Optimal Conditions for Growth

Student Objectives

1. Understand the basic nutritional requirements for microbial growth.
2. Understand and apply methods for measuring microbial growth.
3. Classify microorganisms based on different growth parameters.
4. Record spectrophotometer data, graph data, and interpret bacterial growth curves.

In this laboratory exercise, you will use a spectrophotometer to measure bacterial growth in response to different growth requirements including temperature, oxygen, and salt concentration. You will record the data in your Lab Report and then answer the attached questions.

Name ____________________

Pre-Lab Exercise 5.1

1. How is plant or animal growth different from microbial growth?

2. What does generation time refer to?

3. What do optimal conditions refer to?

4. How are lag time and stationary phase different?

5. What are the common chemical elements required for growth?

6. How can salt inhibit growth?

7. In this lab activity, how will you measure microbial growth?

8. Describe spectrophotometry.

9. What are some limitations to spectrophotometry?

10. During Week 2, when you use the spectrophotometer, during which step do you see potential for error, and how would those errors affect the results?

Determining the Optimal Temperature For Growth: Week One

Week One Materials

Broth culture per table: *Escherichia coli.*

Materials per table: Six TS broth tube inoculating loops, igniter, and Bunsen burner.

1. Each lab table will experiment with six different temperatures: 5°C, 25°C, 30°C, 37°C, 45°C, and 60°C.
2. Obtain a TS broth tube and label the tube with the temperature, your initials, and table number. Repeat for all temperatures.
3. Aseptically inoculate your broth with one loop of *Escherichia coli.*
4. Place your inoculated tube onto the classroom test tube rack for that specific temperature.

Comparing Growth in the Presence and in the Absence of Oxygen: Week One

Week One Materials

Broth culture per table: *Escherichia coli.*

Materials per table: Two TS broth tubes, inoculating loops, igniter, and Bunsen burner.

1. Obtain two TS broth tubes and label one tube "with oxygen" and the second tube "without oxygen", your initials, and table number.
2. Aseptically inoculate your broth with one loop of *Escherichia coli.*
3. Place your inoculated tube onto the appropriate classroom test tube rack for incubation.

Determining the Optimal Salt Concentration for Growth: Week One

Week One Materials

Broth culture per table: *Escherichia coli.*

Materials per table: Seven TSB+salt tubes of the following concentrations: 0%, 0.5%, 1%, 2%, 3%, 4%, 5%, inoculating loops, igniter, and Bunsen burner.

1. Label the salt tubes with the salt concentration, your initials, and table number.
2. Aseptically inoculate each salt tube with one loop of *Escherichia coli.*
3. Place your inoculated tube onto the classroom test tube rack for incubation.

Determining the Optimal Temperature for Growth: Week Two

Week Two Materials

Materials per table: Incubated broth tubes, one un-inoculated Tryptic Soy Broth control tube, two cuvettes, 10 ml pipette, 10 ml pipetter, UV-Vis Spectrophotometer

1. Turn on the spectrophotometer five minutes before use to allow it to warm up.
2. The number on the left is the wavelength (measured in nm) being tested. Use the "nm" up and down buttons to set the wavelength to 600nm.

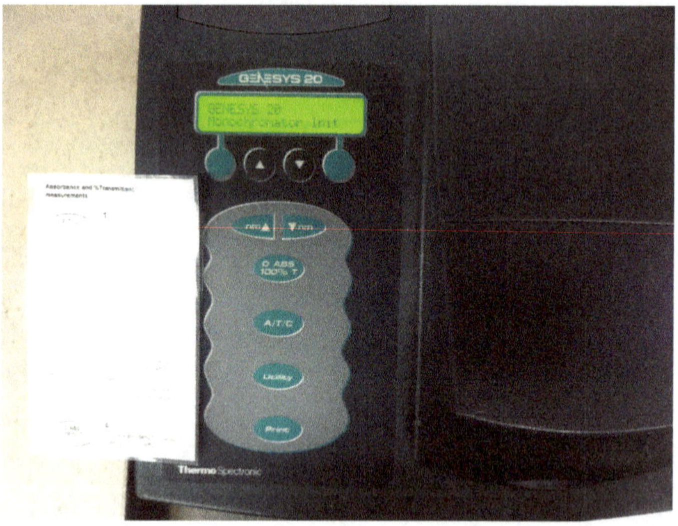

Courtesy of Authors

3. The number on the right is the absorbance (A) or % transmittance (T) reading. If it is not displaying a "T", hit the "A/T/C" button until "T" is on the right.
4. Make sure the cuvettes are clean and dry. Be careful not to touch the cuvettes with your bare hands, as oils from your fingers can interfere with the spectrophotometer.
5. Carefully pipette 3 ml of the un-inoculated Tryptic Soy Broth into the first cuvette.

6. Place the cuvette in the spectrophotometer and close the lid.

Courtesy of Authors

7. Press the "0Abs/100%T" button to set the TSB as the control blank reading.

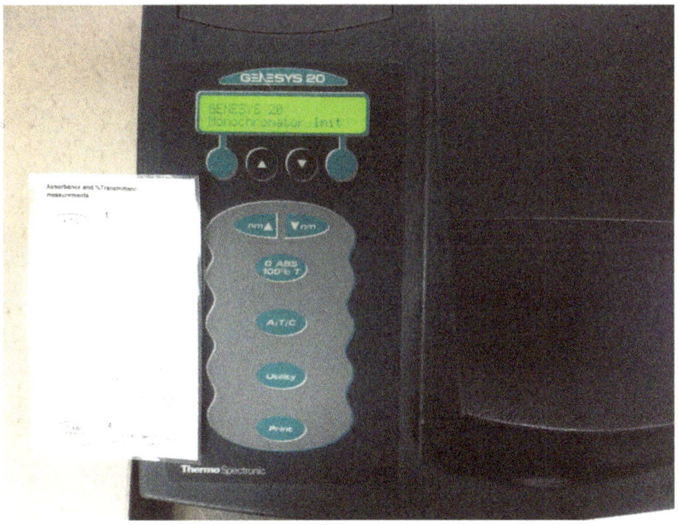

Courtesy of Authors

8. Carefully re-suspend the bacterial sediment and pipette 3 ml of the incubated broth tube (the tube with the bacteria) into the second cuvette.

9. Place the cuvette in the spectrophotometer and record the data as both %T and ABS. To cycle between %T and ABS, press the A/T/C button.

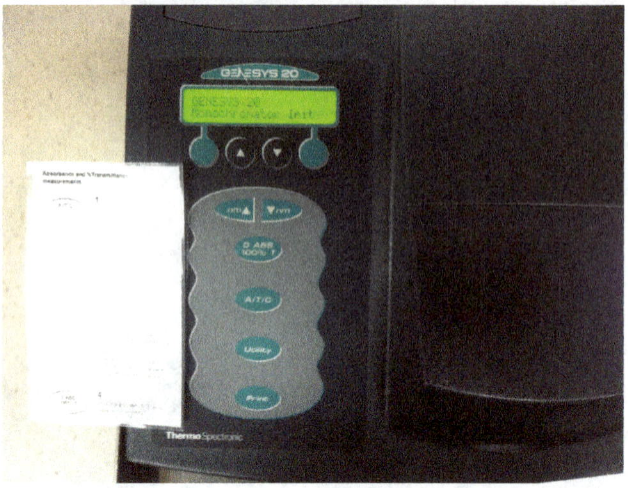

Courtesy of Authors

10. Gather all % transmittance and absorbance data for all temperatures and plot the data on the graph in your Lab Report using the absorbance data.
11. Discard the broth from the cuvette into the waste container, rinse the cuvette with water and repeat for remaining samples.

Comparing Growth in the Presence and in the Absence of Oxygen: Week Two

Week Two Materials

Materials per table: Two incubated broth tubes, two cuvettes, 10 ml pipette, 10 ml pipetter, one un-inoculated Tryptic Soy Broth control tube, UV-Vis Spectrophotometer

1. The Spectrophotometer should still be calibrated. If calibration is needed, repeat Step 7 from the previous protocol.
2. Carefully re-suspend the bacterial sediment and pipette 3 ml of the incubated broth tube (oxygen) into the second cuvette.
3. Place the cuvette in the spectrophotometer and record the data as both %T and ABS. To cycle between %T and ABS, press the A/T/C button.
4. Repeat Steps 2 and 3 for the 'without oxygen" tube.

Determining the Optimal Salt Concentration for Growth: Week Two

Week Two Materials

Materials per table: Incubated broth tubes, one un-inoculated Tryptic Soy Broth control tube (make sure the control you are using is the correct salt concentration), two cuvettes, 10 ml pipette, 10 ml pipetter, UV-Vis Spectrophotometer

1. The Spectrophotometer should still be calibrated. If calibration is needed, repeat Step 7 from the previous protocol.
2. Carefully re-suspend the bacterial sediment and pipette 3 ml of the incubated broth tube (the tube with the bacteria) into the second cuvette.
3. Place the cuvette in the spectrophotometer and record the data as both %T and ABS. To cycle between %T and ABS, press the A/T/C button.
4. Discard the broth from the cuvette into the waste container, rinse the cuvette with water and repeat for remaining samples.
5. Gather all %transmittance and absorbance data for all concentrations and plot the absorbance data on the graph in your Lab Report.

Name ______________________________

Exercise 5.1 Lab Report

Week Two—Determining the Optimal Temperature for Growth

1. Record the absorbance for each of the experimental conditions in the chart below:

Temperature	% Transmission	Absorbance
5°C		
25°C		
30°C		
37°C		
45°C		
60°C		

2. Plot the temperature data from above on the graph below:

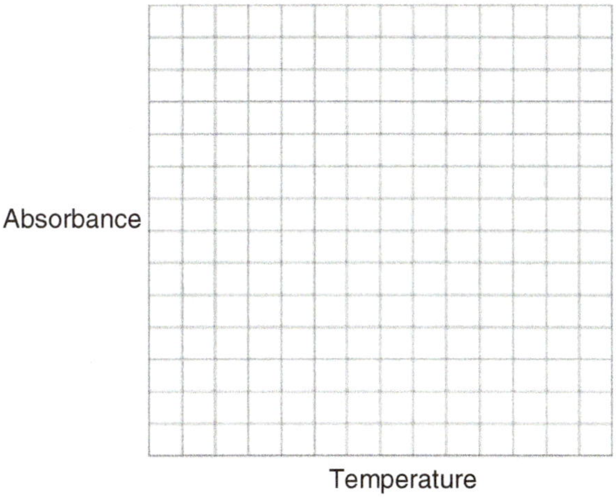

Courtesy of Authors

3. What is the optimal temperature for growth for *E. coli*? How did you determine this?

____

____

4. How would you classify *E. coli* based on temperature requirements?

____

____

Week Two: Comparing Growth in the Presence and in the Absence of Oxygen

5. Record the absorbance for each of the experimental conditions in the chart below:

Absorbance with Oxygen	Absorbance without Oxygen

6. How would you classify this organism based on oxygen requirements? How did you determine this?

____

____

Week Two: Determining the Optimal Salt Concentration for Growth

7. Record the absorbance for each of the experimental conditions in the chart below:

Salt Concentration	% Transmission	Absorbance

8. Plot the salt concentration data above on the graph below:

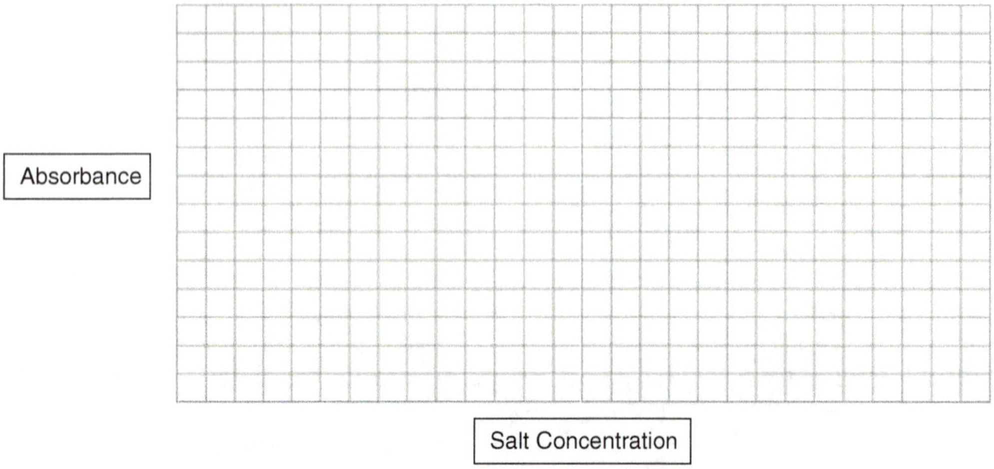

Courtesy of Authors

9. What is the optimal salt concentration for the growth of *E. coli*? How did you determine this?

____

10. How would you classify this organism based on salt requirements (obligate halophile, Facultative halophile, or non halophilic)? Explain your reasoning.

____

____

11. You measure an inoculated sample and the spectrophotometer gives you an output of 85%. What is the absorbance of this sample? (You will need to use the equation in the lab manual.)

____

____

Use the growth curve to answer the questions 12–15.

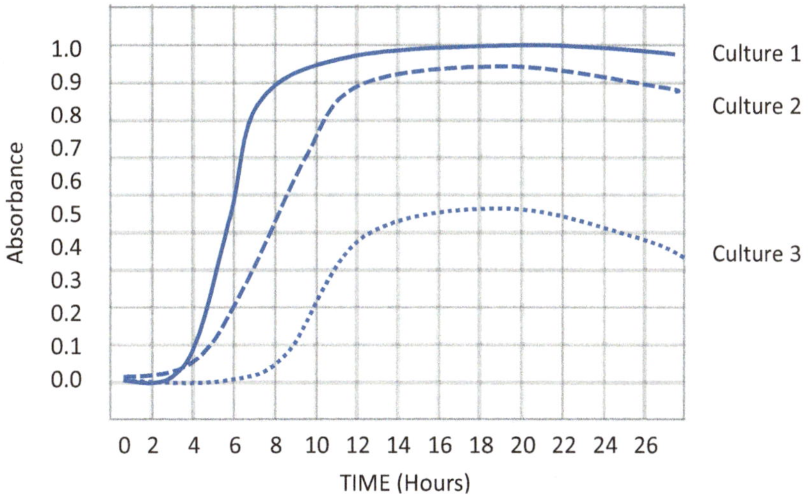

Courtesy of Authors

12. Which culture has the longest lag time?______________________

13. Which culture has the fastest growth rate? ______________________

14. Which culture has the highest cell density? ______________________

15. Assume that the generation time is two hours for Culture 1. What is the growth rate?

____

____

Experimental Growth Curves

To generate the growth curve below, you transferred cells from an old plate in the refrigerator to a flask of sterile medium.

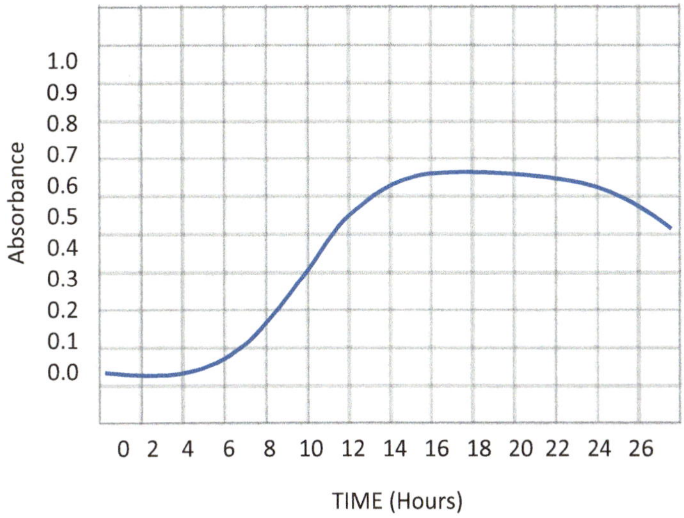

Courtesy of Authors

For each experiment (questions 16–21), predict how the growth curve would change relative to the curve shown.

16. Assume that the curve above was generated by growing cells in their optimum pH range. Predict what would happen if you repeated the experiment at two pH units less than optimal. **Explain your answer in terms of lag time, growth rate and cell density.**

____

____

____

____

17. Assume that the curve above was generated by growing cells in undiluted Tryptic Soy Broth (TSB). Predict what would happen to the curve if you diluted the TSB by 50% with distilled water. **Explain your answer in terms of lag time, growth rate and cell density.**

____

____

____

18. Assume that the curve above was conducted in media containing 7.5% salt. What would happen if you repeated the experiment using using a non halophile? **Explain your answer in terms of lag time, growth rate and cell density.**

19. Assume that the curve above was generated by growing cells at 15°C. These cells are mesophiles. Predict what would happen if you repeated the experiment 37°C. **Explain your answer in terms of lag time, growth rate and cell density.**

20. Predict what would happen to the growth curve if you were to repeat the experiment, but omit all nitrogen from the growth medium. **Explain** your reasoning.

21. Predict what the curve would look like if you repeated the experiment under anaerobic conditions. **Explain** your answer for each scenario:

    A. If the cells are aerobes:

    B. If the cells are facultative anaerobes:

The Ecological Importance of Microbes

Microorganisms are found everywhere on earth. The adult human body, for example, contains three pounds or more of microbial biomass. The discovery of environmental microorganisms that invade the human body resulted in the development of the field of environmental microbiology. **Environmental microbiology** can be defined as the study of microorganisms within all ecosystems and their beneficial or detrimental effects on human welfare, and is more of an applied field of microbiology. The field of environmental microbiology is different from, although related to, **microbial ecology**, which involves the interactions between microorganisms and other organisms in the air, soil, or water. In this chapter we will explore how humans rely on microorganisms for life, as we know it, to exist.

Microorganisms play the majority role in bio-geochemical cycles that occur on earth. Imagine the earth as having a living skin, the soil, which is a thin layer of organic and inorganic material that covers a large portion of earth. This soil is crucial for human life and can contain billions of microorganisms in a very small amount of material. In this medium is where many of the bio-geochemical cycles take place. Cycles such as the carbon, nitrogen, phosphorus, and sulfur cycles are microbial driven and occur in the soil and water. As you know, these elements are required for life, as they are found in the macromolecules in cells. Since the goal of all organisms is to pass on genetic material, organisms must acquire or make nutrients in order to reproduce.

The Carbon Cycle

Carbon is the fundamental element to sustain life; this element is found in all organic molecules in the body. Primitive earth (before life evolved) did not have much, if any, carbon available in a fixed form to sustain life. The majority of carbon was found as inorganic carbon dioxide in the atmosphere. How do we go from an earth filled with inorganic carbon to one rich in organic carbon? The answer to that is due to the role of microorganisms; early microbes began removing carbon dioxide from the atmosphere through a carbon fixation process called photosynthesis. Cyanobacteria, a photosynthetic bacteria, and algae, a photosynthetic eukaryote, along with plants (which have symbiotic cyanobacteria in the cells that do the photosynthesis) begin the carbon cycle and are considered primary producers. You will see that every human depends on the primary producers for life to exist. Primary producers are also called autotrophs, which basically means "to make their own food". Autotrophic organisms convert carbon dioxide into organic molecules through photosynthesis; these organic molecules are used for building cell components such as cell walls. In fact, what many people do not realize is that the material you observe in a plant is largely carbon-based and was once carbon dioxide in the atmosphere. When you burn a plant or any other carbon-based material, by-products of combustion include carbon dioxide, and therefore you are in a sense completing the carbon cycle. There are other types of autotrophy; however, we will focus on photosynthesis as our primary example in this textbook. As a result of carbon fixation through photosynthesis, the organism releases oxygen as a product,

Fig. 6.1 Fossilized Banded Iron Formation
Paulo Afonso/Shutterstock.com

which we and other aerobic respiring organisms rely on for life. In fact, the majority of oxygen we use was produced by cyanobacteria beginning approximately 2.5–3 billion years ago when they first evolved. For about 200 million years these cyanobacteria produced oxygen, which was quickly captured by dissolved iron. Evidence of this resulted in banded iron formations (BIFs) in fossilized sediments. (Figure 6.1).

During that 200 million-year time period the earth remained largely anaerobic. Following this period, oxygen began accumulating in the atmosphere, thus changing earth forever; it is thought that huge populations of anaerobic bacteria went extinct at this time. As you can see, we rely on autotrophs for oxygen; however, humans rely on autotrophic plants as well. Even if you are a strict meat-eater, that cow that you eat will eat plants which fix carbon from the atmosphere. The so called "meat-eater" is considered to be a consumer. Consumers or heterotrophs rely on the activities of autotrophs and utilize the organic materials produced by the autotroph for growth. The organic carbon we consume is incorporated into our cells and will largely remain with us until we die (we really are what we eat!). If you think about it, the carbon in our bodies has been through countless organisms throughout the evolution of earth. As a result of consumption by the consumers, we release carbon dioxide as a by-product of metabolism, and thus we complete another carbon cycle! When an organism does indeed perish, a decomposition process takes place—this does not happen on its own, but microorganisms are involved. Fungi and bacteria are largely considered decomposers, which digest and utilize the carbon from the remains of primary producers and consumers. As a result of decomposition, bacteria and fungi release carbon dioxide as a product (Figure 6.2). By now you should have a better appreciation for microorganisms and other autotrophs as they set the stage for human existence. Consider the earth without decomposers—how would the world look?

Nitrogen Cycle

Nitrogen is an important element for all organisms. Nitrogen makes up approximately 14% of the dry weight of a cell and is found in amino acids, which make up proteins and nucleic acids, which in turn make up a cell's DNA and RNA. Nitrogen is said to be a growth-limiting nutrient, that is, if a cell does not have nitrogen available, then the cell cannot grow. The majority of nitrogen on earth is found in the atmosphere as a gas, forming roughly around 78% of atmospheric gases. Nitrogen gas (N2) is not a usable form for organisms to incorporate into their cells.

Nitrogen Fixation

Nitrogen is made available to organisms through lightning strikes, which cause nitrogen gas and water (H_2O) to react to form ammonia (NH_3) and nitrates (NO_3). Precipitation would then carry these molecules to the ground, which organisms can then utilize. This process of nitrogen fixation through lightning strikes only makes up a small amount of available nitrogen on

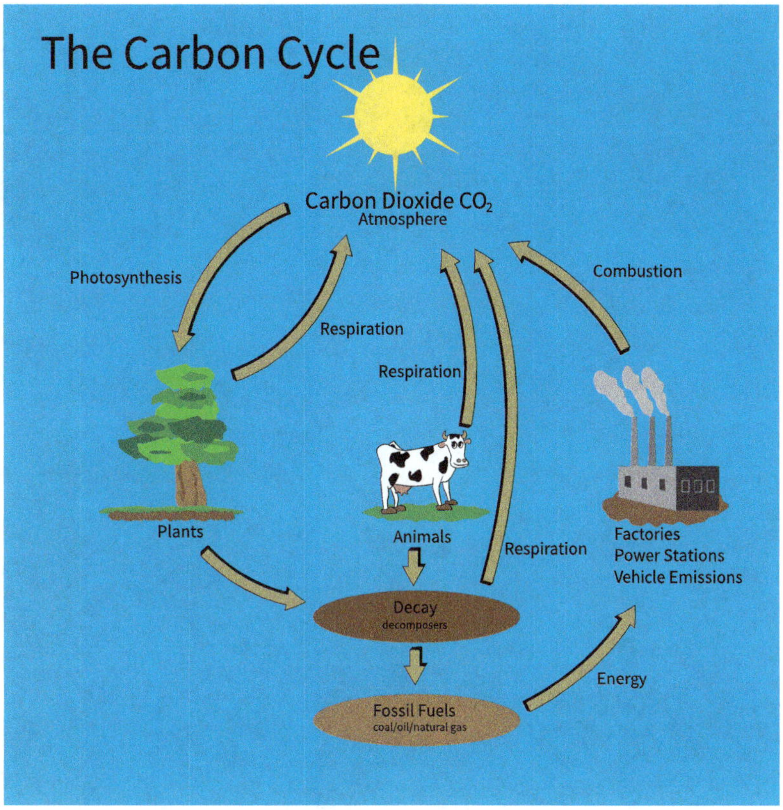

Fig. 6.2 The Carbon Cycle
Photoiconix/Shutterstock.com

earth. Prokaryotic microbes called diazotrophs complete the majority of nitrogen gas fixation. Diazotrophs can be found free, living in the soil, in the water (cyanobacteria), or with a symbiotic relationship with certain plants. Plants such as legumes—that include beans, alfalfa, and peas—have nodules in their root system that house diazotrophs for nitrogen fixation. Nitrogen fixation is a very energy-intensive process performed by a special enzyme called nitrogenase; this enzyme converts N_2 gas into ammonia NH_3 and requires approximately 16 molecules of ATP for every 1 molecule of nitrogen that is fixed from the atmosphere. The free-living diazotrophs were extremely important for the evolution of all life forms. If it were not for these nitrogen-fixing organisms, the amount of nitrogen available would not support life, as we know it (Figure 6.3).

Ammonification

Not all microbes are capable of fixing nitrogen from the atmosphere. The majority of microbes obtain their nitrogen from a decomposition process called ammonification. Ammonification is the process of decomposing organic nitrogen (usually amino acids) to obtain nitrogen. The proteins of dead organisms are degraded to remove the amine group from the amino acid creating ammonia. In moist environments ammonia is converted to ammonium (NH_{4+}), both ammonia and ammonium are available forms of nitrogen for organisms to utilize. Have you ever wondered why your cat litter box smells like ammonia or perhaps your chicken coop? The ammonia smell is from bacteria decomposing proteins in the animal waste.

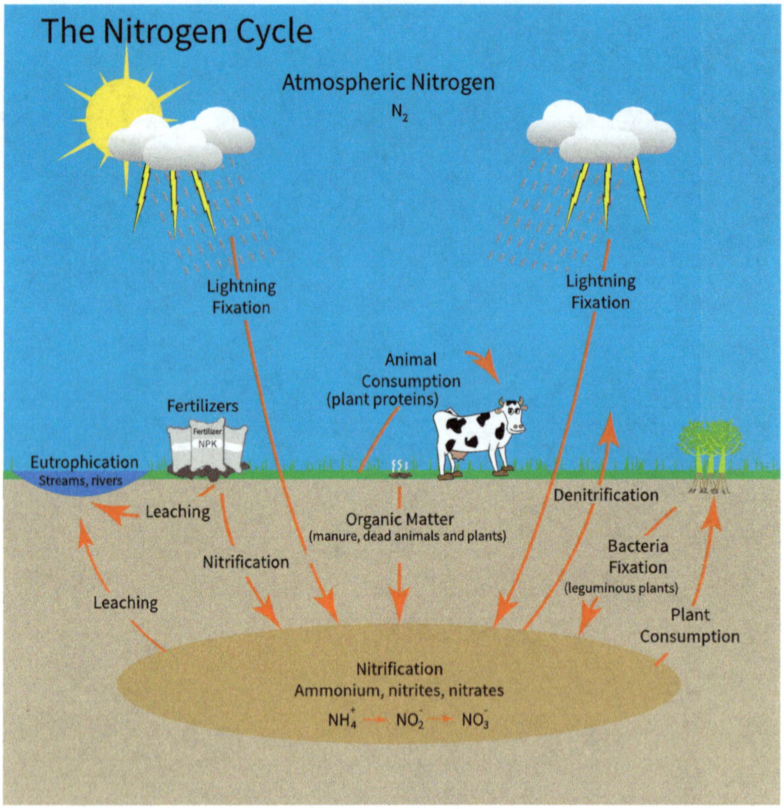

Fig. 6.3 The Nitrogen Cycle
Photoiconix/Shutterstock.com

Nitrification

Nitrification occurs when bacteria called nitrifiers oxidize ammonium (NH_{4+}) to form nitrite (NO_{2-}) and ultimately form Nitrate (NO_{3-}). Nitrate is an available source of nitrogen used by plants for growth. Take a look at a bag of fertilizer and usually you will see nitrogen in the form of potassium nitrate or ammonium nitrate, which you add to your garden to support plant growth. Industrial processes are capable of fixing nitrogen into forms plants can utilize, before modern fertilizers farmers relied (and still do) on bacterial processes to support plant growth.

You might already see a cycle developing: nitrogen gas was in the atmosphere, then fixed into a form available to a microbe, microbes degrade proteins from dead organisms, nitrifiers convert ammonium to nitrate.

Denitrification

The nitrogen cycle is completed when available nitrogen is returned to a gaseous form. This is completed through an anaerobic process when nitrate is used as a terminal electron acceptor. Denitrifiers convert Nitrate (NO_{3-}) to Nitrite (NO_{2-}); some organisms such as those from the genus *Pseudomonas* can then convert nitrite to nitrogen gas, therefore completing the nitrogen cycle. These processes largely take place in the soil. Now imagine that you fertilize your lawn with a nitrogen-rich fertilizer. The grass will surely absorb some of the nitrogen; however, a large amount of the fertilizer is utilized by bacteria and converted back to nitrogen gas. Now imagine if you bag your lawn clippings and have a truck pick up your yard waste as many cities do; that money you spent to fertilize your lawn is now being carted away to a dump that turns that into compost. What microbial processes would occur in your soil if you mulched your lawn clippings? Would you need to fertilize very often?

Applied Environmental Microbiology

Understanding bacterial growth requirements and microbial adaptation has allowed scientists to use microbes for beneficial processes. One field that has emerged using microbes for beneficial purposes is the field of **bioremediation**, which uses microbes to degrade or detoxify harmful pollutants. The pollutants can be introduced to the environment by accident (oil spill), on purpose (insecticides), or by convenience of disposal. Certain man-made (synthetic) compounds are similar to what is found naturally in the environment and can be easily degraded, since organisms in the environment have enzymes capable of "recognizing" the compound. Other synthetic compounds such as certain herbicides are referred to as **xenobiotics**, which is a term used for compounds that persist in the environment for a long period of time (Figure 6.4). Xenobiotics persist in the environment since microorganisms do not recognize the synthetic compound as "food", and therefore they do not have enzymes that will degrade the compound.

The most well-known application of microbes is to use them to clean up oil spills. Since oil is a hydrocarbon and microbes need carbon for growth, we can stimulate microbes to "eat" the oil by adding nutrients to a contaminated site, referred to as **biostimulation**. The nutrients added would be the growth-limiting element nitrogen (no nitrogen = no growth) and other elements such as potassium and phosphorus; therefore natural bioremediation is basically applying fertilizer to a contaminated area (Figure 6.5). Other types of bioremediation involve the *addition* of microbes not normally present in a population; this type of remediation is referred to as **bioaugmentation**.

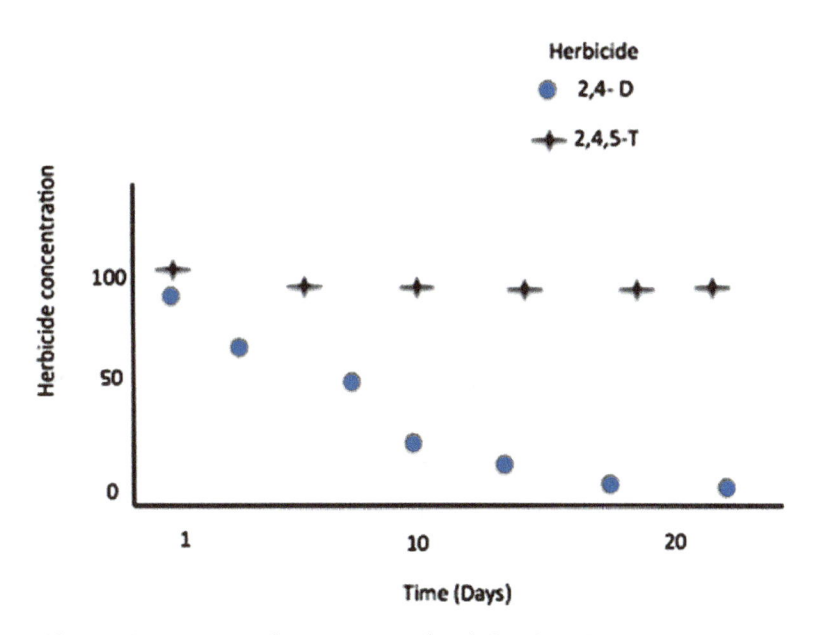

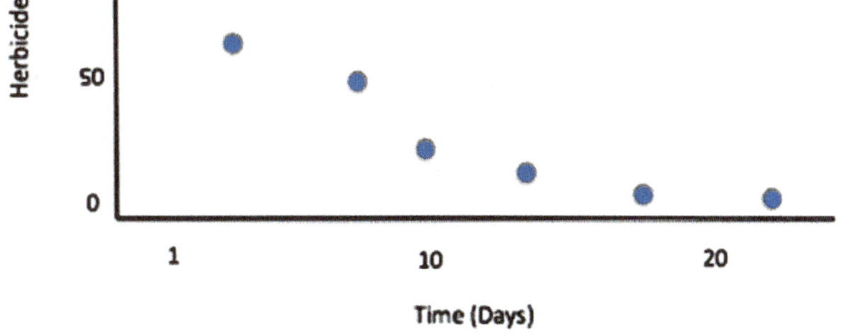

Fig. 6.4 2,4-D and 2,4,5-T are both herbicides; 2,4-D is readily broken down in the environment; 2,4,5-T is considered a xenobiotic since it will persist over time in the environment.
Courtesy of Author

Fig. 6.5 Bioremediation of a contaminated shoreline following the Exxon Valdez oil spill. This is an example of biostimulation; by the addition of nutrients, the amount of oil degraded is easily observed.

Waste Water Treatment

Townships with large enough populations rely on underground sewer systems to flow waste water from the home to a treatment facility. Rainwater and snow melt will also make their way to the same treatment plant. The plant will remove most of the pollutants and microbes and release the treated water back into the environment. However, some pollutants can be released in very low concentrations such as prescription medications. Occasionally, there are lapses in the removal of harmful microbes such was the case in Milwaukee, WI in 1993; 400,000 people were infected with an intestinal parasite *Cryptosporidium parvum* which causes watery diarrhea. Diarrhea caused by this parasite can result in the loss of 10–15 liters of fluid a day! The wastewater at the treatment facility is collected in large underground wells. The remaining steps are outlined below:

Primary Treatment (Figure 6.6): This is a process to physically remove large waste materials that will settle out of the water.

- The water is pumped from the wells to the treatment facility where it is sprayed with aluminum sulfate or ferric chloride, which are coagulants. The coagulants help remove phosphorus from the wastewater for removal.
- Upon entry into the treatment facility, the liquid is screened to remove large solid material.
- Large skimmers remove scum and floating waste that was not removed by the screens.
- The remaining solids are allowed to settle out of solution forming sludge and are removed from the water and then the water flows to the secondary treatment tanks.

Secondary Treatment (Figure 6.6): There are four different methods that can be utilized, and all of them are biological processes used to convert the solids suspended in sewage into inorganic compounds. Microbial growth is encouraged, allowing aerobes to degrade organic compounds to carbon dioxide and water. Toxic or hazardous materials can drastically affect this process.

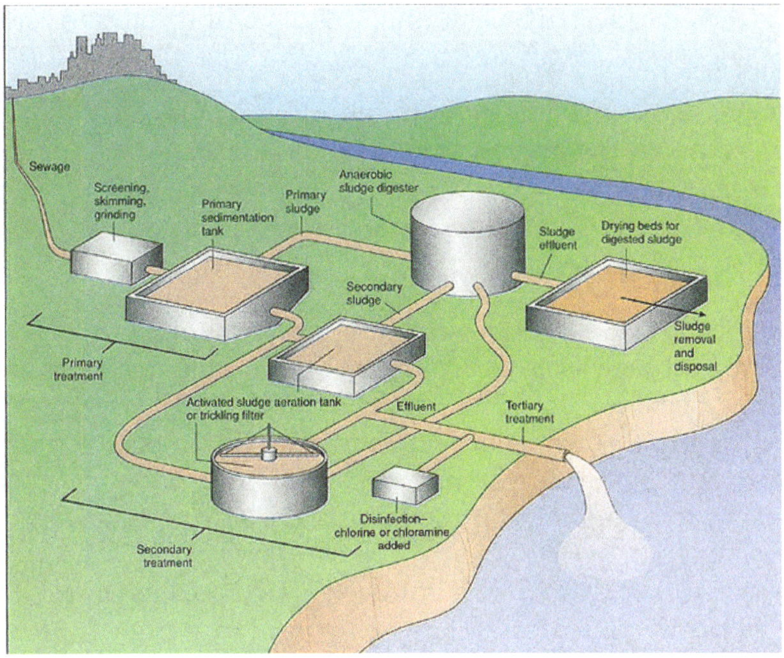

Fig. 6.6 Basic Schematic of Sewage Treatment
© Kendall Hunt Publishing Company

Activated sludge treatment:

- Large numbers of microbes are inoculated into the wastewater from previous sludge treatment applications.
- The sludge is aerated delivering large amounts of oxygen to stimulate aerobic growth.
- Following aeration the suspended particles are allowed to settle, from where it is then removed.
- The water is then disinfected by chlorine, UV light, or ozone and released into a nearby river or lake.

Trickling Filtration: (Used at small treatment facilities.)

- A large rotating arm will spray sewage over a bed of gravel and rocks or plastic.
- The surfaces develop biofilms that degrade organic materials.
- Filtered water is sent to a sedimentation tank to remove sludge.
- The water is then disinfected by chlorine, UV light, or ozone and released into a nearby river or lake.

Lagoons:

- Contaminated wastewater is deposited into shallow ponds or lagoons.
- The water remains for several days.
- Cyanobacteria and algae grow and provide oxygen for aerobic bacteria to degrade the sewage.
- The water is then disinfected by chlorine, UV light, or ozone and released into a nearby river or lake.

Artificial Wetlands:

- This follows the same principle as lagoons with more advanced designs providing habitat for birds and other wildlife while treating sewage.

- The water is then disinfected by chlorine, UV light, or ozone and released into a nearby river or lake.

Advanced Treatment: This involves the removal of ammonia, nitrates, and phosphates. All three chemicals can stimulate the growth of algae and cyanobacteria leading to large masses of surface scum.

Municipal Drinking Water Treatment

- Water is pumped into large reservoirs and solids are allowed to settle.
- Water is filtered through a thick bed of gravel. This removes many microorganisms such as protozoa and bacteria.
- Additional filtration through activated charcoal will remove organic chemicals. These chemicals may be harmful or give the water a bad taste.
- Microorganisms form biofilms on the filter materials and remove carbon and nitrogen from the flowing water.
- Lastly the water will be treated with disinfectants such as chlorine, UV light, or ozone to kill remaining harmful microbes.
- The treated water makes its way to your home.

Treatment of Solid Wastes

The sludge mentioned during the treatment of wastewater is largely sent to landfills where it is buried along with other household and industrial waste. Landfills pose a large problem: run off of rain water can contaminate ground water, take up large amounts of space, pollute the nearby air with methane and other noxious smells, and decrease property values, to name a few. One way to combat growing landfills is by composting. **Composting** is a natural decomposition process that takes household food waste and turns it into a natural fertilizer. Microbial metabolism inside of the piled waste creates heat. Temperatures will reach upwards of 60° Celsius and pathogens are killed. Thermophilic bacteria will not be harmed and will continue to decompose the organic waste. If the pile is frequently aerated by mixing the pile, the compost can be complete in about 6–7 weeks. (Figure 6.7)

Fig. 6.7
Marina Lohrbach/Shutterstock.com

Control of Microbial Growth

As a healthcare professional, you will be placed into many situations that require you to disinfect working areas. You may be the dental hygienist preparing the chair and instruments between patients. You may be the nurse at the doctor's office preparing the examining room between patients. You may be the surgical technician, the respiratory therapist, or the home healthcare professional; therefore it is extremely important that you learn the different methods to control the growth of microbes in order to prevent infection and disease.

Before an object is to be sterilized the objects must be cleaned. **Cleaning** is the removal of all foreign material, such as soil or organic material, from an object. Cleaning is normally accomplished with water, mechanical action, detergents or enzymatic products. The cleaning process is extremely important, since many chemicals are inactivated by the presence of organic material. Once the items are cleaned they can be sterilized. **Sterilization** is the complete elimination or destruction of all forms of microbial life. It is accomplished by chemical or physical means.

Chemical Control Methods

Chemicals that control the growth of microorganisms are generally divided into two main groups, antiseptics and disinfectants. **Antiseptics** are chemicals used on living tissue in an attempt to reduce the population of microbes present; **disinfectants** are chemicals used on inanimate objects and are potentially harmful to living tissue. Objects must first be cleaned before they can be successfully disinfected. Disinfection is essential for controlling the spread of microorganisms. A good disinfectant should kill microorganisms on a surface, but it also must have low toxicity to humans, have a substantial shelf life, be useful in dilute form, and be inexpensive and easy to obtain. Currently, the U.S. Department of Agriculture (USDA) lists over 8,000 disinfectants for hospital use and thousands more for general use. An important thing to note is that of these 8,000 disinfectants, no single disinfectant is appropriate for all scenarios. Disinfectants are used to disinfect inanimate objects, and disinfection is the process that eliminates many or all pathogenic microorganisms, except for bacterial spores, from inanimate objects. Antiseptics, on the other hand, are chemicals that are safe to use on human skin and tissue to kill some, but not all, microbial life. They should not be used to disinfect inanimate objects. The FDA (Food and Drug Administration) regulates antiseptics.

Levels of Disinfection

There are three levels of disinfection:

1. High-level disinfection destroys all microorganisms and some bacterial endospores. These are chemicals that act as **sterilants,** which are used to kill all microbial life on inanimate objects. The FDA (Food and Drug Administration) regulates sterilants as well. Sterilants are high-level disinfectants, which are being used in such a way as to create sterilization,

usually by increasing the exposure time to the chemical and/or increasing the concentration of the chemical.

2. Intermediate-level disinfection destroys mycobacteria, vegetative bacteria, most viruses, most fungi, but not bacterial endospores.
3. Low-level disinfection destroys most vegetative bacteria, some viruses, some fungi, but not mycobacteria and not bacterial endospores.

Disinfectants should state on their label which level of disinfection they achieve.

Common Antiseptics and Disinfectants

Alcohol

The maximum microbicidal activity of alcohol is achieved when its concentration is between 60%–80%. Alcohol can be used as an intermediate-level disinfectant. Its use is limited by its disadvantages. Alcohols can be used effectively as antiseptics for skin preparation of injection sites. Alcohol-based hand gels are recommended for handwashing in healthcare settings, in between washing with plain, non-antimicrobial soaps.

Alcohol is inactivated by organic matter, such as bodily secretions and excretions (blood, urine, feces, etc.). Hands need to be washed whenever there is exposure to such substances. Alcohol functions by denaturing proteins. They leave no residue, are relatively nontoxic, are easy to obtain, and are inexpensive. However, alcohols are volatile, they evaporate quickly (not allowing for extended contact time), and do not kill endospores.

Chlorine

Chlorine is an intermediate-level disinfectant. It functions by oxidizing proteins and nucleic acids. It is effective, convenient, and inexpensive. It may be corrosive and can be inactivated by organic molecules. Its high pH decreases its effectiveness.

Solutions containing liquid bleach used in healthcare must be made up fresh daily. It can be used in a 1:100 solution with water for ordinary cleaning. If high organic load is present, it should be used in a 1:10 solution. More stable forms of chlorine, such as sodium dichloroisocyanurate and chloramines, are used in hospitals.

Its common uses include glassware and surface disinfection, water and wastewater treatment, to disinfect food-processing equipment, and as household and industrial bleaches to maintain and improve hygiene standards. Pharmaceuticals and medicines which either contain chlorine, or are made using it, include those for treating leukemia, malaria, pneumonia, whooping cough, hay-fever, typhoid fever, meningitis, Hodgkin's disease and peritonitis.

Chlorine is one of the most commonly manufactured chemicals in United States. Its most important use is as a bleach in manufacture of paper and cloth, but it is also used to make pesticides (insect killers), rubber, and solvents. Chlorine is used in drinking water and swimming pool water to kill harmful bacteria. It is used as part of the sanitation process for industrial waste.

Recommended Procedure for Disinfection of Small Quantities of Drinking Water

Occasions arise which require the disinfection of small quantities of drinking water. Disinfection may be necessary because of temporary contamination of a supply which is satisfactory under normal conditions.

Four items must be considered if chemical disinfection is to be effective: (1) the water must be free of turbidity or dirt, (2) the chemical must be applied in a sufficient amount to guarantee disinfection and must be uniformly distributed to contact every particle of water, (3) at least

30 minutes of contact time must be provided between the chemical and water to allow time for the chemical to destroy any disease-causing bacteria or viruses, and (4) the water after treatment must be protected from further contamination.

The most widely used chemical for water disinfection is chlorine. This chemical is used by most municipal water systems and effectively destroys disease-causing bacteria and viruses when applied in dosages far below the amount harmful to humans. The chemical can be purchased in the form of chlorine gas, chlorine powder, or liquid chlorine. Household bleaches such as *Clorox* and *Hilex* are liquid chlorine. This form is best for treatment of small quantities of water because no equipment is required and liquid can be easily measured by a dropper, spoon, or measuring cup.

Typically household bleaches will consist of 5.25% available chlorine, although it is possible to have slightly different solution strengths depending on the particular product. The first step in disinfecting drinking water is to determine the percentage of available chlorine in the solution that can be obtained from grocery stores.

The next step is to determine the amount of chlorine to use in properly treating a quantity of water. The following table gives the amount of chlorine necessary to treat given quantities of water. By knowing the amount of water to be treated, the amount of solution can be determined from the table.

Strength of Solution	Brand Name of Solution	Amount of Solution for Disinfection of Drinking Water				
		1 gallon	2 gallons	3 gallons	4 gallons	5 gallons
5.25%A	*Clorox, Hilex, Purex*	3 drops	6 drops	9 drops	12 drops	15 drops

The water to be treated should be place in a clean container. After the chlorine solution is introduced into the water, it should be agitated slightly to mix the chlorine and water. The container should then be covered and the water allowed to stand for at least 30 minutes before using. An odor and taste of chlorine should be apparent. These will disappear with time, are entirely harmless to humans, and are your guarantee of safe drinking water.

Procedure for Chlorine Disinfection of Water Wells

A water well should be thoroughly cleaned and disinfected with a strong chlorine solution after original construction; any repair or maintenance; flooding; a period of non-use; or when two or more 'unsafe' bacteriological water samples are traced to the well.

Adequate chlorine requires a certain chlorine dosage for a minimum contact time—100 parts per million for 2 hours, or 50 parts per million for 8 hours, or 25 parts per million for 24 hours.

Chlorine for disinfection for these water systems can be either 5.25% sodium hypochlorite solution or 65% calcium hypochlorite powder. A 5.25% hypochlorite solution is common household bleach such as *Hilex*, *Clorox*, or *Purex*, available at grocery stores and supermarkets. The 65% calcium hypochlorite powder is available from chemical supply houses and is known commercially as *HTH*, *Perchloron*, or *Pittchlor*.

Household chlorine bleach can release chlorine gas if it is mixed with other cleaning agents. Chlorine can react with many organics to form carcinogenic (cancer-causing) compounds.

People working in facilities that use chlorine to manufacture other chemicals have the highest risk of being exposed to chlorine. Chlorine gas can cause irritation of the eyes, skin, and respiratory tract. Exposure to high levels can result in corrosive damage to the eyes, skin, and respiratory tissues, and could lead to pulmonary edema and even death in extreme cases. Chlorine gas has been found in at least 60 of the 1,591 National Priorities List sites identified by the Environmental Protection Agency (EPA).

The emergence of the protozoan *Cryptosporidium parvum* as a major cause of recreational waterborne disease has prompted public health workers to reevaluate existing recommendations and regulations for water quality and use. Frequent fecal contamination of recreational water, the high level of resistance to chlorine by *C. parvum* oocysts, the low oocyst dose required for infection, and high numbers of bathers make it imperative that we understand how oocyst inactivation is affected by recreational water conditions, including fecal contamination.

Routine use of recreational waters (swimming pools, etc.) by diapered children from day-care facilities, who are known to have an elevated prevalence of *C. parvum* infection, increases the potential for waterborne disease transmission. Prevention plans that combine engineering changes (improved filtration and turnover rates, separate plumbing and filtration for high-risk 'kiddie' pools), pool policy modifications (fecal accident response policies, test efficacy of barrier garments such as swim diapers), and patron and staff education should reduce the risk for waterborne disease transmission in public recreational water venues. Education efforts should stress current knowledge about waterborne disease transmission and suggest simple prevention measures such as refraining from pool use during a current or recent diarrheal episode, not swallowing recreational water, using proper diaper changing and handwashing practices, instituting frequent timed bathroom breaks for younger children, and promoting a shower before pool use to remove fecal residue.

Iodine Tincture

Iodine tincture is an intermediate-level disinfectant. It functions by oxidizing proteins. It is not as quickly inactivated by organic molecules as is chlorine. It is effective at a wide pH range. It may be corrosive and can be inactivated by organic molecules. It stains instruments, clothing, and the skin. It is painful on non-intact skin (remember the sting!). It is used for small-scale drinking water treatment (i.e., by campers, hikers, or in emergency situations) and as a skin antiseptic.

Iodophors

Because iodine causes irritation and discoloring of skin, iodophors have largely replaced iodine as the active ingredient in antiseptics. Iodine molecules rapidly penetrate the cell wall of microorganisms and inactivate cells, resulting in impaired protein synthesis and alteration of cell membranes.

Iodophors are composed of elemental iodine, iodide or triiodide complexed with a solubilizing agent, such as a surfactant or povidone. The result is a water-soluble material that releases free iodine when in solution. Iodophors are prepared by mixing iodine with the solubilizing agent; heat can be used to speed up the reaction. The amount of molecular iodine present ("free" iodine) determines the level of antimicrobial activity of iodophors. Typical 10% povidone-iodine formulations contain 1% available iodine and yield free iodine concentrations of 1 ppm.

Iodophor formulations are designed to make finished topical solutions more gentle to the skin and increase its shelf-life when compared to other iodine-based applications (such as tincture of iodine). The povidone complex is known as Povidone-Iodine (polyvinylpyrrolidone or PVPI).

Betadine, *Iofec, Isodyne, Losan, Tamed Iodine, Prepodyne, Weladol, Wescodyne* and *Vetadine* are iodophor solutions.

Iodine and iodophors have bactericidal activity against gram-positive, gram-negative, and certain spore-forming bacteria (e.g., *Clostridia* and *Bacillus* spp.) and are active against *Mycobacteria*, viruses, and fungi. In concentrations used in antiseptics, iodophors are not sporicidal.

The antimicrobial activity of iodophors is affected by pH, temperature, exposure time, concentration of total available iodine, and concentration of emollients.

Iodophors are rapidly neutralized in the presence of organic material such as blood or sputum.

Povidone iodine absorption has been a concern in the treatment of pregnant and lactating mothers because of the possibility of induced transient hypothyroidism.

Phenolics

Phenolics are low to intermediate-level disinfectants. They function by destroying cell membranes and denaturing proteins. They remain active in the presence of organics. They are irritating to tissue, leave a toxic residue, and create toxic vapors. Phenolics are used to clean surfaces.

Phenolics are the active ingredient in most bottles of 'household disinfectant'. They are also found in some mouthwashes, disinfectant soaps and handwashes. Phenol is probably the oldest disinfectant (used by Lister) and was called carbolic acid in the early days of antiseptics. Phenol is rather corrosive to the skin and sometimes toxic to sensitive people, so the somewhat less corrosive substitute phenolic o-phenylphenol is often used as part of a disinfectant formula. Hexachlorophene is a phenolic which was once used as a germicidal additive to some household products but was banned due to suspected harmful effects.

Phenols are effective against bacteria (especially gram-positive bacteria) and enveloped viruses. Phenols are not effective against non-enveloped viruses and spores. Enveloped viruses include: Coronavirus, Pox and Rabies virus. Non-enveloped viruses include: Papilloma, Parvovirus and Rotavirus.

Phenols maintain their activity in the presence of organic material, and are therefore more useful in foot baths and areas in which organic material cannot be completely removed. Phenolic disinfectants (including cresols and pine oil) are generally safe but prolonged exposure to the skin may cause irritation. Phenolic disinfectants include: *O-Syl*, *Matar*, *Septicol*, *Hexachlorophene*, *Environ*, *One-Stroke*, *Lysovet*, *Tek-Trol*, *Pantek*, *Discan*, and *Staphene*.

Most phenolics have a relatively high toxicity rating and are usually skin irritants, especially so in the concentrations in which they are present in the typical formulation.

Chlorhexidine

Chlorhexidine gluconate is cationic bisbiguanide. Chlorhexidine is a low to intermediate-level disinfectant. It is a **phenol derivative** commonly used as a skin cleanser for hand scrubbing or washing by operating room personnel; for handwashing by medical personnel; for pre-operative skin preparation; and for skin wound and general skin cleansing.

Antimicrobial activity of chlorhexidine is due to disruption of cell membranes, resulting in precipitation of cellular contents.

Chlorhexidine has good activity against *Staphylococci* and other gram-positive bacteria, less activity against gram-negative bacteria and fungi, and only minimal activity against tubercle bacilli (TB). Chlorhexidine is not sporicidal. It has activity against enveloped viruses (e.g., herpes simplex virus, HIV, cytomegalovirus, influenza, and RSV) but substantially less activity against non-enveloped viruses (e.g., rotavirus, adenovirus, and enteroviruses).

Addition of low concentrations (0.5%–1.0%) of chlorhexidine to alcohol-based preparations results in greater residual activity than alcohol alone.

Chlorhexidine gluconate products are used in hand-care and wound-care applications under the brand names *Hibitane*, *Hibiclens*, *ChloraPrep*, *Betasept*, *Biopatch*, and *Exidine*. As stated above, these products are used for general skin cleansing, as a surgical scrub and a pre-operative skin preparation. These cleanser solutions are not suitable for use as mouthwashes.

Chlorhexidine is often used as an active ingredient in mouthwashes designed to kill dental plaque and other oral bacteria, thus improving bad breath. It is sometimes marketed under the brand names *Peridex*, *Periochip*, *Periogard Oral Rinse*, *Corsodyl* or *Chlorohex*. Chlorhexidine gluconate-based products are utilized to combat or prevent gum diseases such as gingivitis.

Known to be deactivated by fluoride, mouth rinses containing chlorhexidine gluconate should be taken at least 30 minutes after using fluoridated oral products (i.e. toothpaste and mouthwash). For best effectiveness, food, drink, and mouth rinses should be avoided for at least one hour after use. Recently, mouthwash formulations containing both chlorhexidine digluconate and fluoride have been developed and found to be effective.

Chlorhexidine is used to help treat periodontal disease, a disease of the gums caused by bacteria growing beneath the gum line. Chlorhexidine works by killing the bacteria. Chlorhexidine implants are placed after the teeth have been thoroughly cleaned. For adults, one implant is inserted into each gum pocket that is too deep. Up to eight chlorhexidine implants are placed between the teeth and gums in places where the gum has a deep pocket. Treatment may be repeated every three months. For children, use and dose will be determined by the dentist.

It is not necessary to remove the implants; they will dissolve on their own. The dentist will check the depth of the pockets in the gums every three months to see if they need to be treated again.

Products containing chlorhexidine gluconate in high concentrations must be kept away from eyes (corneal ulcers) and the inner ear (deafness). It is used in minute concentrations in some contact lens solutions. In some countries it is available by prescription only.

QUATS (Quaternary Ammonium Compounds)

QUATS (Quaternary Ammonium Compounds) are low-level disinfectants. They function by destroying cell membranes. They are inexpensive. However, they are readily inactivated by detergents, fibers, and other compounds. *Pseudomonas aeruginosa* can grow in and contaminate QUAT solutions. They are used to clean surfaces, in the food industry, and in general housekeeping.

Quaternary ammonium compounds (QUATs) are composed of nitrogen atom linked directly to four alkyl groups. Alkyl benzalkonium chlorides are the most widely used as antiseptics.

Antimicrobial activity involves adsorption to cytoplasmic membrane, with subsequent leakage of low molecular weight cytoplasmic constituents. QUATs are primarily bacteriostatic and fungistatic. They are more active against gram-positive bacteria than against gram-negative bacilli. QUATs have relatively weak activity against *Mycobacteria* and fungi and have greater activity against lipophilic viruses.

Because of weak activity against gram-negative bacteria, QUATs are prone to contamination by these organisms. Several outbreaks of infection or pseudoinfection have been traced to quaternary ammonium compounds contaminated with gram-negative bacilli.

Quats are pulmonary and ocular irritants and systemic poisons if ingested. They are the principal cause of occupational dermatitis and asthma in healthcare and janitorial workers. Studies have documented that approximately 2% of healthcare workers are allergic to QUATs. Most are sold as concentrates that are extremely toxic, irritating and corrosive. Errors in dilution can result in highly toxic or, alternatively, ineffective use dilutions.

Quats are extremely sensitive to hard water and usually require a chelant to work in these conditions.

Classification of the 'generation' of quaternary ammonium compounds is as follows:

- First Generation: Benzalkonium chlorides (example: Benzalkonium chloride). First generation quats have the lowest relative biocidal activity and are commonly used as preservatives.
- Second Generation: Substituted benzalkonium chlorides (example: alkyl dimethyl benzyl ammonium chloride). The substitution of the aromatic ring hydrogen with chlorine, methyl and ethyl groups resulted in this second generation quat with high biocidal activity.
- Third Generation: 'Dual Quats' (example: contain an equal mixture of alkyl dimethyl benzyl ammonium chloride + alkyl dimethyl ethylbenzyl ammonium chloride). This mixture of two specific quats resulted in a dual quat offering increased biocidal activity, stronger detergency, and increased safety to the user (relative lower toxicity).
- Fourth Generation: 'Twin or Dual Chain Quats'—dialkylmethyl amines (example: didecyl dimethyl ammonium chloride or dioctyl dimethyl ammonium chloride). Fourth generation quats are superior in germicidal performance, have lower foaming, and have an increased tolerance to protein loads and hard water.
- Fifth Generation: Mixtures of fourth generation quats with second-generation quats (example: didecyl dimethyl ammonium chloride + alkyl dimethyl benzyl ammonium chloride) Fifth generation quats have an outstanding germicidal performance, they are active under more hostile conditions and are safer to use.

Glutaraldehyde

Glutaraldehyde is a colourless liquid with a pungent odor used to sterilize medical and dental equipment. It is also used for industrial water treatment and as a chemical preservative. But it is toxic, causing severe eye, nose, throat and lung irritation, along with headaches, drowsiness and dizziness.

Glutaraldehyde is a dialdehyde that is slightly acidic in its natural state. In a buffered (pH balanced) alkaline solution, it is a highly effective microbicidal agent. Glutaraldehyde is corrosive to metal instruments and is not to be used as a holding solution prior to instrument cleaning and sterilization. It is widely used in the cold sterilization of dental items, such as suction hoses. Glutaraldehyde is usually a clear liquid that turns green when activated. It has a strong odor and is an irritant to the skin, eyes, and respiratory system. It can cause allergic reactions in some individuals. The chemical should be used in separate areas where there is control over ventilation and occupational exposure. Unused glutaraldehyde solution should be stored in tightly closed, properly labeled containers in a cool, secure, properly marked location. For disposal of solutions, consult with local and regional regulations.

Widely available from a number of manufacturers, glutaraldehyde is found in products with properties, e.g., buffered, potentiated, alkaline, acidic, etc. The percentage of concentration along with the temperature determines the time needed to achieve sterility. Glutaraldehydes are EPA registered as steriliant/disinfectant chemicals, effective at penetrating blood and bioburden. Most glutaraldehydes must be activated before use by adding an appropriate buffer. Activated solutions have an effective life of 14 to 30 days, depending on the product. Newer acidic potentiated (pH neutral) glutaraldehydes do not require activation. To measure appropriate strength of solutions, glutaraldehyde test strips are available. Manufacturers' instructions should be closely followed.

Glutaraldehydes, such as *Metricide*, *Cidex*, etc., can be used as sterilants, if a treatment time of 10–12 hours is employed. It has intermediate to high activity level. It is commonly used for high level disinfection at a 1.0–1.5% MEC (minimum effective concentration).

Glutaraldehyde creates toxic vapor. Its ceiling limit cannot exceed 0.2 ppm with 7–15 air exchanges required per hour to provide proper ventilation. Workers should wear PPE, including nitrile rubber, butyl rubber, or polyethelyne gloves and goggles to minimize skin or mucous membrane contact.

Hydrogen peroxide

Hydrogen peroxide functions by creating hydoxyl-free radicals which destroy cell membranes, DNA, and other cell components. 3% hydrogen peroxide is used as a skin antiseptic. 3–6% hydrogen peroxide is a stable and effective disinfectant. 6–25% hydrogen peroxide concentrations can be used as chemical sterilants.

The STERRAD 200 Sterilization System uses low-temperature hydrogen peroxide gas plasma technology. This combination process allows medical instruments to be sterilized.

Hydrogen peroxide is used as a 6% solution for bleaching hair. Some disinfectant solutions for contact lenses contain 3% hydrogen peroxide. Chlorine free bleaches contain 6% hydrogen peroxide. Some newer fabric stain removers/bleaches contain 5–15% hydrogen peroxide.

Peracetic Acid

Peracetic Acid is characterized by rapid action against all microorganisms, including bacterial endospores, and can be used as a chemical sterilant. It remains effective in the presence of organics. It functions by destroying cell membranes and denaturing proteins. It can corrode copper, brass, bronze, plain steel, and galvanized iron. It is unstable when diluted.

Ethylene oxide (ETO)

Ethylene oxide is a gaseous sterilizing agent used for commercial and medical applications. ETO can be explosive and must be mixed with an inert gas, such as carbon dioxide. Gas leaks are considered a medical emergency. ETO is affected by temperature and humidity. It requires 3–12 hours to achieve sterilization. Toxic residue must be removed for objects, which may require an additional 8–12 hours.

Physical Control Methods

Physical control methods include temperature, radiation, and osmotic pressure. **Temperature** is a very common and effective way of controlling microorganisms. Examples of temperature control are autoclaving, boiling, oven, pasteurization, and refrigeration. **Ultraviolet (UV) radiation** is used as a physical control method, because it has microcidal activity against microorganisms; UV light is absorbed by microbial DNA and causes DNA mutation. **Filtration** is another common physical control method which physically separates microbes from liquid.

Temperature

Vegetative microorganisms can generally be killed at temperatures from 50°C to 70°C with moist heat. Bacterial endospores, however, are very resistant to heat and extended exposure to much higher temperature is necessary for their destruction. High temperature may be applied as either moist heat or dry heat.

Autoclave

The **autoclave** uses moist heat under pressure to achieve sterilization. The most common setting for use is: 121°C for 15–20 minutes at 15 pounds per square inch (psi). The autoclave can damage heat-sensitive materials. Autoclaving must allow for drying of items at the end of its cycle. If not, moist items could again become contaminated. Autoclaving with a 1N NaOH solution is thought to destroy prions.

Immediate-use sterilization (also known as flash sterilization) with the autoclave uses a setting of 135°C for 10 minutes at 15 psi. This procedure is to be used in emergency situations only. It is often used for operating rooms where sterile equipment is continually needed. Increasing the temperature decreases the required time needed to achieve sterilization.

Biological indicators are used to ensure that autoclaves are sterilizing. A tube containing the endospores of ***Bacillus stearothermophilus*** is placed in the autoclave with a regular load. The tube also contains growth media in a separate container. After autoclaving, the container is crushed, exposing the endospores to the growth media, and then allowed to incubate. **If the media changes color after incubating**, that means the endospores were able to survive in the autoclave, then germinated and grew in the incubator. This would mean that **the autoclaving did NOT work!**

Chemical indicators involve the use of heat-sensitive tape. Tape is attached to the outside of loads, turning color during autoclaving. A changed indicator gives a visual sign that heating occurred. A changed chemical indicator does not ensure sterilization of items. Use of tape should **never** replace the periodic testing of the autoclave using biological indicators.

Boiling

Boiling is not an effective means of achieving sterilization because it does not kill endospores, but it is an excellent high-level disinfection process. This method is used in home and emergency situations, where other methods are not available.

Dry Heat

The most common setting for **dry heat sterilization** is 200°C for 1.5 hours. It requires increased temperatures and times as compared to moist heat sterilization.

Pasteurization

Pasteurization is the process of heating food or other substances under controlled conditions—for a short amount of time and under controlled temperatures—to kill pathogens and reduce the total number of microorganisms without damaging the substance.

Low Temperature

Refrigeration/Freezing is not a sterilization technique but it does control the growth of microbes by slowing or stopping growth. Refrigeration and freezing are most commonly used for preservation of food.

Radiation

The ultraviolet portion of the light spectrum includes all radiations with wavelengths from 100 nm to 400 nm. This wavelength does not penetrate glass or plastic; therefore, UV light is used to control microorganisms on the surfaces of objects or in the air. **UV radiation** involves the use of ultraviolet light to kill bacteria by damaging the bacterial DNA. Thymine dimers are formed within the DNA from adjacent thymine molecules. UV light cannot penetrate grease, plastic, glass, etc. and, therefore, is not an effective method of sterilization. UV light is harmful to the eyes and skin and can promote the development of skin cancer.

Biotransfer cabinets have germicidal UV lights that aid in disinfecting objects contained within them. However, because this UV radiation is dangerous to all forms of life, it is only turned on when the cabinet door has been closed. An important consideration when using UV light is that it has very poor penetrating power. Only microorganisms on the surface of a material that are exposed directly to the radiation are susceptible to destruction. UV light can also damage the eyes, cause burns, and cause mutation in cells of the skin.

Ionizing radiation, such as X-rays and gamma rays, has much more energy and penetrating power than ultraviolet radiation. Ionizing radiation ionizes water and other molecules to form **radicals** (molecular fragments with unpaired electrons) that can disrupt DNA molecules and proteins. Gamma radiation is often used to sterilize pharmaceuticals and disposable medical supplies such as syringes, surgical gloves, catheters, sutures, and Petri plates. This method can also be used to retard spoilage in seafood, meats, poultry, and fruits. The United States Post office uses gamma radiation for certain classes of mail.

Filtration

Filtration is an effective method used to sterilize heat-sensitive liquids and reduce the microbe levels in drinking water. Air can also be filtered using HEPA filters which remove airborne particles as small as 0.3 microns. Outdoor enthusiasts may be familiar with drinking water filtration systems, which are used to remove bacteria and protozoa from environmental water sources such as rivers and lakes. The membrane filters used in filtration methods have varying pore sizes depending on the application; the most common size used to remove most bacteria and protozoa are 0.45 micron and 0.22 micron pores. Filtration methods are not effective for removing viruses unless the pore sizes are small enough to capture viral particles, since most viral particles range in size from 5–300 nm (0.005–0.30 microns). The influenza virus is estimated to have a 125–150 micron diameter and the even smaller rhinovirus responsible for the common cold range in size from 0.018–0.028 microns.

Osmotic Pressure

In the natural environment, microorganisms are constantly faced with alterations in osmotic pressure. Water tends to flow through semipermeable membranes, such as the cytoplasmic membrane of microorganisms, towards the side with a higher concentration of dissolved materials (solute). In other words, water moves from greater water (lower solute) concentration to lesser water (greater solute) concentration. When the concentration of dissolved materials or solute is higher inside the cell than it is outside, the cell is said to be in a **hypotonic environment** and water will flow into the cell. The rigid cell walls of bacteria and fungi, however, prevent bursting or plasmolysis. If the concentration of solute is the same both inside and outside the cell, the cell is said to be in an **isotonic environment**. Water flows equally in and out of the cell. Hypotonic and isotonic environments are not usually harmful to microorganisms. However, if the concentration of dissolved materials or solute is higher outside of the cell than inside, then the cell is in a **hypertonic environment**. Under this condition, water flows out of the cell, resulting in shrinkage of the cytoplasmic membrane or plasmolysis. Under such conditions, the cell becomes dehydrated and its growth is inhibited.

Order of Resistance of Microorganisms

Bacterial endospores are the most resistant to disinfection (i.e., the hardest to destroy!). *Mycobacteria* (TB bacteria) are the next most resistant. This is due to the bacterium's waxy lipid (mycolic acid) in its cell wall. Nonlipid (non-enveloped or naked) viruses come next, followed by fungi. Vegetative bacteria are easier to destroy. Least resistant are the lipid (enveloped) viruses. The lipid of the virus' envelope is easily destroyed by most disinfectants. Once the envelope is destroyed, the virus is not able to survive.

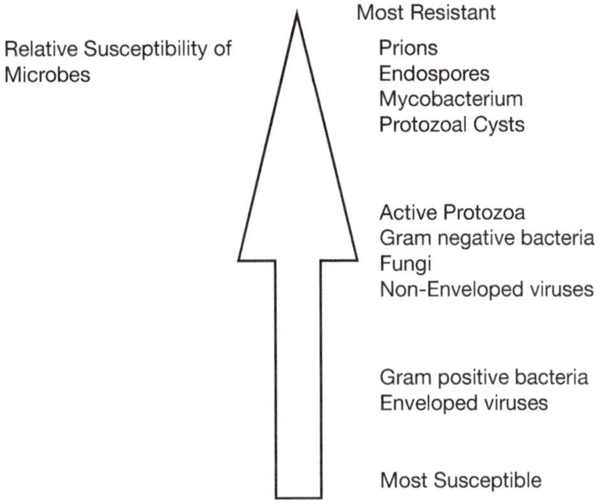

Courtesy of A. Swarthout

Factors Affecting the Success of the Disinfection and Sterilization Process

- Previous cleaning of the object. This is a must! An object cannot be disinfected or sterilized without first being cleaned to remove excess organic material.
- Organic load on the object. Is it coated in blood or other body fluids?
- Type and level of microbial contamination (see above: order of resistance). Was the object (piece of equipment, table surface, etc.) in contact with a patient with an endospore-forming infection (*Clostridium difficile*)? Was the patient in end stages of AIDS?
- Concentration of and exposure time to disinfectant. Every disinfectant comes with a label stating the concentration at which it must be used, if dilution is allowed, and the exposure time required to achieve disinfection.
- Condition of the object (grooves, crevices, hinges, lumens, etc.). Tubes with lumens, such as endoscopes, are notorious for being contaminated!
- Temperature and pH of disinfection process. Again, every disinfectant comes with a label stating information concerning the temperature and pH at which it will work. If we do not consider all of these factors in our use of a disinfectant, then it's very possible that we are not really disinfecting!

Spaulding's Classification System of Healthcare Items

All healthcare-related items are classified into one of three groups. Their classification determines the level of either sterilization or disinfection that is required.

Critical care items are items which enter sterile tissue or the vascular system. These include implants, scalpels, needles, cardiac and urinary catheters, and other surgical instruments. Critical care items must be processed by sterilization.

For those of you studying for the dental hygiene program, please note that all items used in the oral cavity in dentistry are considered to be critical care items and must be processed by sterilization! This has to do with the frequency of bleeding that can occur during dental procedures.

Semicritical care items are items which come in contact with mucous membranes or non-intact skin. These include respiratory therapy and anesthesia equipment, flexible endoscopes, cervical diaphragm fitting rings, laryngoscopes, and endotracheal tubes. Semicritical care items must be processed by high-level disinfection. It is recommended that semicritical items be rinsed with sterile water after disinfection. Semicritical items, such as thermometers and hydrotherapy tanks, only require intermediate-level disinfection.

Noncritical care items are items which come in contact with intact skin but not mucous membranes. These include bedpans, blood pressure cuffs, crutches, bed rails, linens, some food utensils, bedside tables, wheelchairs, and patient furniture. Noncritical care items are processed by low-level disinfection.

Lab Exercise 7.1

Control Methods: Effectiveness of Chemical and Physical Controls

Student Objectives

1. Define the following terms: antiseptic, disinfectant, sterilization, cleaning, and active ingredient of a chemical.
2. Interpret a chemical sensitivity test.
3. Evaluate the zone of inhibition of specific chemicals.
4. Evaluate the relative effectiveness of common disinfectants used in the health care setting.
5. Analyze the various disinfection methods.
6. Perform the UV test on select bacteria.
7. Interpret the effectiveness of UV radiation on bacteria and correlate time to death rate.
8. Correlate temperature and growth rate.
9. Describe the various physical methods used to control microbial growth.
10. Interpret the effectiveness of filtration on removing bacteria from liquid.

Name ______________________

Pre-Lab Exercise 7.1

1. What is the difference between disinfectants and antiseptics?

2. Describe the levels of disinfection listed below.

   Sterilization:

   High-level:

   Intermediate-level:

   Low-level:

3. List the sterilization methods used in healthcare settings.

4. Describe the types of microorganisms that are most resistant to control procedures.

5. List the factors affecting the disinfection process.

6. List the various methods that use temperature to control growth.

7. Describe the temperature, pressure, and time most commonly used with an autoclave.

8. Describe UV radiation, when UV radiation is used, and the limitations involved with UV radiation.

9. Describe ionizing radiation.

10. How is osmotic pressure used to control microbial growth?

11. Why is membrane filtration not effective at removing viral particles from a solution or from the air?

In this lab exercise, we will evaluate the activity of chemical and physical controls on specific bacteria. The first exercise will test three disinfectants and three antiseptics using sterilized filter discs impregnated with a given chemical, placed upon a lawn culture of four different types of bacteria. The second exercise will test the effectiveness of temperature on a bacterium. The third exercise will test the effectiveness of ultraviolet radiation on two bacteria.

Effectiveness of Chemicals—Week One

Week One Materials

Laboratory materials: Three disinfectants and three antiseptics from home and in the lab, sterile filter discs, beakers, forceps, Bunsen burners, and safety glasses.

Broth cultures per table: *Staphylococcus*, *Bacillus*, *E. coli*, and *Pseudomonas*.

Materials per student: 1 Mueller-Hinton agar plate, china marker, 1 sterile hockey stick, 1 sterile 1-ml pipette, and mechanical pipetter.

1. At the lab tables, each student will use one of the four different bacteria (so that all four bacteria are used at each table). **Record** the name of the bacterium you are using.
2. Obtain a Mueller-Hinton plate and label the top with your initials and the name of your bacterium.
3. Collaborating as a class, we will assign a number to each disinfectant.
   Label the numbers on the BOTTOM of the Petri plate, spaced evenly apart.
   Record the name of the disinfectant with the corresponding number.
4. Your instructor will pour a small amount of each test chemical into different beakers for the entire class to use.
5. Using a sterile 1-ml pipette, transfer **0.1 ml** of your assigned bacterium onto the surface of the agar. Do NOT open the lid until you are ready to transfer. **Dispose** of the used pipette in the **Nalgene container**.
6. Use a sterile hockey stick to spread the inoculum over the entire surface of the agar. The inoculum MUST evenly cover the entire agar surface to create a bacterial lawn. *Cover the Petri plate*. Dispose of the used hockey stick in a biohazard container.
7. Use the Bunsen burner and sterilize the forceps; pick up a single sterile disc, and touch the edge of the disk to the disinfectant. *DO NOT place the disc more than halfway into the solution. Allow the solution to absorb into the entire disc.*
8. Open the lid and place the disk on the lawn culture, in the numbered area.
9. Press the disk lightly on the agar. Close the lid immediately.
10. Continue this preparation for the remaining two disinfectants and three antiseptics being tested. The disks should be in a circle pattern, spaced apart evenly.
11. Tape the plate, place in a storage can (do not invert), and put on the cart to be incubated at 37°C for 24 hours.
12. Go to the **lab report** section and **record** the names of the disinfectants and antiseptics and their active ingredients. Be sure to include the number assigned to the test chemical.

Effectiveness of Temperature—Week One

Week One Materials

Broth culture per table: *Alcaligenes viscolactis*

Materials per table: Five TS broth tubes, inoculating loops, igniter, and Bunsen burner.

1. Each lab table will experiment with six different temperatures: 5°C, 25°C, 30°C, 37°C, and 45°C. Each student will be assigned one or two temperatures.
2. Obtain a TS broth tube and label the tube with the temperature you are assigned, your initials, and table number.
3. Aseptically inoculate your broth with one loop of *Alcaligenes viscolactis*.
4. Place your inoculated tube onto the classroom test tube rack for that specific temperature.

Effectiveness of UV Radiation—Week One

Week One Materials

Laboratory materials: UV light, cardboard disks with numerical cut-outs, safety glasses, and Nalgene containers.

Broth cultures per table: *Staphylococcus* and *Bacillus.*

Materials per student: 1 tryptic soy agar (TSA) plate, 1 sterile hockey stick, 1 sterile 1-ml pipette, and mechanical pipetter.

At the start of this lab activity, each student will be assigned one of the two bacteria and each table will be assigned a time to expose the bacteria to UV light. You will receive a cardboard disk with a numerical cut-out that corresponds to the time assigned.

Simo988/
Shutterstock.com

Table 1: *Staphylococcus* = 20 seconds and 40 seconds.
Bacillus = 20 seconds and 40 seconds.
Table 2: *Staphylococcus* = 1 minute and 2 minutes.
Bacillus = 1 minute and 2 minutes.
Table 3: *Staphylococcus* = 3 minutes and 4 minutes.
Bacillus = 3 minutes and 4 minutes.
Table 4: *Staphylococcus* = 5 minutes and 7½ minutes.
Bacillus = 5 minutes and 7½ minutes.
Table 5: *Staphylococcus* = 10 minutes and 15 minutes.
Bacillus = 10 minutes and 15 minutes.

1. At the lab tables, two students will handle *Staphylococcus* and two students will handle *Bacillus* (a spore-forming bacterium). Record the name of the bacterium you are using and the time you are assigned.
2. Obtain a tryptic soy agar (TSA) plate and **label** with your initials, the name of your bacterium, and the time on the bottom of the plate.
3. Using a sterile 1 ml pipette, transfer **0.1 ml** of your assigned bacterium onto the surface of the agar. **Dispose** of the used pipette in the **Nalgene container**.
4. Use a sterile hockey stick to spread the inoculum over the entire surface of the agar. The inoculum MUST evenly cover the entire agar surface to create a bacterial lawn. Cover the Petri plate. Dispose of the used hockey stick in a biohazard container.
5. Remove the Petri plate lid and place downward on a clean area of the lab table. Cover the exposed plate with the instructor-assigned cardboard disk with numerical cut-out.
DO NOT ALLOW THIS CARD TO COME IN CONTACT WITH THE BACTERIAL LAWN.

6. Carefully take the covered Petri plate to the germicidal UV light and place the plate inside the UV box. Make sure your hand, wrist, and arm are covered.
7. Allow the lawn to be exposed to the germicidal effects of the UV light for the assigned time. WATCH THE CLOCK CAREFULLY.
8. When the exposure is completed, remove your plate immediately. Remove the cardboard disk and leave the cardboard inside the UV light box.
9. Bring your Petri plate back to your lab table and place the lid on the plate.
10. Tape the plate and place it upside down on the incubation tray to be incubated at 37°C for 24 hours.

Effectiveness of Filtration—Week One

Week One Materials

Laboratory materials: 1 *E. coli* culture, 1 99-ml bottle, sterile 1-ml micropipette tips, micropipetter, 3 EMB agar plates, 1 sterile 0.45 micron filter, 2 sterile hockey sticks, sterile forceps, sterile membrane filtration apparatus, sterile liter flask, vacuum pump, and paper towels.

Set-Up:

1. Set up the filtration apparatus as demonstrated by your instructor.
2. Place the filtration apparatus onto the 1-liter flask. Ensure the rubber stopper is snug inside the opening of the flask. Ensure your apparatus is secure. Apply paper towels around the flask in case there is a leak.

Inserting the Filter:

3. Sterilize the blunt end forceps by briefly passing them through the Bunsen burner flame.
4. Use the sterile forceps to remove the sterile 0.45-micron filter from its package.
   - Remove the blue cover from the filter and throw away.
   - The filter is white with a grid on one side.
5. Temporarily remove the top part of your filtration apparatus: separate the funnel from the base. Using the sterile forceps, place the sterile 0.45-micron filter on the filter holder of the apparatus, the grid side up. Ensure that the filter is centered exactly on the filter holder.
6. Set the funnel back on the base (filter holder) and fasten it in place.
7. Connect the hose from the flask to the vacuum pump.

Diluting the Sample:

8. Gently re-suspend the *E. coli* sample by rolling the test tube between your hands.
9. Using a sterile micropipette tip, aseptically transfer **0.1-ml** of the *E. coli* sample into a 99-ml bottle.Place the pipette back in its package and **discard** in the **biohazard waste container**.
10. Secure the top of the bottle, then shake the bottle up and down 25 times.

Plate a Pre-Filtration Control:

11. Use a new sterile micropipette tip, transfer 0.1 ml of the diluted *E. coli* sample to an EMB plate (labeled Pre-Filtration).
12. Use a sterile hockey stick to spread the inoculum over the entire surface of the agar. The inoculum MUST evenly cover the entire agar surface to create a bacterial lawn. *Cover the Petri plate*. Dispose of the used hockey stick in a biohazard container.

Filtration:

13. Turn on the vacuum.
14. Slowly pour the remaining contents of the diluted *E. coli* sample into the funnel of the filtration apparatus.
    Leave the vacuum on until the entire sample goes through the membrane.
15. Turn off the vacuum and disconnect the hose.

Plating:

16. Sterilize the forceps as before.
17. Aseptically remove the filter from the filter holder with the sterile forceps. Transfer the filter onto the center of the **EMB agar** with the grid side up.
18. Label the plate with your table number. Tape the plate, place in a storage can (do not invert), and put on the classroom tray to be incubated.
19. Place all glassware of the filtration apparatus on the incubation cart. Place the clamp and vacuum pump on the demo cart.

Plating a Post-Filtration EMB Plate

20. Using a new sterile pipette tip, aseptically transfer 0.1 ml of the filtration effluent onto a new EMB plate.
21. Using a new sterile hockey stick spread the inoculum over the entire surface of the agar.

End of week one procedures

Effectiveness of Chemicals—Week Two

1. Examine your plates. Compare your results with those at your lab table, and then with everyone in the lab.
2. Analyze the zones of inhibition. Use a metric ruler to measure the diameter of the zones of clearing surrounding each of the disks. Be certain to measure in millimeters, not centimeters. Compare zones of each test chemical to one another.
3. **Record** the zone of inhibition data in the **lab report**. Share your analysis with others at your table.
4. You can analyze the effectiveness of one antiseptic or disinfectant by comparing the size of the zone to that of a different organism i.e. bleach, Pseudomonas = 22mm and Staphylococcus = 30mm; this would indicate that bleach is more effective at killing Staphylococcus. You cannot compare the size of the zones to other chemicals i.e. bleach, Pseudomonas = 22mm and Lysol, Pseudomonas = 10mm; this does not necessarily mean bleach is more effective than Lysol since the two chemicals have different chemical properties allowing them to diffuse into the media differently.
5. Discard the plates on the autoclave cart.

Effectiveness of Temperature—Week Two

1. Line-up the five broth tubes with *E. coli* in a test tube rack. Examine the turbidity (growth) of each broth.
2. Correlate temperature with the amount of turbidity. **Record** the results in the **lab report**.
3. Remove any tape from the test tubes and place tubes on the autoclave cart.

Effectiveness of UV Radiation—Week Two

1. Examine your plate. Compare your results first with those at your lab table, then with everyone in the lab.
2. If you can see your number imprinted in the bacterial lawn of your Petri plate, record your experiment as a "kill" (K). If you are unable to discern the number, record your exposure time as a "no kill" (NK). **Record** your data in the data table of the **lab report**.
3. Complete the **chart** that is created on the chalkboard for the class. Fill in the chart in the **lab report** section and analyze the results.
4. Discard the plates on the autoclave cart.

Effectiveness of Filtration—Week Two

1. Examine the pre-filtration plate. Draw and record your observations in the lab report.
2. Examine the EMB plate with the membrane filter. Draw and record your observations in the lab report.
3. Examine the post-filtration plate. Draw and record your observations in the lab report.
4. Discard the plates on the autoclave cart.

Name ______________________________

Exercise 7.1 Lab Report

Disinfectant Susceptibility Test

1. Disinfectant results. Record the zone of inhibition in millimeters.

Name of Disinfectant and its Active Ingredient	*Staphylococcus*	*Bacillus*	*E. coli*	*Pseudomonas*

2. Which **disinfectant** seemed to have the best action? (See number 4 under 'Week two—Effectiveness of chemicals, to determine effectiveness)

____

3. What conclusions can you make from the data on the **disinfectants** tested? What have you learned through this activity regarding the use of various disinfectants in the healthcare setting?

____

____

____

4. Which **antiseptic** seemed to have the best action? (See number 4 under 'Week two-Effectiveness of chemicals' to determine effectiveness)

5. What conclusions can you make from the data on the **antiseptic** tested? What have you learned through this activity regarding the use of various **antiseptics** in the healthcare setting?

Effectiveness of Temperature

6. Rank the order of least turbid (no growth) to most turbid (growth).

	5°C	25°C	30°C	37°C	45°C
A. viscolactis					

7. Correlate temperature and microbial growth. What are your conclusions?

____

____

Effectiveness of Ultraviolet Radiation

8. Record your results. Label **K** for "kill". Label **NK** for "no kill".

Organism	Exposure Times						
Staphylococcus	20 sec	40 sec	1 min	2 min	3 min	4 min	5 min
Survival							
Staphylococcus	7½ min	10 min	15 min				
Survival							
Bacillus	20 sec	40 sec	1 min	2 min	3 min	4 min	5 min
Survival							
Bacillus	7½ min	10 min	15 min				
Survival							

9. Which bacterium seems to be the most resistant to UV radiation?
Correlate the time of UV exposure to the bactericidal rate.

10. A student performing this lab forgot to remove the Petri plate lid prior to exposing *Staphylococcus* to germicidal UV radiation. After incubating the plate for 24 hours, he/she found no killing had occurred. Explain what went wrong here.

Effectiveness of Filtration

Draw (using correct colors) observations of the filtration plates.

PRE-FILTRATION PLATE	MEMBRANE FILTER PLATE	POST-FILTRATION PLATE

Record observations from each of the three plates above; note any colony morphology differences and change in abundance of microorganisms:

Pre-filtration plate:

Membrane filtration plate:

Post-filtration plate:

11. Was the membrane filtration method effective at removing *E. coli* from the water sample? Why or why not?

____

____

12. If your post-filtration plate had microbial growth, were the colony morphologies the same as the pre-filtration plate? If they were the same, provide some reasons as to why the filtration was not successful in removing all of the microorganisms. If your post filtration plate had no growth answer the question as if the plate did have growth.

____

____

____

13. In the chart below consider the situation and list the ***specific*** physical or chemical method most commonly used for each example.

Situation Requiring Microbial Control	**Specific Method Used (specific)**
Milk	
Surgical tools	
Operating room	
Drinking water	
Swimming pool	
Kitchen counter	
Hospital room	
Injection site on skin	

Lab Exercise 7.2

Water Quality Analysis

Student Objectives

1. Perform proper water sample collection techniques.
2. Explain the kinds of bacteria that serve as indicators of contamination of water by human pathogens.
3. Explain the principle of the membrane filtration method for the analysis of water.
4. Perform a coliform count using the membrane filtration method for the analysis of water.
5. Aseptically transfer a bacterial sample using a sterile pipette.
6. Understand coliforms and how they are used to assess water quality.
7. Evaluate the EMB medium for selective and differential results.
8. Understand the standards of drinking and recreational water.
9. Evaluate disinfection and sterilization techniques of water.
10. List and describe the types of waterborne infections.

Name ______________________

Pre-Lab Exercise 7.2

1. What is the purpose of this laboratory exercise?

2. Explain the EPA's standards regarding drinking water and coastal recreational water.

3. Explain what *E. coli* and *Enterobacter aerogenes* may indicate about your water sample.

4. Describe the techniques used during emergency disinfection of drinking water.

5. List the criteria to be a good indicator organism for water quality.

6. Examine the "Common Causes of Gastroenteritis" table. What do the waterborne infections have in common regarding type of diarrhea, site of infection, and clinical signs?

7. Describe *E. coli* and *Cholera* as waterborne infections. Why is *E. coli* more abundant?

8. Describe *Cryptosporidium parvum* and *Giardia intestinalis* as waterborne infections.

INTRODUCTION

Drinking water, polluted water from natural sources, and treated effluent water must be examined regularly for determining their quality. Because water can be a carrier of pathogenic microorganisms capable of causing diseases in humans, water supplies used for drinking and cooking should be checked for possible contamination on a regular basis. It is necessary to establish maximum limits or standards for microbial contamination to ensure a pure, wholesome water supply and sufficient protection from waterborne diseases. Water quality standards have been established by every state.

Most waterborne pathogens find their way into the water from the feces of infected individuals. Unless there is a massive epidemic, those pathogens will only be found in very low concentrations that are difficult to isolate. Therefore, microbiologists have adapted the practice of identifying **indicator microorganisms** that are normally found along with pathogens in the intestinal tract. Their presence in water indicates possible fecal contamination and, therefore, an increased probability of waterborne disease. The coliform bacteria such as *E. coli*, *Enterobacter*, and *Klebsiella* are most frequently used as indicators.

To be indicator organisms, the bacteria must meet the following criteria:

1. They are normal microbiota of the human intestinal tract and their presence in high numbers in water would, in all likelihood, indicate human contamination.
2. Coliforms are very hardy bacteria able to survive for long periods outside the host, making it possible to isolate and identify them long after they have been released.
3. They are relatively easy to culture, requiring only the most basic microbiological materials and expertise.
4. They are found in high enough numbers in contaminated water to give a statistically valid count.

Escherichia coli and *Enterobacter aerogenes* are **coliforms** we will test for in this laboratory exercise. High numbers of *E. coli* may indicate that the water is unsafe for drinking, swimming, or bathing; therefore *E. coli* is a major indicator of fecal contamination. *Enterobacter aerogenes* is an indicator of soil run-off; this organism is commonly found in soil and grain samples. This experiment will use a bacteriological examination technique called the membrane filter method. The membrane filter technique is widely used for the examination of bacteria from a water sample. In this method the bacterial cells are filtered through a membrane. Using a vacuum for suction, the sample passes through the filtration apparatus. The excess water is passed through the filter and collected into a 1-liter flask. The bacteria, along with the membrane, are then placed on an EMB plate (see Chapter 4 for details regarding this media). Depending on your instructor, you may perform the serial dilution method which is an accurate method used to calculate the number of bacteria in the original water sample. In both procedures it is important to correctly determine the number of CFUs (colony forming units) on the media. Using the serial dilution method you are more likely to obtain a "countable plate", since microbial load levels can vary based on where the sample is from. **A countable plate has between 30–300 CFUs on the agar surface**. Upon incubation, coliforms produce visible colonies, which can be counted and their color analyzed.

Standards for drinking water are set by the EPA (Environmental Protection Agency).

The current standard with regard to *drinking water* and total coliforms:

1. Contaminant level goal = zero
2. Contaminant level maximum = when more than 40 samples are collected in one month, no more than 5% of the samples be coliform-positive
3. Contaminant level maximum = when less than 40 samples are collected in one month, no more than one sample collected be coliform-positive

This is the current standard with regard to *coastal recreational waters*:

States have the flexibility in determining the appropriate risk, as long as they do not exceed the EPA's acceptable risk levels when adopting their water quality standards of bacteria. The EPA established the 'Beach Act', a water quality criteria corresponding to an illness rate of 0.8% for swimmers in freshwater and 1.9% for swimmers in marine waters. The EPA developed these criteria values for *E. coli* and *Enterococci* concentrations that corresponded to the estimated illness rates associated with these coliforms. Most states determined that the coliform number could be between 70–200 coliforms per 100 ml.

Should the *drinking* water sample that you test in the lab exceed the above standards, it is recommended that you reevaluate how you acquired your sample. First consider where your drinking water sample is from (well water or municipal source). Coliforms in a municipal water sample would be extremely rare. The high colony count may be the result of sampling error and not contamination by fecal coliforms. Therefore, don't jump to any conclusions. You should be consulted as to how you obtained your sample and advised to resample your water source and repeat the test. Should the numbers remain high, you should contact the water and wastewater treatment facility of your city or county to have your water source professionally evaluated. Finding coliforms in well water is not uncommon; typically, as long as *E. coli* is not found, the water is considered safe to drink. If coliforms are found, then your well can be disinfected. Information regarding well disinfection procedures can be found earlier in this chapter.

Waterborne Infections

The presence of fecal coliforms may indicate the presence of more harmful pathogens such as *Cryptosporidium parvum* and *Giardia intestinalis*, which are two intestinal protozoans most commonly responsible for waterborne infections. *Giardia intestinalis* cysts and *Cryptosporidium parvum* oocysts can survive through the usual levels of chlorination used in water treatment plants; therefore, water filtration is necessary to remove them. Due to the small size of the oocysts of *C. parvum*, refined water filtration is needed.

Noroviruses, also known as Norwalk-like viruses (NLV), are responsible for waterborne viral infections. Norwalk-like virus is a highly contagious, acute GI infection that shows symptoms of the "stomach flu."

Escherichia coli and *Shigella sonnei* are the two most common bacteria responsible for waterborne bacterial infections. *Escherichia coli* is very abundant in our environment since it is found in human and animal feces. Cholera is also a waterborne bacterial disease that causes acute diarrheal illness, with severe water loss. Cholera is caused by the bacterium *Vibrio cholera*. In the United States, cholera has been virtually eliminated by modern sewage and water treatment systems. Cholera is problematic in non-industrialized nations.

To avoid waterborne diseases, hikers, campers, and hunters must not drink or use water from streams. Water should not be used to brush teeth, to rinse fruits or vegetables, to make ice cubes, or for any other activity that would accidentally allow the water to be swallowed. The best way to treat the water is to bring it to a boil, then vigorously boil for at least one minute. If chlorine or iodine tablets are used, the water must be treated for hours to ensure effectiveness. (Further information is on the next page.)

COMMON CAUSES OF GASTROENTERITIS

	Type of Diarrhea	Site of Infection	Mechanism of Invasion	Clinical Signs of Infection
Entamoeba histolytica	Bloody diarrhea	Intestinal mucosa	Enterocyte invasion, ulcers	High mortality
Giardia intestinalis	Watery diarrhea	Intestinal epithelium	Toxic products	Self-limited or chronic
Cryptosporidium parvum	Watery diarrhea	Intestinal epithelium	Toxic products, intracellular	Asymptomatic, elderly and infants, HIV
Cyclospora	Watery diarrhea	Intestinal epithelium	Enterocyte invasion	Watery diarrhea for weeks or months
E. coli	Watery diarrhea	Intestinal epithelium	Toxic products	Traveler's diarrhea
E. coli O157:H7	Bloody diarrhea	Intestinal epithelium	Shiga-like toxin	HUS, mortality
Norovirus (Norwalk-like Virus)	Watery diarrhea	Intestinal epithelium	Decrease in digestive enzymes, cell death	Self-limited
Shigella	Watery diarrhea	Intestinal epithelium	Toxic products	Day care centers
Shigella dysenteriae	Bloody diarrhea	Intestinal epithelium	Shiga toxin	HUS, mortality

Water Purification

There are two primary ways to treat water: boiling and adding bleach. If tap water is unsafe because of water contamination (from floods, streams, or lakes), boiling is the best method.

- Cloudy water should be filtered before boiling or adding bleach. Disinfectants are less effective in cloudy water.
- Filter water using coffee filters, paper towels, cheese cloth, or a cotton plug in a funnel.

Boiling

- Boiling is the safest way to purify water.
- Vigorous boiling for one minute will kill most disease-causing microorganisms present in water. When boiling is not practical, chemical disinfectants should be used. The two chemicals commonly used are chlorine and iodine.

Purifying by Adding Liquid Chlorine Bleach

- Common household bleach contains a chlorine compound that will disinfect water.
- If boiling is not possible, treat water by adding liquid household bleach, such as Clorox or Purex. Household bleach is typically 5.25 % chlorine. Avoid using bleaches that contain perfumes, dyes and other additives (such as Fresh Scent, Lemon Fresh, or other varieties). Be sure to read the label. To ensure that the bleach is at its full strength, rotate or replace your storage bottle every three months.
- Bleach will kill most, but not all, types of disease-causing organisms that may be in the water (Chlorine and iodine may not be effective in controlling more resistant organisms like *Cryptosporidium* and *Giardia*).
- Place the water (filtered, if necessary) in a clean container. Add the amount of bleach according to the table below.
- Mix thoroughly and allow to stand for at least 30 minutes before using (60 minutes if the water is cloudy or very cold).

Treating Water with a 5–6% Liquid Chlorine Bleach Solution

Volume of Water to be Treated	Treating Clear/Cloudy Water: Bleach Solution to Add	Treating Cloudy, Very Cold, or Surface Water: Bleach Solution to Add
1 quart/1 liter	3 drops	5 drops
½ gallon/2 quarts/2 liters	5 drops	10 drops
1 gallon	⅛ teaspoon (8–10 drops)	½ teaspoon
5 gallons	½ teaspoon	1 teaspoon
10 gallons	1 teaspoon	2 teaspoons

Purifying by Adding Iodine

- If boiling water and adding liquid chlorine bleach are not possible, chemical disinfection with iodine (e.g., Globaline, Potable-Aqua, or Coghlan's, found in pharmacies and sporting goods stores) is another method for making water safer to drink.
- Cloudy water should be strained through a clean cloth into a container to remove any sediment or floating matter, and then the water should be treated with iodine.
- *Cryptosporidium* (a parasite that can cause diarrhea) and other coccidian parasites (e.g., Cyclospora, Toxoplasma) might not be killed by this method.

Iodine tablets

- Follow the tablet manufacturer's instructions.
- If water is cloudy and cannot be filtered, double the number of tablets.
- If water is extremely cold, less than 5° C (41° F), an attempt should be made to warm the water, and the recommended contact time (standing time between adding a chemical disinfectant to the water and drinking the water) should be increased to achieve reliable disinfection.

Note: Be sure the tablet size is correct for a liter of water.

Water Analysis Activity—Week One Membrane Filtration Method

Water Specimen Collection

1. Your instructor will discuss collection of water samples the week before the lab begins. Bring water samples from your well, lakes, ponds, rivers, and streams. Avoid municipal tap water if you want to see potential growth on your media; as mentioned, it is extremely rare to find coliforms in municipal water.
2. Collect your water sample in a sandwich baggie, half-full. Place the filled baggie into a storage container to provide support to the water-filled bag on its journey to Delta.

Materials

Materials per table: Your water sample, 99-ml bottle, one sterile 1-ml pipette, mechanical pipetter, EMB agar plate, blunt end forceps, sterile 0.45 micron filter, sterile membrane filtration apparatus, 1-liter flask, vacuum pump, and paper towels.

Set-Up:

1. Set up the filtration apparatus as demonstrated by your instructor.
2. Place the filtration apparatus onto the 1-liter flask. Ensure the rubber stopper is snug inside the opening of the flask. Ensure your apparatus is secure. Apply paper towels around the flask in case there is a leak.

Inserting the Filter:

3. Sterilize the blunt end forceps by briefly passing them through the Bunsen burner flame.
4. Use the sterile forceps to remove the sterile 0.45-micron filter from its package.
   - Remove the blue cover from the filter and throw away.
   - The filter is white with a grid on one side.
5. Temporarily remove the top part of your filtration apparatus: separate the funnel from the base. Using the sterile forceps, place the sterile 0.45-micron filter on the filter holder of the apparatus, the grid side up. Ensure that the filter is centered exactly on the filter holder.
6. Set the funnel back on the base (filter holder) and fasten it in place.
7. Connect the hose from the flask to the vacuum pump.

Preparing the Sample:

8. Gently shake your water sample in order to re-suspend all material.
9. Using a sterile pipette, aseptically transfer **1-ml** of your water sample into a 99-ml bottle. Place the pipette back in its package and **discard** in the **Nalgene container**. *Put the pipetter back in the storage bin.*
10. Secure the top of the bottle, then shake the bottle up and down 25 times.

Filtration:

11. Turn on the vacuum.
12. Slowly pour the contents of the 100ml bottle into the funnel of the filtration apparatus. Leave the vacuum on until the entire sample goes through the membrane.
13. Turn off the vacuum and disconnect the hose.

Plating:

14. Sterilize the forceps as before.
15. Aseptically remove the filter from the filter holder with the sterile forceps. Transfer the filter onto the center of the **EMB agar** with the grid side up.
16. Label the plate with your table number. Tape the plate, place in a storage can (do not invert), and put on the classroom tray to be incubated.
17. Place all glassware of the filtration apparatus on the incubation cart. Place the clamp and vacuum pump on the demo cart.

Water Analysis Activity—Week One
Serial Dilution Method

Materials

Materials per table: Your water sample, 1 99-ml bottles, micropipette tips, micropipetter, 1 EMB agar plates, 3 TSA plates, sterile hockey sticks.

1. Use a china marker to label the one 99-ml water dilution bottles: "10^{-2}".
2. Use a marker to label the bottom of each TSA plate with your table number. Label one plate "10^{-1}", a second plate "10^{-2}", and a third plate "10^{-3}".
3. Dilute your water sample by transferring 1.0 ml of your sample into 99 mls of water (10^{-2} bottle). Shake the water bottle vigorously for 20 seconds.
4. Using the same pipette aseptically transfer 0.1 ml of your original water sample onto your 10^{-1} TSA plate. Have someone else at your table spread the sample on the agar surface using a sterile hockey stick.
5. Using a new pipette tip aseptically transfer 0.1 ml of water from the 10^{-2} dilution bottle to the 10^{-3} TSA plate. Have someone else at your table spread the sample on the agar surface using a NEW hockey stick.
6. Using the same pipette tip aseptically transfer 1 ml of water from the 10^{-2} dilution bottle to the 10^{-2} TSA plate. Have someone else at your table spread the sample on the agar surface using THE SAME hockey stick.
7. **Allow the liquid to absorb into the agar before inverting plates for incubation.**
8. To inoculate your EMB plate transfer either 1 ml or 0.1 ml onto your EMB plate. If you transfer 0.1 ml you will label that plate 10^{-1}. Have someone else at your table spread the sample on the agar surface using a NEW hockey stick. **Allow the liquid to absorb into the agar before inverting plates for incubation.**

Water Analysis—Week Two
Membrane Filtration Method

1. With the Quebec colony counter, count the number of colony-forming units (CFU) on the eosin methylene blue (EMB) plate.
2. Place the plate on the **colony counter**, lid side up (not inverted).
3. Adjust the lighting and magnifying lens to provide an optimum view.
4. Use a hand tally counter. Be sure that it is set on zero before proceeding.
5. Count the colonies that are separated from one another on the membrane filter of the Petri plate. You are counting the total number of **CFU/ml (colony-forming units per ml)** of the original water sample.
6. **Record** the number of **CFU/ml** in the original water sample in the Lab Report.
   If the number of colonies is too few to count, record as "TFTC."
   If the number of colonies is too numerous to count, record as "TNTC."
7. Examine the **EMB** plate for the presence of specific coliforms. Determine if you have *E. coli* and/or *Enterobacter aerogenes*. Report your analysis in the lab report.

***EMB NOTE*: Since the bacteria will grow on a filter paper, and not directly on the EMB plate, *E. coli* will NOT have a green metallic sheen. Instead, *E. coli* colonies will have a purple color on the filter paper.**

Water Analysis—Week Two
Serial Dilution Method

1. Observe each of the four EMB plates and determine which plate is countable.
2. Count the colonies that are separated from one another on the Petri plate. You are counting the total number of **CFUs** on the plate. You must use a calculation to determine how many **CFU/ml** were in the original water sample.
3. To determine the number of **CFU/ml**, use the following calculation:

$$\text{CFUs/ml} = 1/\text{dilution factor} \times \text{number of colonies}$$

4. **Record** the number of **CFU/ml** in the original water sample in the Lab Report.
   If the number of colonies is too few to count, record "TFTC."
   If the number of colonies is too numerous to count, record "TNTC."
5. Examine the **EMB** plate for the presence of specific coliforms. Determine if you have *E. coli* and/or *Enterobacter aerogenes*. Report your analysis in the lab report.

Name ______________________________

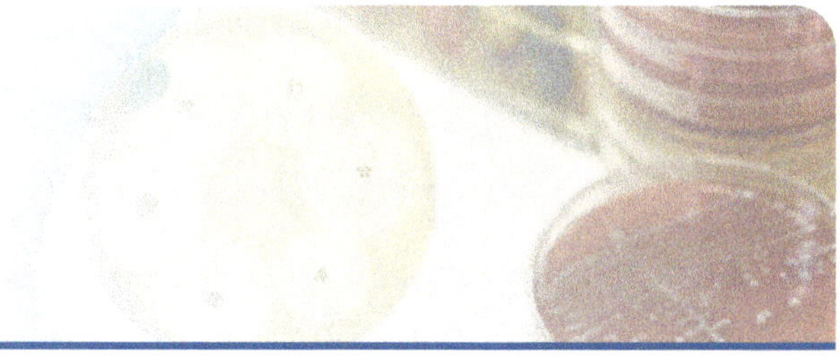

Exercise 7.2 Lab Report

Week Two—Membrane Filtration Analysis

1. Source of the water = ______________________________

   Sketch your EMB plate. Label the different colonies and describe them below.

   Describe the appearance of the colonies on the plate:

Week Two—Serial Dilution Analysis

2. Source of the water = ______________________________

   Sketch your **COUNTABLE EMB** plate. Label the different colonies and describe them below:

3. Sketch your **COUNTABLE TSA** plate. Describe them below:

4. Describe the presence of *E. coli* and/or *Enterobacter aerogenes* on your EMB Plate.

5. Record the *final analysis* of your water here. State where your water sample was collected from. Is your water sample safe to drink?

6. Did anyone's sample of drinking water test below the acceptable standards? If so, mention where they got the water (home, city, etc.) and how the sample was collected.

7. Sarah Goodbar, the nurse's aid, had a bumper crop of melons and strawberries from her garden. She decided to share the fruits of her labor with her patients at the Geriatrics Facility. In order to keep the fruits cool and fresh she decided to place them in the ice-making machine. An outbreak of dysentery occurred among the patients of the facility. The Infection Control Professional determined that the infection was "waterborne." Explain what happened here and how it was confirmed!

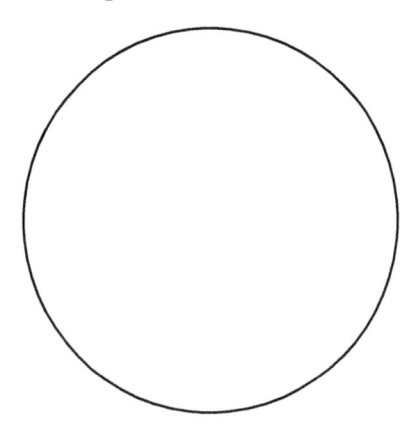

8. A water sample is taken from the whirlpool at Physical Therapy and tested for coliforms. A dilution of 10^{-3} was used before using the membrane filtration method. The next day, the EMB plate showed 50 colonies. The colonies are small with a green metallic sheen. (CFU/ml = 1/dilution factor × number of colonies)

   a. What is the estimated number of bacterial contaminants in the water sample (CFU/ml)?

   ____

   b. Describe the type of bacteria in the water.

   ____

   c. What is the most likely source of these bacteria?

   ____

9. How will this exercise influence your daily life?

____

____

____

Lab Exercise 7.3

Food Quality Analysis and Bacterial Population Counts in Food

Student Objectives

1. Aseptically transfer a bacterial sample using a sterile pipette.
2. Inoculate a Petri plate with a bacterial sample using a sterile pipette and pour melted media in an aseptic manner.
3. Perform a standard plate count (SPC): a serial dilution, plating out the samples, incubating, and performing the count.
4. Use the term CFU/ml (Colony-Forming Units per ml).
5. Understand that microbial load is used to determine the conditions under which ground beef needs to be stored and its expected shelf-life.
6. Understand the use of microbes for food production.
7. Evaluate foodborne infections and foodborne intoxication.
8. Use and analyze the serial dilution method with different volumes.
9. Apply microbial spoilage of food as a public health concern.

Name ____________________

Pre-Lab Exercise 7.3

1. List examples of how microbes are used in food production.

2. What conditions might cause an undesirable or spoiled food product?

3. You discover that a food item contains 10^{10} bacteria per gram or ml. Explain the possible cause of this condition.

4. Under what circumstances can a pipette be reused when performing a serial dilution?

5. When should a used pipette be replaced with a new, sterile pipette?

6. What is a standard plate count (SPC)?

7. How are *Salmonella* and *E. coli* infections similar?

8. Describe the complications of infection related to *E. coli* O157:H7.

INTRODUCTION

Microbiology is important to all aspects of food production and quality control. Many microbial activities provide us with foods. Many of the foods produced by microbes are the result of fermentation of carbohydrates with the production of acids and alcohols that give foods their characteristic aromas and tastes. **Fermentation** refers to the metabolic processes that release energy from the carbohydrate or other organic molecules, do not require an electron transport system, and use an organic molecule as a final electron acceptor.

Acid production may serve several purposes. By lowering the pH, acid protects food from the growth of and spoilage by detrimental bacteria. Acidity can also produce chemical changes such as the coagulation of proteins, as in yogurt production. In yogurt production, microorganisms ferment lactose and produce lactic acid; eating significant numbers of lactose-fermenting organisms may actually be a benefit by aiding in the digestion of dairy products and increasing the population of "good" bacteria to out-compete potentially pathogenic organisms. In alcoholic fermentations (such as wine production) yeast ferment sucrose to produce ethyl alcohol and carbon dioxide. Alcohol also acts as a preservative, and it may be oxidized to produce acetic acid (this is what happens to wines when they get a vinegary taste).

Microbiology is also important with regard to **food spoilage**, which is caused by undesirable and/or disease-causing organisms that replicate in foods. Bacteria and fungi occur in most foods in varying quantities. Generally, numbers are less than 100,000 per gram or ml. The number of microbes can be abnormally high if growth has occurred either in the case of fermented food (desirable) or food spoilage (undesirable). Spoiling food may contain greater than 10^7 bacteria per gram or ml. To determine the numbers of bacteria in a food item, the number of **colony forming units (CFU)** is determined by homogenizing a food sample, making a dilution series, and plating the dilution on agar. The plates are then incubated and the CFU per gram or ml in the original food item are determined. An evaluation can then be made of the quality of the food based on the normal condition of that food. Large numbers (greater than 10^7) can generally (not always) be interpreted to mean that the food has been held under conditions that permit growth of bacteria (good or bad). In many foods, especially previously cooked foods, this suggests that microbes capable of causing foodborne illness may have also grown. Thoroughly cooking food is important in reducing the numbers of pathogens, but some bacteria can produce toxins that will not be destroyed by cooking. This plate count method has been used to establish standards for some foods, such as milk, which must contain less than 15,000 CFUs per ml after pasteurization.

The microbial spoilage of food is an important economic and public health concern. The conditions under which ground beef need to be stored and its expected shelf-life are often determined by its microbial load. The most accurate way to determine this load is by performing viable counts on the meat, a procedure performed by quality control technicians at large meat packing plants and, less frequently, by health inspectors. **Ground beef is generally considered acceptable if the viable counts are less than 10^6 bacteria per gram**.

Foodborne Intoxication and Foodborne Infection

Foodborne intoxication results when a person eats a food product containing an already-produced exotoxin. The exotoxin was produced by bacteria living and reproducing in the food product before it was consumed. The exotoxin is what causes the rapid development of the illness in the person. Symptoms can appear within hours of consumption. *Staphylococcus aureus* and *Clostridium botulinum* produce foodborne intoxication. The toxin produced by *S. aureus* is heat-stable, so is not destroyed by cooking or boiling the food. The toxin produced by

C. botulinum is heat-labile, so boiling the food for 15 minutes immediately before consumption would destroy the toxin.

Foodborne infection results when a person eats a

Standard Plate Count of Food Sample Activity

> **Materials**
>
> **Laboratory materials per Table:** 2.5g food sample, 50 ml falcon tube, sterile hockey sticks, 22.5 ml sterile water, 4 TSA, and 1 EMB plates, sterile 1 ml pipettes, mechanical pipetter, and 1 sterile 99 ml water dilution bottles.

Serial Dilution and Plating Procedure:

1. Carefully mass 2.5 grams of your food sample and add the food to 22.5 ml of sterile water in your 50 ml falcon tube. This is considered a 1:10 dilution = 10^{-1}.
2. Vortex your sample for 45 seconds.
3. Take 1 ml of your vortexed sample and transfer to a 99 ml dilution bottle. This is a 1:100 dilution = 10^{-2}; the final dilution factor is 10^{-3} since you add the exponents from the first 1:10 dilution. Mix well by shaking.
4. To plate the samples start with your 1:10 dilution falcon tube. Transfer 1 ml of the liquid to 1 TSA plate. Use a sterile hockey stick to spread the liquid around the plate evenly. Allow liquid to absorb into the media. Label this plate 10^{-1}.
5. Next take 0.1 ml from the 1:10 dilution falcon tube and transfer to a second TSA plate. Use a sterile hockey stick to spread the liquid around the plate evenly. Allow liquid to absorb into the media. Label this plate 10^{-2}.
6. Next take 1 ml from the 1:100 dilution bottle and transfer to a third TSA plate. Use a sterile hockey stick to spread the liquid around the plate evenly. Allow liquid to absorb into the media. Label this plate 10^{-3}.
7. Next take 0.1 ml from the 1:100 dilution bottle and transfer to a fourth TSA plate. Use a sterile hockey stick to spread the liquid around the plate evenly. Allow liquid to absorb into the media. Label this plate 10^{-4}.
8. To inoculate your EMB plate transfer either 1 ml or 0.1 ml on to your EMB plate. If you transfer 0.1 ml you will label that plate 10^{-1}. Have someone else at your table spread the sample on the agar surface using a NEW hockey stick. **Allow the liquid to absorb into the agar before inverting plates for incubation.**
9. Place all plates on the tray for incubation.

Standard Plate Count of Food Sample Week Two

1. Lay out your plates in the order of dilution. **Countable plates** must have between **30 and 300 colonies**. Determine which plates appear to have TFTC (too few to count/less than 30 colonies) and TNTC (too numerous to count/greater than 300 colonies). Set aside the plates that appear to have fewer than 30 or over 300 colonies.
2. Take the plate (or plates) that appear to fall within our countable range to the Quebec colony counter.
3. Place the plate upside down (inverted) on the **colony counter**.
4. Adjust the lighting and magnifying lens to provide an optimum view.
5. Use a hand **tally counter**. Be sure that it is set on zero before proceeding.

6. Count the colonies that are separated from one another and are on the bottom of the Petri plate; not those that may be growing up the inside edge of the plate.
7. The total number of **CFU/ml** of original culture is determined by multiplying the plate count by the dilution factor of that countable plate. For example, if you counted 160 colonies on the 10^{-6} plate, the final count per ml would be 160×10^6 CFU/ml, or 160,000,000 colony-forming units per ml.
8. **Record** the number of CFU/ml in the original meat sample in the lab report.

Name ______________________________

Exercise 7.3 Lab Report

Week Two—Standard Plate Count Analysis of Food Sample

1. Sketch the dilution plate that was counted. Use an arrow to pick a colony that your group would choose to Gram stain for further analysis.

   Food Sample:____________________________

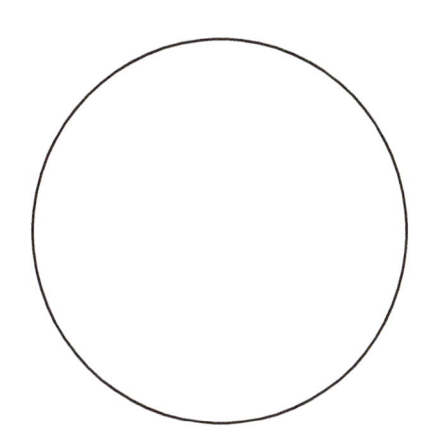

2. Calculate the number of CFU/ml in the original food sample using the countable plate.

   ____

3. Is your food sample acceptable? Explain your answer.

   ____

   ____

4. What is the CFU/ml that is considered acceptable for ground hamburger?

   ____

5. Was there any growth on your EMB plate? Is the plate countable? What does growth on your EMB plate tell you about the types of organisms in your food sample?

   ____

6. Was *E. coli* present on the EMB plate? How can you tell?

   ____

7. When you place one ml of a culture into a 99 ml dilution bottle, what is the final dilution of the sample?

____

8. Write the dilution factor below each tube:

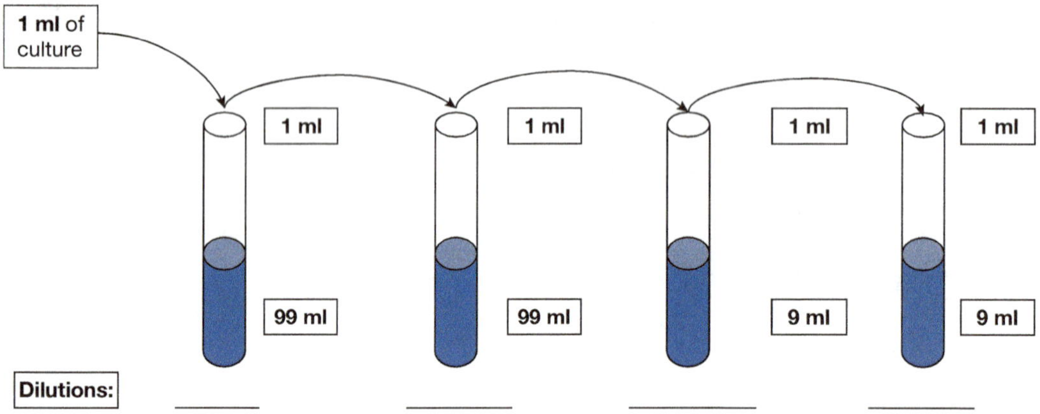

Courtesy of M. Pressler

9. Assume you have an extremely dense culture of bacteria, estimated to be 15,000,000,000 CFU/ml. Determine how many 99 ml dilution bottles, pipettes, and Petri plates would be required to get a countable plate. You may draw it out.

____

____

____

____

10. Describe foodborne intoxication and name the microorganisms that cause it.

____

____

11. Describe the toxin produced by *Staphylococcus aureus.*

____

____

12. Describe the toxin produced by *Clostridium botulinum.*

____

____

13. Describe foodborne infection and name the microorganisms that typically cause foodborne illnesses.

14. Describe the transmission, infection, and symptoms of the disease caused by *Campylobacter*.

Microbial Metabolism

Understanding microbial metabolism is important for a wide variety of reasons. Microbiologists can study bacterial metabolic pathways as a model for human pathways. Scientists can answer fundamental questions such as: How do cells gain energy to form cell structures? How do pathogens acquire energy and nutrients at the expense of the health of a patient? How does yeast turn grape juice into alcohol? Not only can we answer these questions by understanding microbial metabolism, we can also identify unknown microorganisms through biochemical testing. Biochemical tests allow microbiologists to identify end products from metabolic pathways, since not all bacterial species produce the same enzymes given the same substrate or "food" source, because a particular bacterium may or may not be able to utilize the substrate to grow. In this chapter you will learn how microbes use enzymes in metabolic pathways and how microbial cells metabolize glucose and produce energy for the cell's needs.

Metabolism is the sum of all chemical reactions within a cell. These chemical reactions can either be catabolic (energy harvesting) or anabolic (biosynthetic) reactions. **Catabolic** reactions are processes used by cells to break down complex organic molecules into simpler compounds. During this process energy is released, meaning that the cell gains energy (Figure 8.1). An example of a catabolic reaction would be the required steps a cell must complete in order to break down glucose and gain Adenosine Triphosphate (ATP). **Anabolic** reactions within a cell use the ATP energy in order to build complex molecules from simpler molecules. An example of an anabolic reaction would be the steps a cell takes to build cell structures such as cell membranes or cell walls.

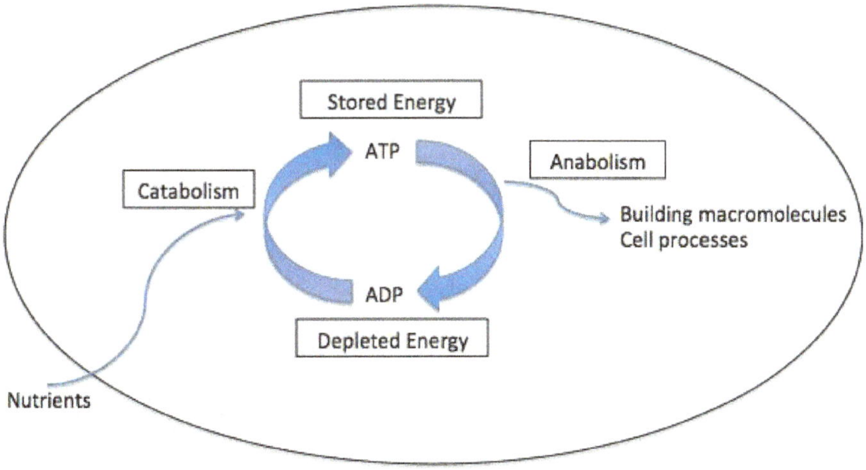

Figure 8.1 Metabolic Action of a Cell, Showing the Cyclical Reaction between ATP and ADP. Nutrients are Catabolized to make ATP while other Cell Processes use ATP

Courtesy of Authors

Figure 8.2 Example of a Linear Biochemical Pathway. Some Pathways may be Branched or Circular. Note that where each Arrow is Located in the Pathway, an Enzyme would be Involved in the Formation of that Product

Courtesy of Authors

As mentioned above, in order for a cell to break down glucose and make ATP, there are a series of steps a cell must complete These so called "steps" are referred to as **metabolic pathways** or **biochemical pathways** (Figure 8.2).

Enzymes

In a metabolic pathway there are **substrates** (which are the starting compounds), intermediates, and end-products. Enzymes are involved in each step of a pathway, which facilitate the formation of intermediates and end-products. **Enzymes** are composed of protein and act as biological catalysts that lower the activation energy of a chemical reaction and therefore speed up the reaction (Figure 8.3). Some substrates are too large to be transported into the cell. Therefore certain cells can produce **exoenzymes** which are enzymes that are secreted from the cell and break down the substrate outside of the cell into simpler molecules, which can then be transported inside the cell. Amylase is one such exoenzyme secreted by various bacteria. Amylase hydolyzes starch (polysaccharide) into monosaccharides. Once the starch has been broken down outside of the cells the monosaccharides can then be utilized by the cell for growth. In the lab we can observe changes in media enriched with starch and determine which bacteria can produce the exoenzyme amylase.

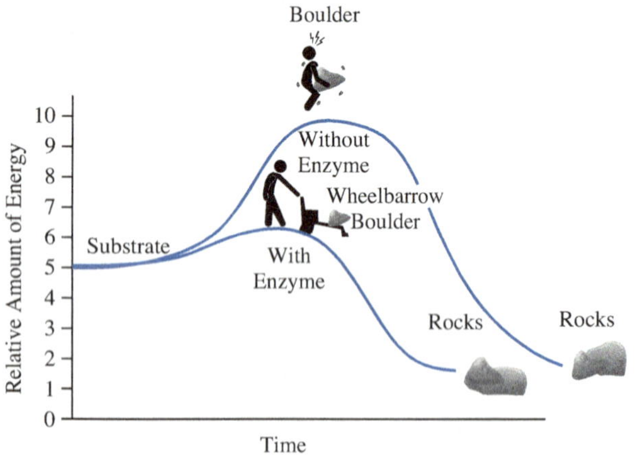

Figure 8.3 Example of how the use of an Enzyme Lowers the Amount of Energy for a Reaction to take Place. Note that the Time is also Decreased for the same Product to be Made. Biochemical Reactions within a Cell could take Place without Enzymes; However, Life as we know it would not exist without them

Enzymes are substrate-specific, meaning that a different substrate requires a different enzyme. Typically enzymes are named by adding "ase" to the end of the substrate, for example lipase would be a substrate-specific enzyme that breaks down lipids. Enzyme specificity can be explained by a 'lock and key' theory, now commonly called the **induced fit model**. The key is the substrate, and as you know, the keys on your key chain are specific for only one lock. Likewise here, the lock is the enzyme and the substrate has a shape that fits into the enzyme's **active site**. Once the substrate binds to the active site, this is refered to the enzyme-substrate complex (Figure 8.4).

Many enzymes are complete on their own; however, some enzymes called **apoenzymes** have non-protein components called cofactors (Figure 8.5). Cofactors can be inorganic elements such as iron, magnesium, or zinc, or they

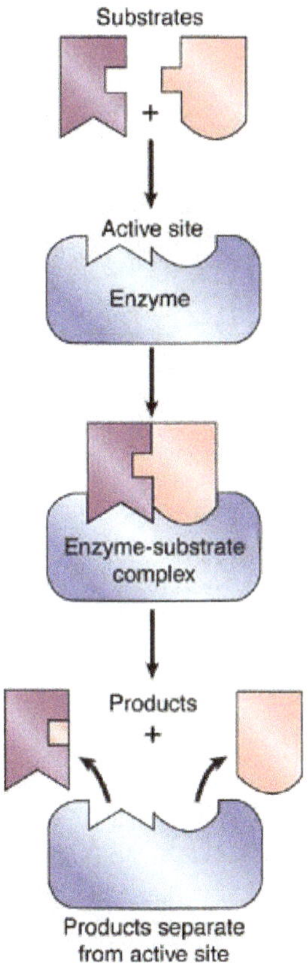

Figure 8.4 Schematic Representation of Enzyme Action inside a Cell. Would this be a Catabolic or Anabolic Reaction? Note that after the Products are Released, the Enzyme is Recycled in the Cell and can Bind to more Substrate. (Kendall Hunt Image Figure 6.3 Energy and Metabolism)

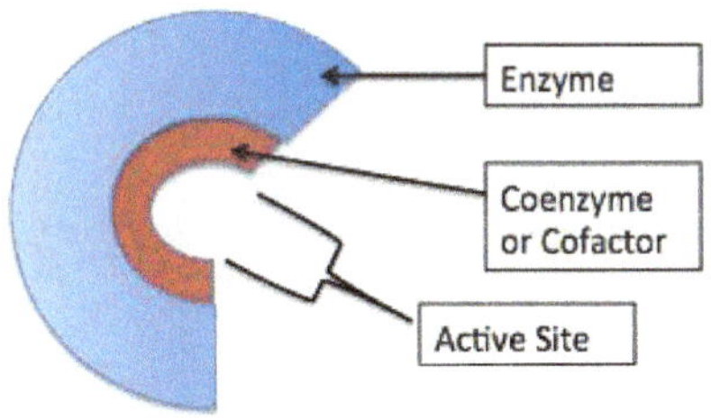

Figure 8.5 Anatomy of an Apoenzyme.

Courtesy of Author

can be coenzymes which are derived from vitamins which cannot be synthesized by certain organisms. For example, *Escherichia coli* can synthesize most of its own vitamins and convert them to coenzymes; however, humans must consume vitamins in order to have proper cell metabolism. Since *E. coli* has a symbiotic relationship with humans, Vitamin K can be obtained from the *E. coli* living within our GI tract. The binding of the apoenzyme with its coenzymes and cofactors is refered to as a **holoenzyme**.

Some enzymes have sites separate from the active site called an **allosteric site**. Depending on the enzyme, certain molecules can bind to these sites, which results in a change in the active site. There are some biochemical pathways in which the molecule that binds to the allosteric site would increase the performance of that enzyme in the pathway. However, we will focus on how enzyme activity can be inhibited from molecules binding to an allosteric site. When an enzyme is inhibited by a molecule binding to the allosteric site, the molecule will actually change the shape of the active site (Figure 8.6a). When the active site shape is changed, the substrate can no longer bind to the active site, and therefore, no products will be made. Cells can actually take advantage of these types of enzymes to regulate a metabolic pathway. A process called **feedback inhibition** allows a cell to shut down an entire biochemical pathway when the end product of the pathway acts as an allosteric inhibitor on the first enzyme in the pathway (Figure 8.6b). *Escherichia coli* can control the synthesis of isoleucine by this mechanism. The presence of isoleucine allosterically inhibits the first enzyme in the pathway, which will prevent the synthesis of isoleucine. Once isoleucine is depleted, *E. coli* can resume production of the amino acid.

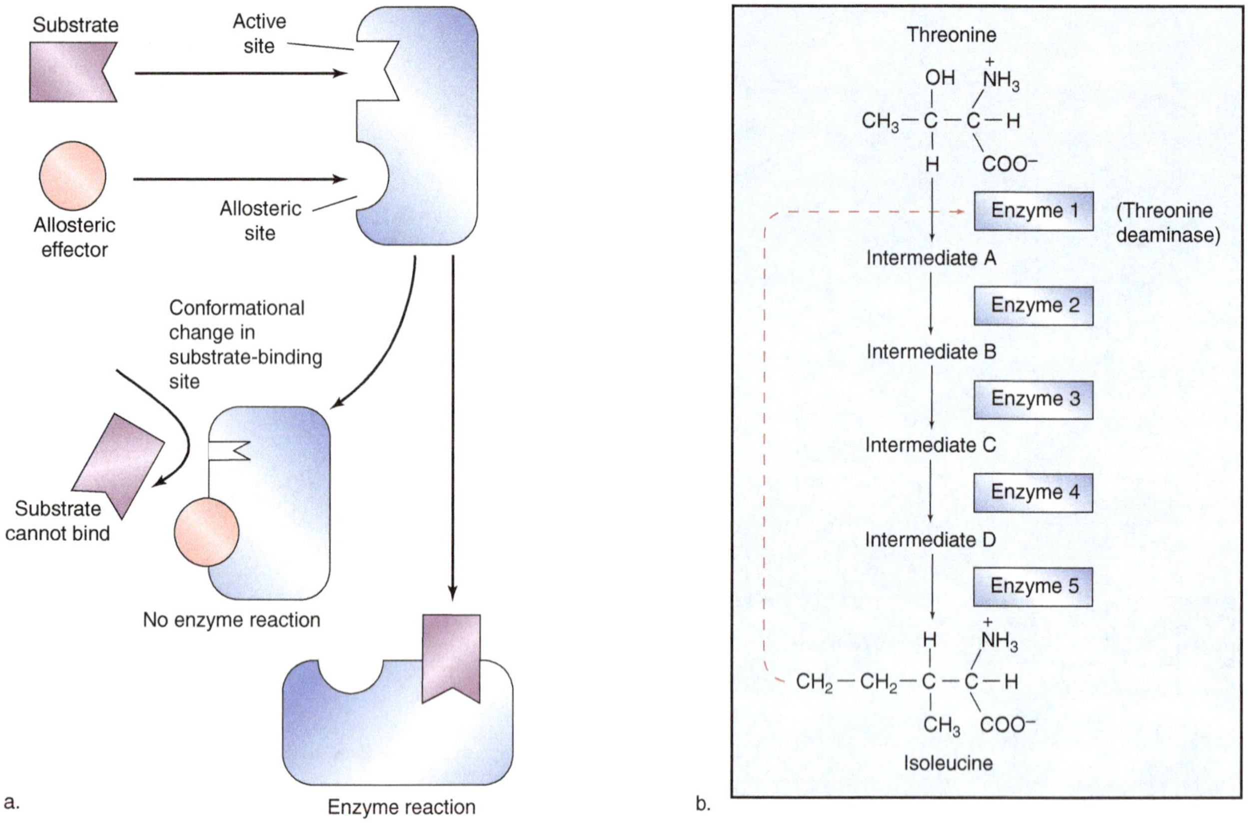

Figure 8.6a Image showing how Allosteric Inhibitors change the Active Site, thereby Making the Enzyme Nonfunctional.

Figure 8.6b Feedback Inhibition. Once the Concentration of Isoleucine is High Enough Inside the Cell, isoleucine acts as an Allosteric inhibitor on Enzyme 1 in the Metabolic Pathway to Create Isoleucine. The Enzyme would be Distorted as seen in 8.6a, thereby Shutting the Pathway Down.

Enzyme Inhibition

Enzymes can be inhibited in a variety of ways, as you will see here. Microbiologists can take advantage of understanding enzyme anatomy and function to develop antibiotics. If certain biochemcal pathways are shut down, the growth of the organism may cease. There are two main ways in which inhibitory molecules act on enzymes: competitively and noncompetitively. Competitive inhibiton refers to molecules that compete with the substrate for the active site. Once bound, a competitive inhibtior prevents the substrate from binding and prevents the formation of end products (Figure 8.8). An example of a competitive inhibitor is the antibiotic sulfanilamide (commonly called sulfa). Sulfanilamide is a competitive inhibitor that competes for the active site that normally binds with a molecule called PABA. PABA is converted into folic acid within the cell and is required for the synthesis of nucliec acids (DNA and RNA). If there is no folic acid, then the cell cannot undergo cell replication. Sulfa drugs are selectively toxic; since humans do not synthesize folic acid, we must absorb our folic acid from the foods we eat. Noncompetitive inhibitors attach to the allosteric site on enzymes thereby altering the shape of the active site (Figure 8.9). There is not a specific example of an antibiotic that acts in this way;

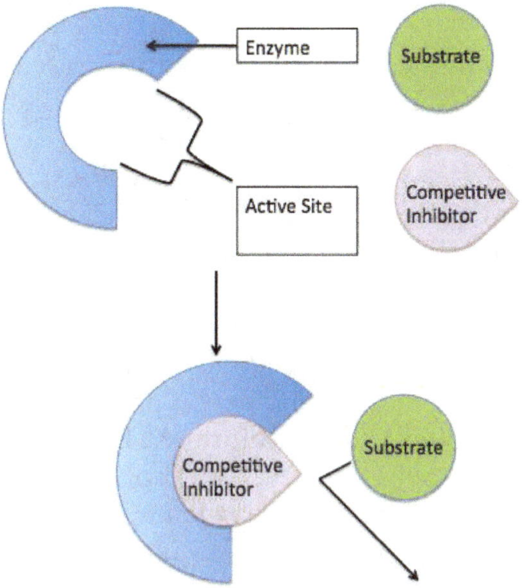

Figure 8.7 This image shows how a Molecule with a Similar Chemical Structure or "Shape" can Bind to the Active Site of an Enzyme. This is Refered to as Competitive Inhibition since the Inhibitior is Competing with the Substrate. Once the Inhibitior is Bound to the Active Site, the Reaction is Stopped

Courtesy of Author

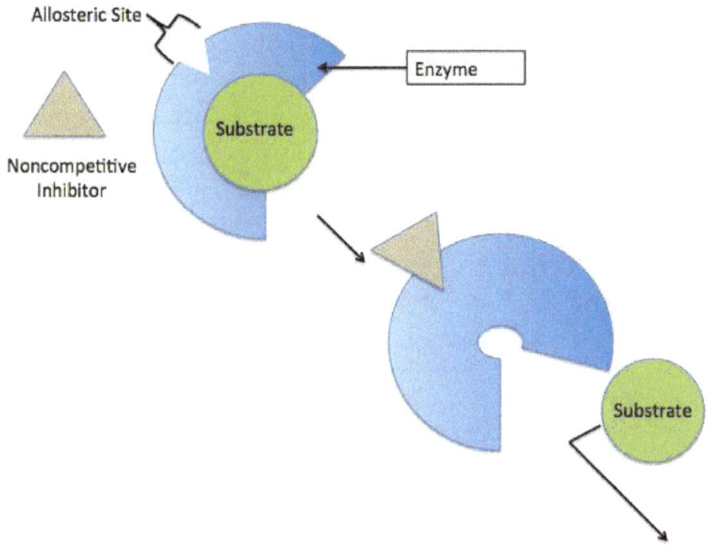

Figure 8.8 Noncompetitive Inhibitiors Bind to an Allosteric Site on the Enzyme. Once a Molecule is Bound to the Allosteric Site the Active Site is Distorted and the Enzymatic Pathway is Shut Down

Courtesy of Author.

however, heavy metals may bind to allosteric sites, which explains some of the toxic effects that metals have on not only bacteria but on humans as well.

Factors that Influence Enzymatic Activity

A cell's ability to survive in extreme temperatures or pH is due to their enzymes' ability to resist those conditions. A thermophile, for example, will have enzymes that are heat-stable and therefore, allow the cell to grow in extreme temperatures. Enzyme activity can be influenced by environmental factors such as pH, temperature, salt, and substrate concentrations and have optimal activity ranges (Figure 8.9).

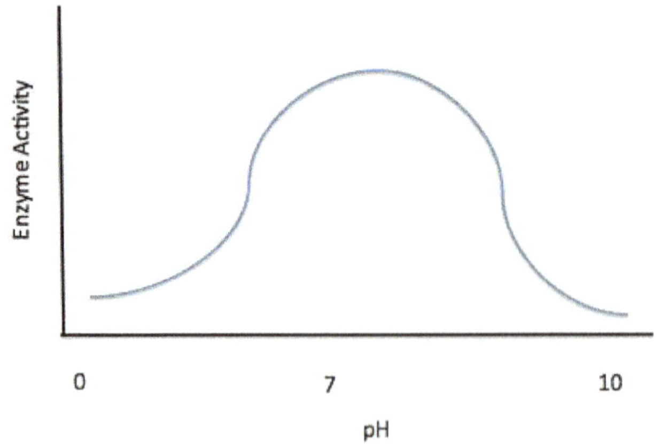

Figure 8.9 The Effect of pH on Enzyme Activity
Courtesy of Author

How Cells Make ATP

As mentioned earlier, ATP is the molecule that cells use to perform cell processes. There is a cyclical role between ATP and ADP (adenosine diphosphate). ATP can be thought of as a "charged" battery and ADP as a "dead" battery. There are two processes used by heterotrophic bacteria to make ATP: substrate-level phosphorylation and oxidative phosphorylation. Phosphorylation refers to the addition of a phosphate to ADP (2 phosphates) to form ATP (3 phosphates). Substrate-level phosphorylation, as the name suggests, is when a cell uses a substrate or "food source" to phosphorylate ADP. Glycolysis and the Krebs cycle are examples of substrate-level phosphorylation, and only a small amount of ATP is made. Oxidative phosphorylation harvests energy from the proton motive force, which will be discussed later, to add a phosphate to ADP.

Glucose Metabolism: The basics

During a catabolic reaction, one molecule (for e.g. Glucose) will act as an energy source or electron donor; when glucose is broken down by a cell to release energy, glucose is oxidized. Oxidation refers to the loss of electrons; when glucose is oxidized, another molecule must be reduced such as NAD+ or gain the electrons glucose lost in the process. What has just been described is a very basic **oxidation reduction reaction** (Figure 8.10).

Oxidation reduction reactions always occur simultaneously. With this example of glucose oxidation, the electrons lost from glucose are transferred to electron carriers in the cell. The two electron carriers involved in glucose metabolism during aerobic respiration in bacteria are NAD+ and FAD. These electron carriers will be reduced to form NADH and $FADH_2$ which will be refered to as reducing power (Table 1).

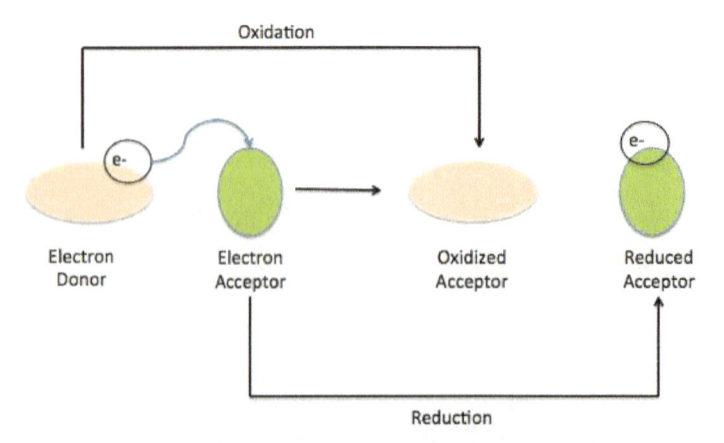

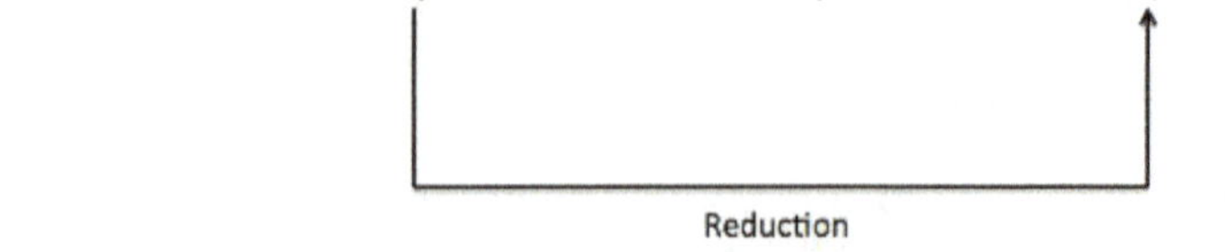

Figure 8.10 Oxidation-Reduction Reaction
Courtesy of Author

Table 1 The two most common electron carriers used by cells. When glucose is oxidized (loses electrons) the electron carriers are there to "grab" them, thereby becoming reduced

Electron Carrier Oxidixed Form	Electron Carrier Reduced Form
NAD+	NADH
FAD	$FADH_2$

This reducing power is used to drive the electron transport system, which in turn will create the proton motive force. Electrons from glucose are transferred to electron carriers and ultimately will combine with a terminal electron acceptor; in **aerobic respiration**, this terminal electron acceptor is oxygen. When oxygen is used as the terminal electron acceptor, the cell can produce the most ATP. During **anaerobic respiration**, inorganic molecules other than oxygen, such as nitrate or sulfate, are used as terminal electron acceptors. Organisms that use an anaerobic process always yield less ATP than if oxygen is used as the electron acceptor. Organisms like facultative anaerobes can either switch their metabolic process depending on what molecules are present in their environment, or they can be strictly fermenters. Certain genera of bacteria such as *Streptococcus* sp. are obligate fermenters and therefore do not respire. In the laboratory we will use thioglycolate media to help us determine an organism's oxygen requirements. However, we can only tell if the organism is a facultative anaerobe or an aerobe using this media. Therefore, in the laboratory activity following this chapter we will use oxidation/ fermentation media (OF) to determine if the organisms ferment glucose or if they use the glucose through cell respiration.

Glucose Metabolism: The Process

Microorganisms oxidize sugars as their primary source of energy for anabolic reactions. Glucose is the most common energy source. However, it should be noted that not all cells can use glucose as an energy source, and may rely on proteins or lipids for energy production. Energy can be obtained from glucose by respiration, which can be aerobic or anaerobic, or through fermentation, which is a process used by cells that cannot respire. Considering glucose metabolism as well, there are three metabolic pathways that cells use to completely oxidize glucose: glycolysis, transition reaction or synthesis of acetyl CoA, and the Krebs Cycle. During these processes, a small amount of ATP is made and reducing power is made, which will be used to drive the proton motive force. We will now look at each of the pathways in more detail.

Aerobic oxidation of glucose chemical equation: $C_6H_{12}O_6 + 6\ O_2 \rightarrow 6\ CO_2 + 6\ H_2O$

$$38\ \text{ADP} \rightarrow 38\ \text{ATP}$$

Glycolysis

As observed in the chemical equation above, glucose is a 6-carbon molecule. During glycolysis, glucose is split into two 3-carbon molecules of pyruvate. In order to split glucose, two ATP molecules are required during what is called the investment phase. The process of glycolysis is actually a 10-step process requiring many different types of enzymes. As a result, the cell produces four ATP for a net gain of two ATP and has also created two NADH. Almost all cells that are capable of using glucose as an energy source use the process of glycolysis. Aerobic and anaerobic respiring bacteria use glycolysis as well as fermenters. However, cell respiration begins with the transition reaction since fermenters do not respire.

Cell Respiration: Transition Reaction

The transition reaction or synthesis of Acetyl-CoA reaction connects glycolysis to the Krebs cycle. From glycolysis we have two 3-carbon molecules of pyruvate which will feed into the transition reaction. During this reaction, each of the pyruvate molecules will lose a carbon in the form of CO_2. Since the pyruvate molecules are oxidized during the transition reaction, two molecules of NAD+ will be reduced to NADH. At the end of the reaction, there will be an end product of two 2-carbon Acetyl-CoA molecules.

Cell Respiration: Krebs Cycle

Acetyl-CoA from the transition reaction will start the Krebs cycle. As Acetyl-CoA is oxidized, CO_2 is given off as a by-product for a total of four CO_2. For each molecule of Acetyl-CoA, the cell will gain one ATP, three NADH, one $FADH_2$ for a total of two ATP, six NADH, and two $FADH_2$. If you recall, we started with a 6-carbon molecule of glucose, lost two carbons in the form of CO_2 during the transition reaction, and lost four carbons during the Krebs cycle as CO_2. Thus the carbon "backbone" of glucose is now completely gone, and yet we have only transformed a total of four ATP (two from glycolysis and two from Krebs cycle). Next we will explore how respiring cells make the majority of their ATP through oxidative phosphorylation.

Cell Respiration: Electron Transport System and the Proton Motive Force

As glucose was oxidized, you noticed that there was a fair amount of reducing power formed (NADH and $FADH_2$). As NAD+ and FAD are reduced, they carry the electrons to the cell membrane, which is the site of the electron transport system (Figure 8.11). The electron carriers NADH and $FADH_2$ will transfer the electrons, thereby becoming oxidized, to proteins in the cell membrane called cytochromes. There are numerous cytochromes involved in the electron transport system; the electrons are passed from one cytochrome to the next, and as a result, the electron energy is used to pump hydrogen ions or protons from the cell membrane. The oxidized electron carriers are shuttled back to glycolysis and the Krebs cycle to pick up more electrons, and are therefore recycled. The electrons eventually make their way back into the cell and combine with O_2 and H+ to form H_2O. As the positively-charged hydrogen ions are pumped out of the cell, they concentrate immediately outside of the cell membrane. There will be a net charge outside of the cell as positive, and therefore the inside of the cell has a net negative charge. The protons are "attracted" to the inside of the cell membrane; however, the cell membrane is not permeable to the protons. This separation of charged ions creates an electrochemical gradient across the membrane. The electrochemical gradient represents potential energy refered to as the **proton motive force**. ATP is harvested when protons flow through a special turbine-like protein called ATP synthase. ATP synthase phosphorylates ADP by oxidative phosphorylation. The theoretical yield for ATP transformed from the proton motive force from the reducing power generated by glycolysis, transition reaction, and Krebs cycle using oxygen as a terminal electron acceptor is 34 ATP. Theoretically, the cell gains three ATP from each NADH and two ATP from each $FADH_2$. A total of 10 NADH and 2 $FADH_2$ from the previous steps are used to drive the proton motive force.

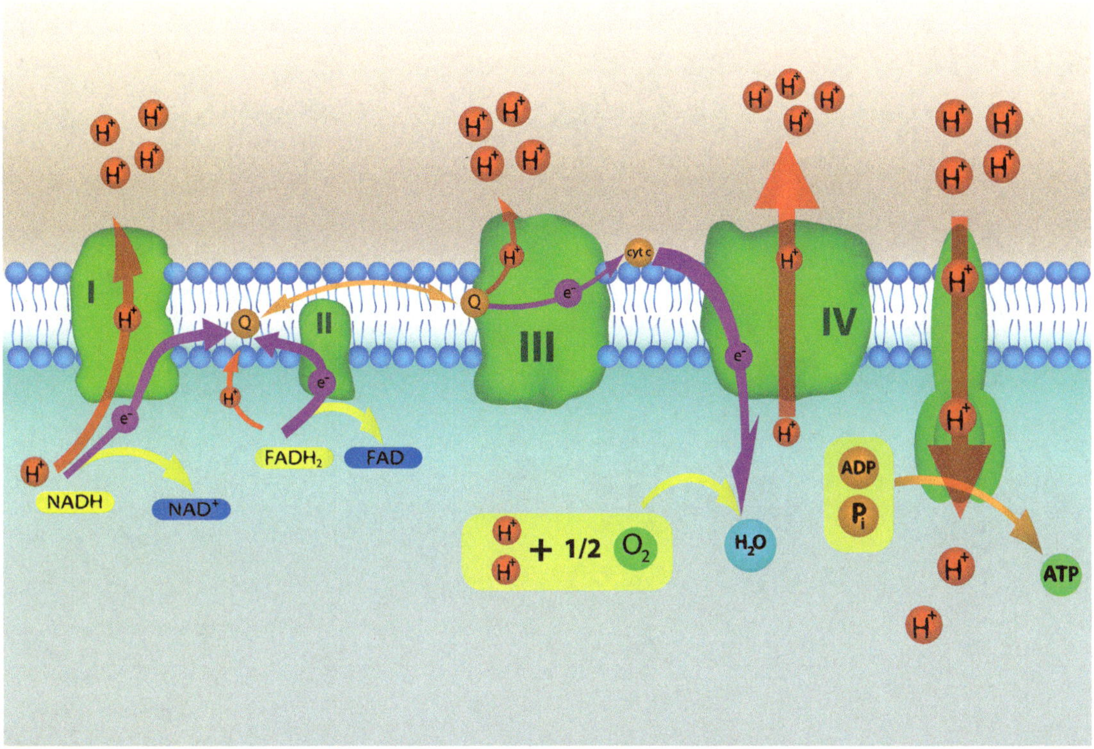

Figure 8.11 Electron Transport System Occurring in the Cell Membrane of Bacteria or Inside the Mitochondria in Animal Cells

Extender_01/Shutterstock.com

Fermentation

The process of fermentation occurs in organisms that cannot respire; therefore, they do not completely oxidize glucose using the transition reaction and Krebs cycle, and do not have an electron transport system. Fermentation is a way cells can recycle NADH in the cell and create useful products for humans. Fermenters are indifferent toward oxygen, meaning that they do not use O_2 to transform energy, nor is O_2 inhibitory toward the fermentation process. During fermentative metabolism, organic molecules act as electron acceptors to recycle NADH. The fermentation process begins with glycolysis in which the cells gain (net) two molecules of ATP, two NADH, and will have two 3-carbon molecules of pyruvate at the end of the process. A popular fermentation process is alcohol fermentation, in which alcohol is the end product of the process. Alcohol fermentation is performed by *Saccharomyces cerevisiae* which is a yeast (eukaryotic). Once the yeast split the 6-carbon molecule of glucose into pyruvate, pyruvate is oxidized to form acetylaldehyde (2-carbon) with CO_2 as a by-product. Acetylaldehyde will then gain electrons from the NADH produced from glycolysis to form alcohol and NAD+. As mentioned earlier, there are many useful products produced by fermentative metabolism, alcohol being one; organisms from the genus *Clostridium* can produce organic solvents such as acetone and isopropanol. *Lactobacillus* sp. produce lactic acid and they are one of the organisms involved in making yogurt.

Lab Exercise 8.1

Metabolism: Enzyme Action, Carbohydrate Catabolism, and Introduction to Biochemical Testing

Student Objectives

1. Define the terms: carbohydrate, catabolism, and hydrolytic enzymes.
2. Explain how anabolic and catabolic reactions are linked.
3. Explain how fermentation and respiration are different.
4. Understand the usefulness of biochemical testing.
5. Compare and contrast aerobic respiration, anaerobic respiration, and fermentation.
6. Understand the transfer of electrons in oxidation/reduction reactions.
7. Use aseptic techniques in order to gain accurate results.
8. Perform and interpret a microbial starch hydrolysis test.
9. Perform and interpret oxidation-fermentation (OF) tests in order to differentiate between the types of catabolism: fermentative and oxidative.
10. Perform nitrate reduction, catalase, and oxidase tests.

Name ______________________________

Pre-Lab Exercise 8.1

1. What is the purpose of this laboratory exercise?

____

____

____

2. Explain what a biochemical test is and why these tests are useful in identifying unknown bacteria.

____

____

____

____

3. How are reduction and oxidation reactions different?

____

____

____

4. Explain the role of electron carriers in ATP formation.

____

____

____

5. Differentiate between aerobic respiration and anaerobic respiration.

____

____

____

6. Differentiate between fermentation and anaerobic respiration.

7. Describe the role of an enzyme.

8. Explain enzyme-substrate specificity in a chemical reaction.

9. Explain what is happening when starch is hydrolyzed by amylase.

10. What color will an OF tube turn if Glucose is metabolized?

11. What is a cytochrome?

12. What test is used to look for the prescence of cytochrome c?

INTRODUCTION

All cells have two fundamental tasks: (1) continually synthesizing new components such as cell walls, membranes, ribosomes, DNA, and surface structures such as flagella or capsules. These actions allow cells to enlarge and eventually divide. (2) Harvesting energy and converting it to a form that is usable in order to power the biosynthetic reactions, transport nutrients and other molecules, and sometime propel or move itself. Cellular metabolism is the sum of all of the chemical reactions within a living organism. **Metabolism** can be divided into either anabolic or catabolic reactions. Biochemical testing is used to determine metabolic processes inside a cell and the results of the testing are extremely important for identifying unknown bacteria.

Anabolism is the metabolic process of building complex organic molecules from smaller components. Anabolic reactions require energy that is gained from catabolism. **Catabolism** is the metabolic process of breaking down complex organic molecules into smaller components which will release of energy. Bacteria use the carbon and energy released in the process. Carbohydrates are often used as a primary source of carbon and energy, a carbohydrate is an organic molecule that contains carbon, hydrogen, and oxygen in a 1:2:1 ratio. These carbohydrates contain individual monomers called **monosaccharides** which are small water-soluble carbohydrates and can be joined together and form a polysaccharide such as starch.

As mentioned earlier in this chapter microbiologists can test for the production of the exoenzyme amylase by using a media enriched with starch. Organisms are inoculated onto a starch media following incubation and, with the addition of Gram's iodine, a halo will be observed around the area of growth if amylase was secreted from the cell. Within the halo amylase "cut" the polysaccharides into monosaccharides, such as glucose, which the cells can easily transport into the cell to use for energy and growth.

Once glucose is transported into the cell, the cell will use a catabolic process to obtain ATP from the molecule. The type of glucose catabolism used by the bacterium can be determined by using **oxidation-fermentation (OF) medium,** which is typically used to determine the carbohydrate catabolism of Gram-negative bacteria. Oxidative catabolism refers to the use of oxygen as an electron acceptor during aerobic respiration. Fermentation does not use oxygen and is the more common type of catabolism used by bacteria. Oxidation-fermentation media contains low concentration of peptone and a high concentration of carbohydrates in a semi-solid agar deep. When glucose is either fermented or undergoes oxidation, acids are produced as by-products. Organisms not capable of using the carbohydrates will rely on the peptone to support growth; these bacteria are referred to as **non saccharolytic** bacteria. Two OF tubes are used for each study organism; one is left open to the air making oxygen available and the other is sealed with mineral oil to keep air out, therefore creating an anaerobic environment. The media contains bromthymol blue, a pH indicator that will turn yellow in the presence of acids, indicating glucose was metabolized by either fermentation or oxidative metabolism. **If the media turns yellow in both the air tube and sealed tube then the organism is capable of fermenting the glucose in the media.** Bacteria not capable of metabolizing carbohydrates (non saccharolytic) give a negative OF result, meaning they are neither oxidizing nor fermenting. **A negative result is indicated by no color change in the oil-covered tube and in some cases an increase in pH or alkaline by-product, changing the bromthymol blue from green to blue in the top of the open tube.** The increase in pH is due to amine production, such as ammonia, by bacteria that break down the peptone (protein) in the medium. Other bacteria give a negative result indicated by no growth or color change in the medium. This result would indicate that the particular bacteria being tested can neither use the protein or the carbohydrate in the media as an energy source. **If only the air tube turns**

yellow then the organism used oxidative catabolism and is therefore not a fermenter of glucose. Inoculation and analysis of OF medium after an incubation period is considered to be a **biochemical test**. Biochemical tests test for metabolic end products which can be used along with colony morphology and the Gram stain in aiding the identification of unknown organisms. Since each bacterial species produces a different set of enzymes the substrate may or may not be utilized.

In aerobic respiration (oxidative), special proteins called **cytochromes** carry the electrons to oxygen, which is the terminal electron acceptor in the electron transport system. There are four classes of cytochromes that have been identified. Testing for the presence of cytochromes is useful in identifying bacteria. The **oxidase test** is used to test for the enzyme cytochrome c oxidase. The presence of cytochrome oxidase is determined by the appearance of pink/maroon precipitate after the addition of the reagent to a bacterial colony. This test provides useful clues when trying to determine the oxygen requirements of a bacterial species as oxidase positive organisms are usually considered non-fermenters.

The **catalase test** is another test used during the identification of bacteria, aerobic respiring organisms will form harmful hydrogen peroxide (H_2O_2) inside the cell. Since oxygen contains two unpaired electrons, oxygen is readily reduced to superoxide. Aerobic respiring organisms produce superoxide dismutase which forms hydrogen peroxide. Hydrogen peroxide is toxic and will kill a cell, therefore the cell must produce catalase which will break down H_2O_2. **Catalase converts H_2O_2 into water and oxygen** thereby neutralizing the toxic hydrogen peroxide. The catalase test is performed by adding hydrogen peroxide to a bacterial colony. The observation of bubbles (oxygen) indicates the organism is catalase positive. This test is one of the main tests to differentiate between *Streptococcus* and *Staphylococcus;* however, this test is performed on all Gram positive bacteria during the unknown lab.

During this lab you will be looking at biochemical characteristics of different microorganisms to gain insights as to how they use oxygen. A test used to determine if a bacterium is capable of anaerobic respiration is the nitrate reduction test. Nitrate reduction is performed by anaerobic respiring bacteria where nitrate is used as a final electron acceptor. Nitrate may be reduced in a step-wise manner during anaerobic respiration, depending on the bacteria involved in the process and depending on the number of accepted electrons. Some bacteria reduce nitrates to nitrites, others reduce nitrate to ammonia, while others reduce nitrate to nitrous oxide or nitrogen gas. During this lab exercise you will use nitrate broth to determine if a bacterium reduces nitrate.

$$\underset{\text{hydrogen peroxide}}{2H_2O_2} \xrightarrow{\text{catalase}} \underset{\text{water}}{2H_2O} + \underset{\text{oxygen}}{O_2}$$

During this lab you will be looking at biochemical characteristics of different microorganisms to gain insights as to how they use oxygen. A test used to determine if a bacterium is capable of anaerobic respiration is the nitrate reduction test. Nitrate reduction occurs when bacteria anaerobically respire using nitrate as a terminal electron acceptor. Nitrate may be reduced in a step-wise manner during anaerobic respiration, depending on the bacteria involved in the process and depending on the number of accepted electrons. Some bacteria reduce nitrates to nitrites, others reduce nitrate to ammonia, while others reduce nitrate to nitrous oxide or nitrogen gas. In this lab you will use **nitrate broth** to determine how a bacterium reduces nitrate, if at all.

$$\underset{\text{nitrate ion}}{NO_3^-} + 2H^+ + 2e^- \longrightarrow \underset{\text{nitrite ion}}{NO_2^-} + H_2O \longrightarrow \underset{\text{nitrous oxide}}{N_2O} \longrightarrow \underset{\text{nitrogen gas}}{N_2}$$

Week One—Enzyme Action And Carbohydrate Catabolism

Part A: Starch Hydrolysis

> **Materials**
>
> **Per table:** Petri plate of nutrient starch agar and inoculating loop.
>
> **Pure Cultures:** *Bacillus subtilis*, *Escherichia coli*, and *Pseudomonas aeruginosa*.

1. Using a Sharpie, draw three sectors on the bottom of your starch agar plate. Label each section with one of the corresponding organisms listed above (label "BS", "EC", and "PA").
2. Streak a short single line of *Bacillus subtilis*, *Escherichia coli*, and *Pseudomonas aeruginosa* near the rim of their corresponding labeled sections of agar (shown below).
3. Invert the plate and incubate at 37°C until the next lab period.

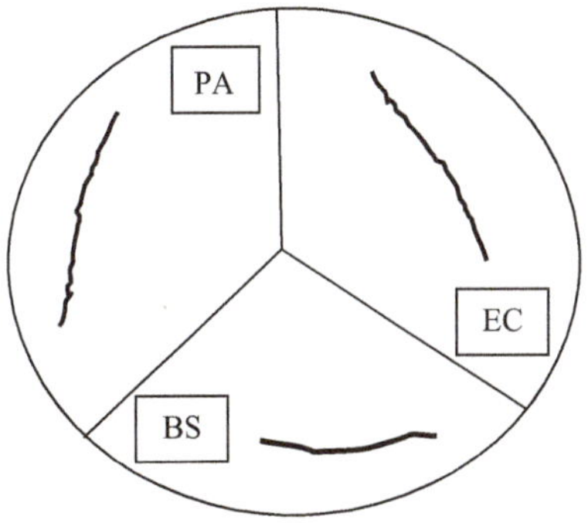

Courtesy of Authors

Part B: OF-Glucose

Materials

Per table: 6 OF-glucose deeps, inoculating needle, and mineral oil.

Pure cultures of *Alcaligenes faecalis*, *Escherichia coli*, **and** *Pseudomonas aeruginosa*.

1. With a Sharpie, label two OF medium tubes with the initials of each of the three organisms listed above. For each organism label one tube "air" and one tube "no-air".
2. Inoculate the OF-medium tubes with the corresponding organisms using an inoculating needle. Dip the flamed, thin wire needle into the culture and stab the agar deep. Try to keep the needle in the center of the tube and penetrate the OF deep for the whole length of the needle. Remove the needle carefully and re-flame.
3. Place drops of mineral oil on top of the agar in each of the three tubes labeled "no-air" (one tube for each organism). Without touching the dropper to the inside of the OF tube, add mineral oil until there is a layer ¼ inch to ½ inch deep on top of the agar.
4. Incubate all tubes at 37° C until the next lab period.

Uninoculated of Media
Courtesy of Authors

Part C: Nitrate Reduction Test

Materials

Per table: Three tubes of nitrate broth and inoculating loop.

Pure cultures of *Escherichia coli*, *Pseudomonas aeruginosa*, **and** *Alcaligenes faecalis*.

1. Label three tubes of sterile nitrate broth "EC", "PA", or "AF" (one tube for each of the cultures listed above).
2. Aseptically inoculate the broth with a loopful of pure culture corresponding to the label.
3. Place in the test tube rack provided for incubation at 37°C.

Part D: Oxidase Test

> **Materials**
>
> **Per table:** One Tryptic Soy Agar (TSA) plate and inoculating loop.
>
> **Pure cultures of** *Escherichia coli* **and** *Pseudomonas aeruginosa.*

1. With your Sharpie, divide the bottom of the agar plate in half. Label one half of the plate "EC" and the other "PA".
2. Aseptically inoculate the plate with one streak of pure culture on the corresponding side of the plate.
3. Invert the plate and incubate at 37°C until the next lab period.

Part E: Catalase Test

Materials

Per table: One TSA plate and inoculating loop.

Pure cultures of *Streptococcus mutans* and *Bacillus subtilis*.

1. With your Sharpie, divide the bottom of the agar plate in half. Label one half of the plate "SM", and the other "BS".
2. Aseptically inoculate the plate with one **streak** of *Streptococcus mutans* on the corresponding half and a **spot** of *Bacillus subtilis* on the other half. *Bacillus subtilis* may spread, so we don't use a full streak.
3. Invert the plate and incubate at 37°C until the next lab period.

Week Two—Enzyme Action and Carbohydrate Catabolism

Part A: Starch Hydrolysis

Materials: Gram's Iodine.

1. Record the presence and appearance of bacterial growth.
2. Flood the plate with Gram's iodine. Clear halos in media surrounding growth represent areas of starch hydrolysis. Areas that stain dark represent starch media that has not changed.

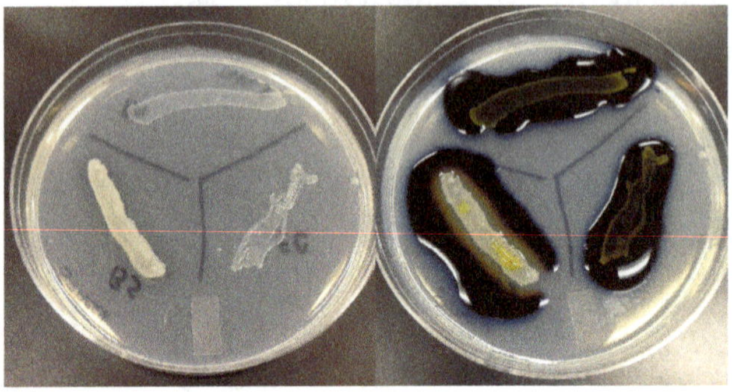

Starch Hydrolysis Results.
Left: After Incubation; **Right:** After Addition of Gram's Iodine to the Plate to Detect the Presence of Starch.
Courtesy of Authors

Part B: OF-Glucose

1. Compare your group's inoculated tubes with the uninoculated OF-media controls located on the instructor lab bench. The control tubes represent the color of the medium prior to the test and allow a basis for comparison.
2. Observe the stab line for growth, and determine whether glucose was catabolized, and if so, the type of glucose catabolism.

OF Media Results: Left: Result for Neither Oxidizer nor Fermenter; **Middle:** Result for Fermentation; **Right:** Result for Oxidizer
Courtesy of Authors

Part C: Nitrate Reduction

Materials: Nitrate reagents A and B, zinc dust, and disposable pipette.

1. Transfer 1 ml of the inoculated nitrate reduction media to 1 well of a spot plate. Add 2–3 drops of nitrate A and 2–3 drops of nitrate B. Appearance of a red color within 30 seconds is a positive test for nitrate reduction. Repeat this for the remaining organisms.
2. If there is no red color, scoop a tiny amount of zinc dust with a toothpick and add it to the wells with no red color and mix.
3. If the broth turns red after the zinc, the test is negative for nitrate reduction.
4. If the broth remains clear after the addition of zinc, this indicates that nitrate was reduced all the way to gas.

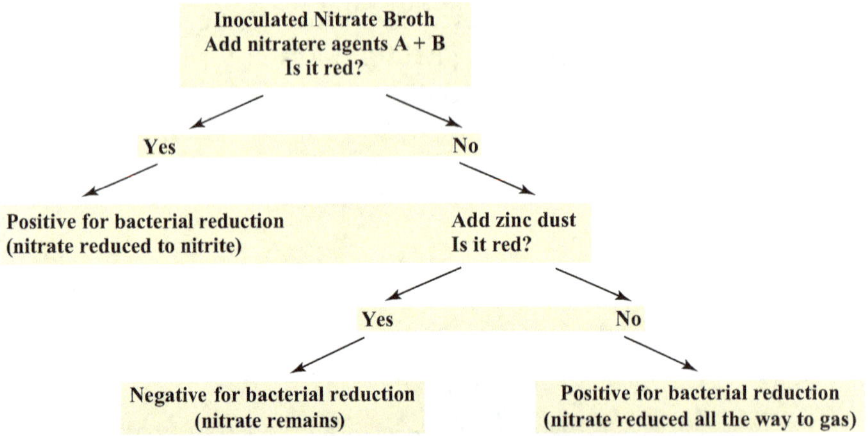

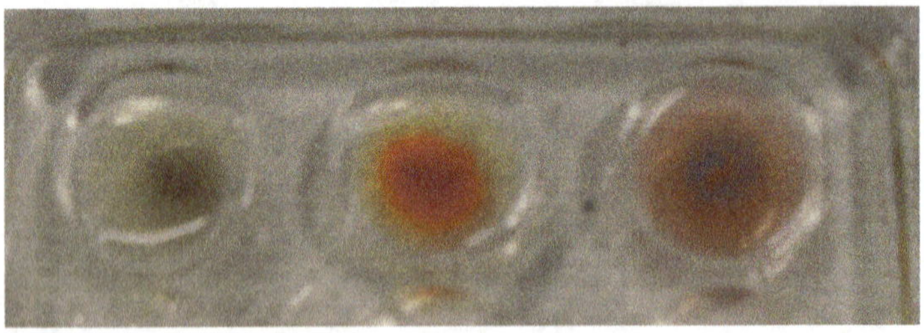

Nitrate Reduction Test Results:

Left: Clear after Addition of Reagents AND Zinc = No Nitrate/Nitrite for Reagents/ Zinc to React with = Nitrate Reduced to Gas

Center: Red after Addition of Reagents = Bacterial Reduction of Nitrate to Nitrite

Right: Pink/Red after Addition of Zinc = No Bacterial Reduction of Nitrate

Courtesy of Authors

Part D: Oxidase

Materials: Oxidase reagent.

1. To test for cytochrome *c* oxidase, drop a small amount of oxidase reagent directly on the colonies.
2. If the bacterium is oxidase-positive, the **colonies** will turn a purple color within two minutes and then slowly change to pink/maroon color. Pay close attention to the amount of time it takes to turn pink/maroon!

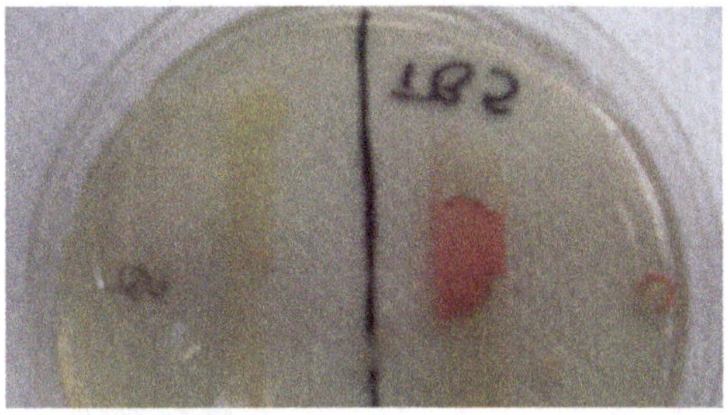

Oxidase Test Results: Left: Oxidase-Negative; **Right:** Oxidase-Positive. **Note:** the Colony must Turn Color, not the Media
Courtesy of Authors

Part E: Catalase

Materials: Hydrogen peroxide (H_2O_2).

1. Place a small amount of hydrogen peroxide directly onto colonies on the NA plate. Make sure not to touch the dropper onto the growth.
2. Observe the colonies for the formation of bubbles. If the bacteria are catalase-positive, bubbles will form on the colonies. Catalase-negative colonies will not release bubbles.

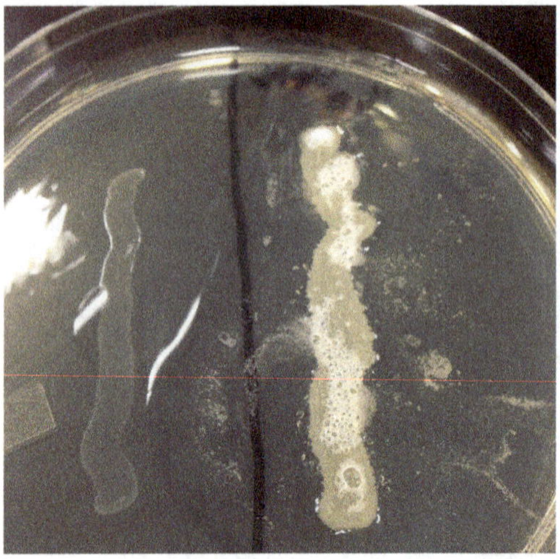

Catalase Test Results:
Left: Catalase-Negative,
Right: Catalase-Positive
Courtesy of Authors

Name ______________________

Exercise 8.1 Lab Report

Part A: Starch Hydrolysis Record with + if there is growth; if there is a lot of growth use +++; and use—for no growth

Organism	Growth (+/–)	Color of Medium Around Colonies after Adding Iodine	Amylase Production: Yes or No
Bacillus subtilis			
Pseudomonas aeruginosa			
Escherichia coli			

1. Based on your observations from your starch hydrolysis plate, after adding iodine, how can you tell amylase is an exoenzyme and not an endoenzyme?

2. What specific types of molecules involved in starch catabolism would be found in the clear area that would **not** be found in the dark area of the starch plate with iodine?

Part B: OF-Glucose Record + for growth; record—for no growth

Organism	Growth (+/–)			Color of Medium			Glucose* Fermenter (F), Oxidizer (O), or Neither (–)
	Open	Oil	Control	Open	Oil	Control	
Alcaligenes faecalis							
Escherichia Coli							
Pseudomonas aeruginosa							

3. How did you determine the metabolic processes of the organisms (oxidizer, fermenter, or neither) in the presence of the glucose substrate? **Explain.**

___

___

4. You observe bacterial growth and acid production in a single tube of OF-glucose medium that is exposed to air. Without the benefit of a tube sealed with oil to use for comparison, can you tell if the organism is oxidizing or fermenting? **Explain.**

___

___

___

___

5. Some microbes turn the pH indicator blue in the OF-glucose medium. What are they metabolizing (utilizing) in the medium? What alkaline by-product is produced?

___

___

___

Part C: Nitrate Test (Anaerobic Respiration)

Organism	Color after addition of reagent A and B	Color after zinc	Gas (+/–)	Bacterial NO_3 reduction? (Yes/No)
Alcaligenes faecalis				
Escherichia coli				
Pseudomonas aeruginosa				

6. If the solution remains colorless after the addition of zinc to a nitrate reduction tube, what does this indicate?

___

___

___

7. Would nitrate reduction in the natural environment occur more often in the presence or absence of oxygen? **Explain.**

___

___

Part D: Oxidase Test

Organism	Color after Reagent	Oxidase Test (+/–)
Escherichia coli		
Pseudomonas aeruginosa		

8. Which type of organism is likely to be oxidase-positive: aerobic, anaerobic, or fermenter? **Explain.**

____

____

____

Part E: Catalase Test

Organism	Reaction to H_2O_2	Catalase Test (+/–)
Bacillus subtilis		
Streptococcus mutans		

9. What do the bubbles represent if you have a catalase-positive organism?

____

____

10. Why does hydrogen peroxide bubble when you pour it on a cut?

____

____

11. Which two groups of bacteria is the catalase test commonly used to differentiate between?

____

Genetics

Everything that happens in a cell is determined by the genetic code of that cell. The genetic code determines what metabolic capabilities the cell has, what conditions the cell can live in, what toxins it produces and what mechanisms it has for invading, colonizing and evading the immune systems of hosts. **Genetics** is the science of heredity; it includes the study of what genes are, how they carry information, how they are replicated and passed to subsequent generations of cells or passed between organisms, and how expression of their information within an organism determines the particular characteristics of that organism. In this laboratory exercise, we will explore the central dogma of molecular biology: DNA → RNA → protein.

Bacterial Genomes

The complete set of genetic material in the cell is the **genome**. The genome includes both chromosomes and extra-chromosomal elements called plasmids (not all bacterial cells have plasmids). **Plasmids** are small circular DNA molecules that carry genes for nonessential cell functions such as antibiotic-resistant genes. **Chromosomes** are cellular structures that physically carry hereditary information. The chromosome contains segments of DNA that code for functional products.

Bacterial Chromosomes

Most bacteria have only one circular molecule of DNA. A few species of bacteria have two or more chromosomes. The chromosome is inherited from the parent cell, and contains all of the information necessary for the cell to survive and reproduce.

Bacterial chromosomes have been extensively studied by scientists, and have been used as models for studying human DNA. Historically it has been easier to study bacterial DNA because the genome of bacteria is so much smaller than the genome of humans. For example, the chromosome of *E. coli* is roughly 4.7×10^6 base pairs (bp) long and is about 1 mm long, which is 1,000 times longer than the entire cell! However, because DNA is very thin and tightly packed inside the cell, this twisted, coiled molecule only takes up about 10% of the cell's volume. The chromosome contains about 3,500 genes, compared to the human genome, which has around 30,000 genes.

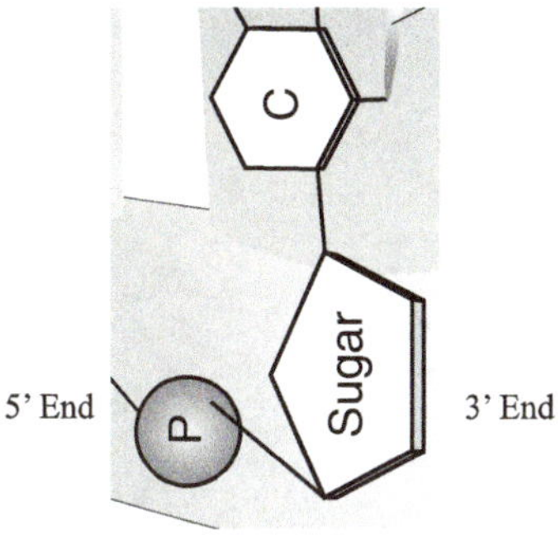

Fig. 9.1 Nucleotide Structure IMAGE from KH Sciences/Biology/ DNA 14.1 DNA Overview © Kendall Hunt Publishing

Structure of DNA

The genome is made of DNA, deoxyribonucleic acid. (Figure 9.2) The building blocks of DNA are nucleotides. Each nucleotide is made of a phosphate group, a deoxyribose sugar and one of four nucleic acids (adenine, thymine, guanine or cytosine). (Figure 9.1)

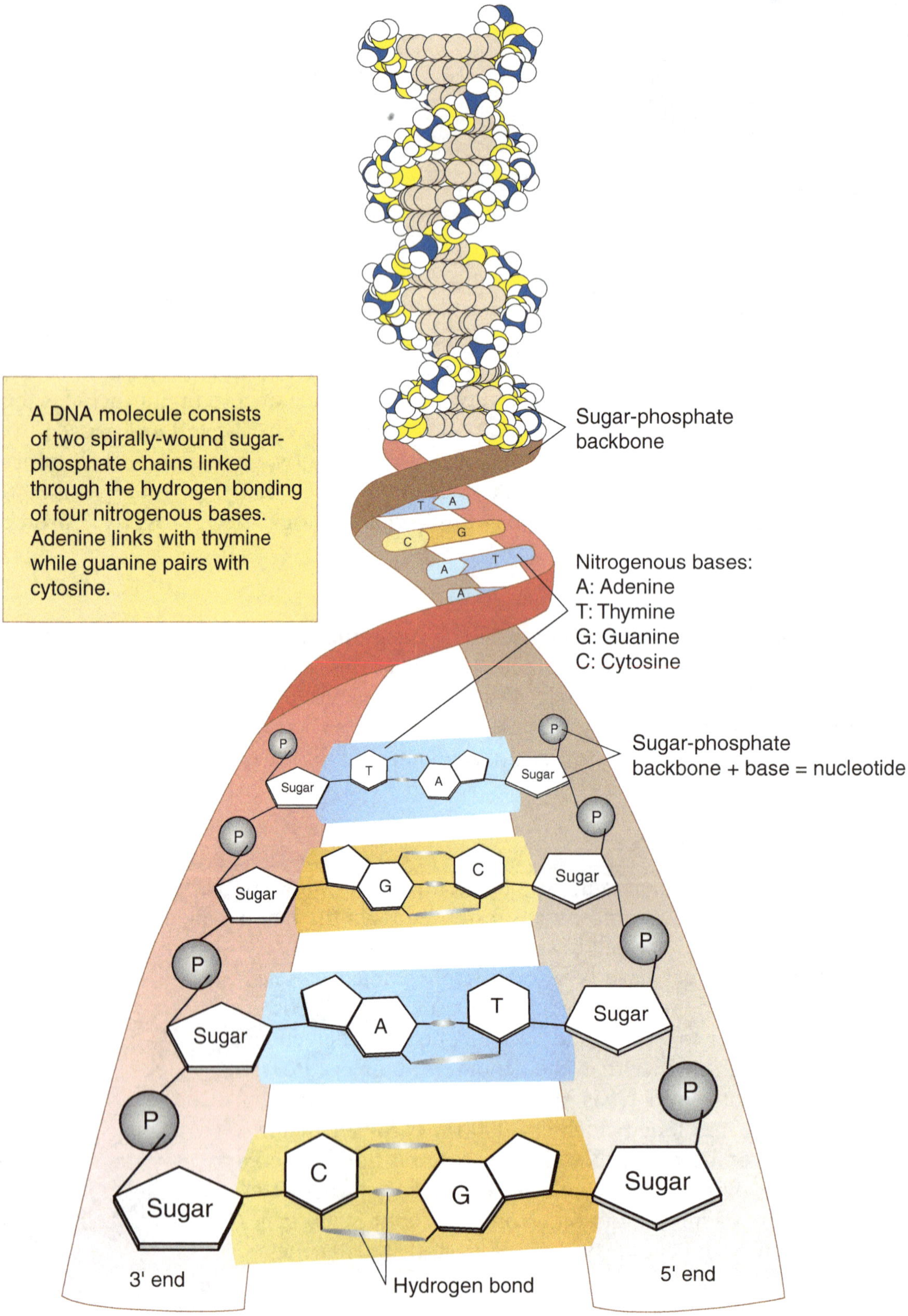
A DNA molecule consists of two spirally-wound sugar-phosphate chains linked through the hydrogen bonding of four nitrogenous bases. Adenine links with thymine while guanine pairs with cytosine.
Sugar-phosphate backbone
Nitrogenous bases:
A: Adenine
T: Thymine
G: Guanine
C: Cytosine
Sugar-phosphate backbone + base = nucleotide
Sugar
P
T
A
G
C
3' end
5' end
Hydrogen bond

Fig. 9.2 Structure of DNA IMAGE from KH Sciences/Biology/DNA 14.1 DNA Overview

The phosphates bind to the sugars of the adjacent nucleotide to create the **sugar-phosphate** backbone of the DNA molecule. Hydrogen bonds between the nucleic acids of the two complementary strands hold the DNA molecule together and allow the molecule to "unzip" for DNA replication and protein synthesis. Adenine is **complementary** to thymine, which means that adenine always binds to thymine on the opposite (complementary) strand. Cytosine is complementary to guanine.

DNA Replication

In order for a cell to reproduce, it must copy its genome so that each of the two daughter cells receives a copy. DNA replication is **semi-conservative**, that is, each "new" strand is made up of half new DNA and half old DNA.

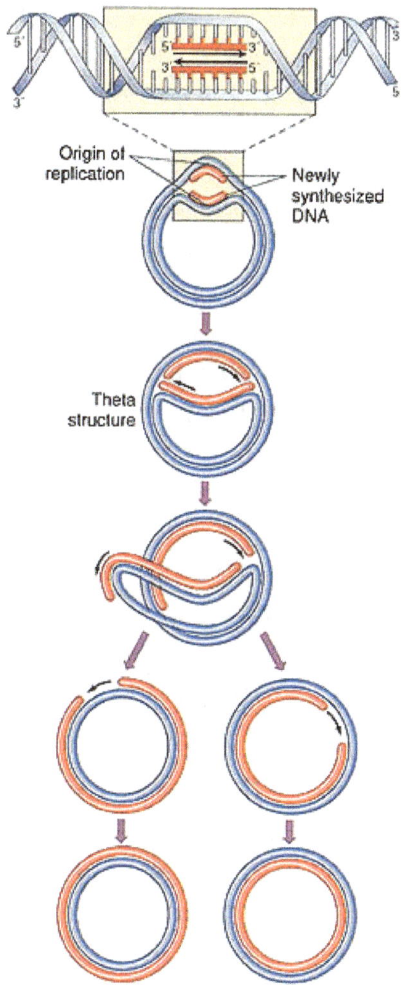

Fig. 9.3 Overview of DNA Replication

Directionality of DNA

DNA is directional—the sugar end of a nucleotide is referred to as the 3' (three prime) end and the phosphate end is the 5' end. The 3' end of a nucleotide binds to the 5' end of the adjacent nucleotide. This is analogous to a person's left and right sides: when several people line up holding hands, the left hand of one holds the right hand of the person next to them. (Figure 9.4)

DNA is synthesized in the 5' → 3' direction. This means that each new nucleotide is added to the 3' end of the previous nucleotide.

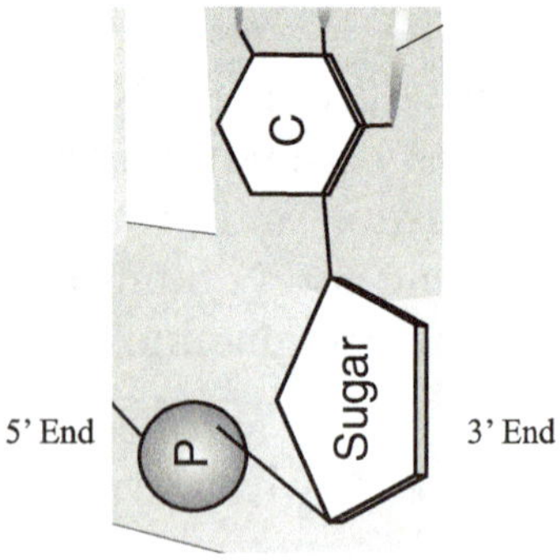

Fig. 9.4 Directionality of DNA IMAGE from KH Sciences/Biology/ DNA 14.1 DNA Overview © Kendall Hunt Publishing

Because of the directionality of DNA, each of the parent strands is synthesized in a slightly different manner. The **leading strand** is synthesized continuously: as the DNA molecule unzips to form the replication fork, synthesis continues in the 5' → 3' direction following along behind the replication fork. On the **lagging strand**, replication is discontinuous because only the short segments of DNA in the proper direction are available at a time for synthesis. Once replication occurs in these segments, the replication enzymes must wait for the replication fork to unzip more DNA before continuing replication. This "hopscotching" creates several short segments of new DNA called Okazaki fragments. (Figure 9.5)

Molecular Events of Leading Strand Synthesis. DNA replication is an anabolic process controlled by enzymes. The events of Leading Strand DNA synthesis are as follows:

1. Specific proteins recognize a region on the chromosome called the origin of replication. The specific sites where DNA replication begins will form replication bubbles, as shown in Figure 9.5.
2. **Helicase** "unwinds" or "unzips" the DNA molecule by breaking the bonds between the complementary base pairs. The separation of strands creates the **replication fork**.

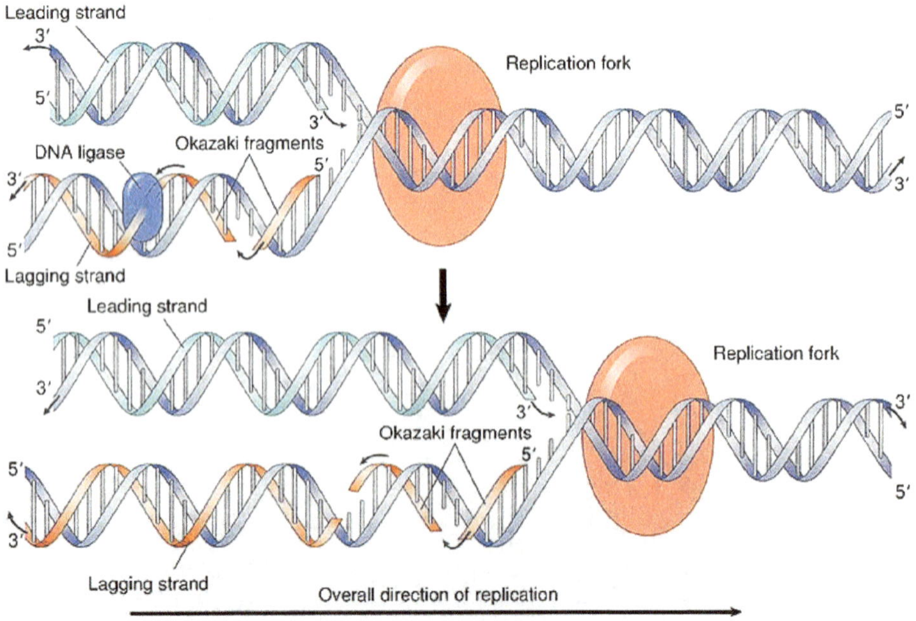

Fig. 9.5 DNA Replication IMAGE: Kendall Hunt © Kendall Hunt Publishing

3. **Single-strand DNA binding proteins (SSBPs)** bind to the DNA strands to stabilize them during replication. Single-stranded DNA is fragile and tends to break.
4. **Primase** synthesizes and adds short segments of RNA called **primers**, which are used to initiate the actual replication of DNA.
5. **DNA Polymerase III** recognizes the primers and binds to the 3` end. DNA polymerase synthesizes one of the strands continuously in the 5` to 3` direction; this is referred to as the **leading strand** (Steps 5 and 6 occur at the same time).
6. The other strand that is replicated is the **lagging strand**; since DNA polymerase cannot add nucleotides to the 5` end, synthesis must be reinitiated periodically as the helicase unwinds the DNA. Each re-initiation is preceded by the synthesis of an RNA primer called an Okazaki fragment.
7. **DNA Polymerase I** removes the RNA primers and replaces them with DNA.
8. **DNA ligase** seals the gaps between fragments on the lagging strand.

Because bacterial chromosomes are circular, two replication forks are formed at the **origin of replication** and DNA replication occurs in a bi-directional fashion around the chromosome. The two forks meet and replication is finished at the **terminus** of replication.

Protein Synthesis

How cells control their growth and regulate their metabolic activities is found on information within the genome on functional units called **genes**. The sequence of nucleotide bases in DNA carries genetic information in units that are referred to as **genes**. Think of the genome as a cookbook in which each individual recipe is a gene and whatever is made from the recipe is the protein.

Cookbook → Recipe → Food

Structural genes encode the information for specific proteins. To create a protein, a gene must first be transcribed into a sequence of nucleotide bases in a messenger RNA molecule (mRNA). The mRNA sequence is then translated into an amino acid sequence of a protein. This sequence of amino acids in a protein molecule determines the shape and chemical characteristics of the protein. Thus, each structural gene specifies a specific protein in the cell that carries out a specific function based on its chemical characteristics and molecular shape.

This flow of information is called **the Central Dogma of DNA** and can be envisioned like this, as shown in Figure 9.6.

DNA → mRNA → Protein

This function of the specific protein gives the cell the specific trait coded for by the gene. The sequence of bases that represents a specific gene is the **genotype**. The specific **trait** (composition, morphology, content and activity of individual enzymes, growth rate, etc.) formed by the genotype of the organism is called the **phenotype**.

Each gene is organized into a promoter, a coding region and a terminator. (Figure 9.7) This is similar to a sentence: the promoter represents the beginning of the gene, the coding region is the actual gene, or "meaning" of the gene, and the terminator signifies the end of the gene, like a period.

Protein synthesis occurs in two phases: transcription and translation.

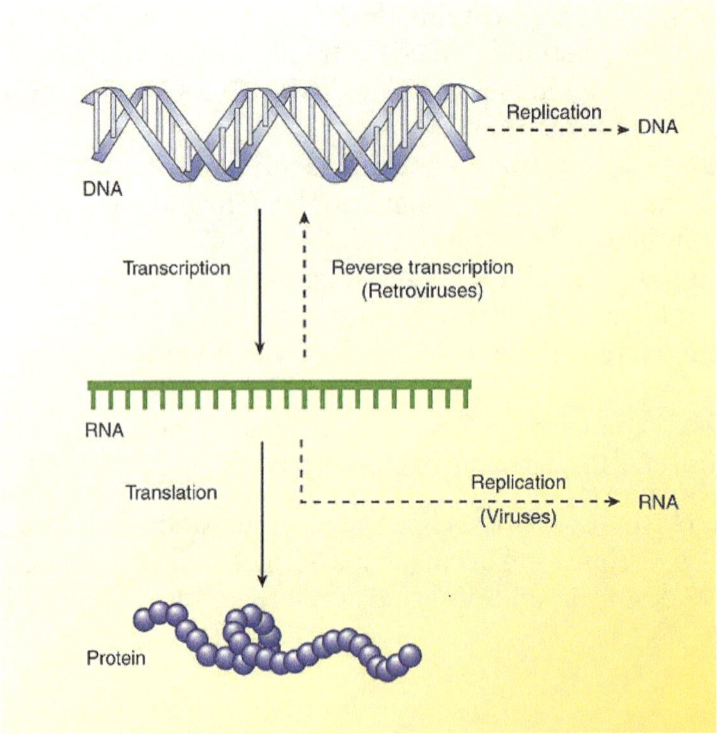

Fig. 9.6 The Central Dogma IMAGE:
© Kendall Hunt Publishing

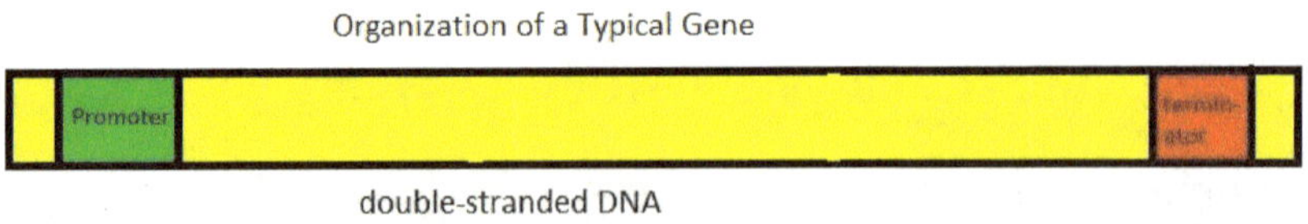

Fig. 9.7 Organization of a Typical Gene
Courtesy of A. Swarthout

Transcription

During transcription, a single gene is copied from DNA into mRNA (messenger RNA). RNA is similar to DNA in structure, but is made of ribonucleic acid rather than deoxyribonucleic acid, and RNA uses the nucleic acid uracil instead of thymine. The base-pairing occurs as it does in DNA except that Adenine is complementary to Uracil.

The entire process is carried out by a single enzyme, **RNA polymerase**, in three steps (Figure 9.8):

1. **Initiation:** The RNA polymerase binds to the DNA at the promoter.
2. **Elongation:** RNA polymerase moves along the template strand of the DNA, synthesizing the complementary single-stranded mRNA. As the RNA polymerase advances along the DNA, RNA nucleotides are added to the growing mRNA.
3. **Termination:** The RNA polymerase reaches the terminator sequence on the DNA and falls off the DNA strand.

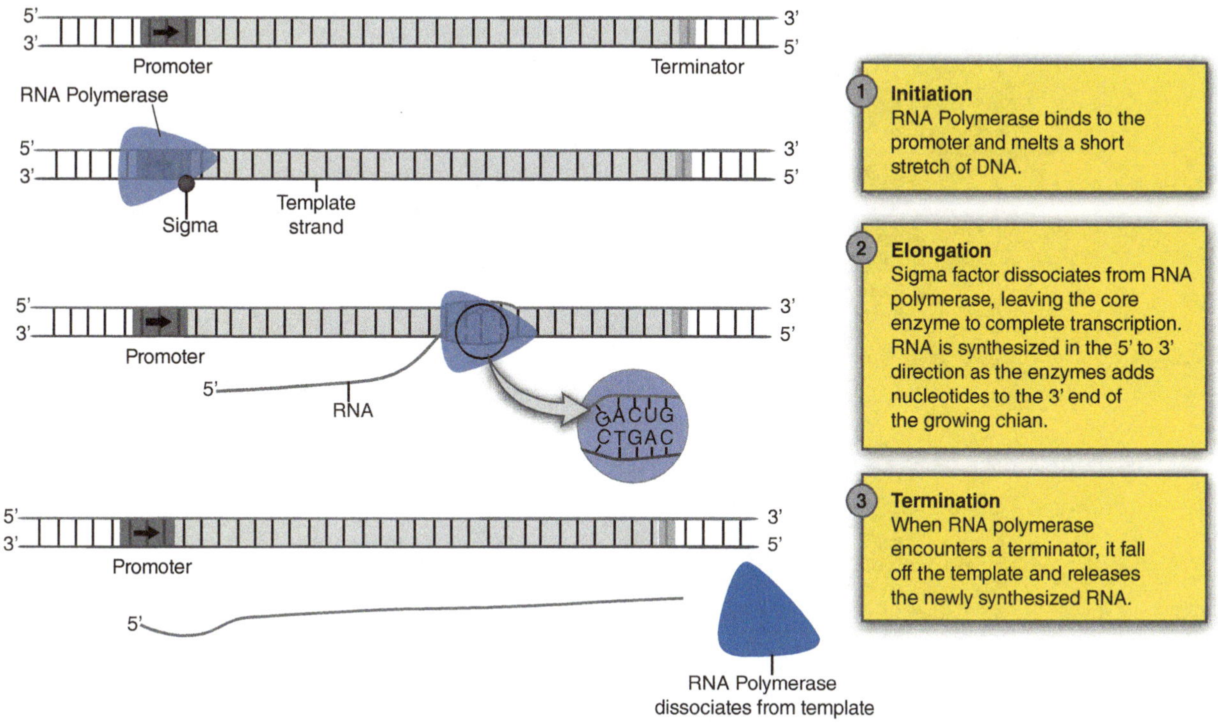

Fig. 9.8 The Process of RNA Synthesis

From Microbiology: A Human Perspective, 7/e by Eugene Nester, Denise Andersen and C. Evans Roberts, Jr. Copyright © 2011 McGraw-Hill Companies, Inc. Reprinted by permisison

Translation

The mRNA created in transcription is used as the template to make protein. The mRNA is organized into three-base long units called **codons.** Each codon codes for a specific amino acid.

Translation begins when mRNA binds to a ribosome, and then a transfer RNA (tRNA)—a molecule that carries an amino acid—attaches to the ribosome. This ribosomal complex positions amino acids in a favorable position so that peptide bonds can be formed between amino acids. Complementary matching of three nucleotides on the tRNA, called an anticodon, and three nucleotides on the mRNA, called a codon, ensures the correct sequence of amino acids. The mRNA passes along the ribosome in short spurts, one codon at a time. As this occurs, the first amino acid forms a peptide bond with the second amino acid. At the same time, a new tRNA enters the ribosome. The first tRNA, which no longer carries an amino acid, leaves the ribosome and is recycled inside the cell. Each time the amino acid is transferred to the growing polypeptide, a new tRNA brings an amino acid. As the ribosome proceeds down the mRNA, eventually a stop codon is encountered. At this point the ribosomal complex falls apart, **protein synthesis** is complete, and the polypeptide is released into the cell where it can be further processed into cell structures.

The codon dictionary:

First letter of code	Second letter of code				Third letter of code
	U	C	A	G	
U	Phenylalanine	Serine	Tyrosine	Cysteine	U
	Phenylalanine	Serine	Tyrosine	Cysteine	C
	Leucine	Serine	STOP	STOP	A
	Leucine	Serine	STOP	Tryptophan	G
C	Leucine	Proline	Histidine	Arginine	U
	Leucine	Proline	Histidine	Arginine	C
	Leucine	Proline	Glutamine	Arginine	A
	Leucine	Proline	Glutamine	Arginine	G
A	Isoleucine	Threonine	Asparagine	Serine	U
	Isoleucine	Threonine	Asparagine	Serine	C
	Isoleucine	Threonine	Lysine	Arginine	A
	Methionine (start)	Threonine	Lysine	Arginine	G
G	Valine	Alanine	Aspartate	Glycine	U
	Valine	Alanine	Aspartate	Glycine	C
	Valine	Alanine	Glutamic acid	Glycine	A
	Valine	Alanine	Glutamic acid	Glycine	G

Fig. 9.9 The Codon Dictionary

Translation also occurs in three steps (Figure 9.10):

1. **Initiation:** This is when the mRNA, first tRNA and two halves of the ribosome all bind together.
2. **Elongation:** This is when new tRNA molecules bring in new amino acids to grow the polypeptide chain.
3. **Termination:** This is when the stop codon is reached and protein synthesis is complete.

Regulation of Gene Expression

In the language of genetics, "expressed" means "on". Most people are surprised to learn that not all genes are "on" all the time. Think of the cookbook example above. Can you image if your "chocolate chip cookie" gene were on all the time? All you would do is bake chocolate chip cookies!

Gene regulation is a method that cells use to cope with changing conditions of the environment. Cells use control mechanisms to synthesize the maximum amount of cell material from a limited supply of nutrients.

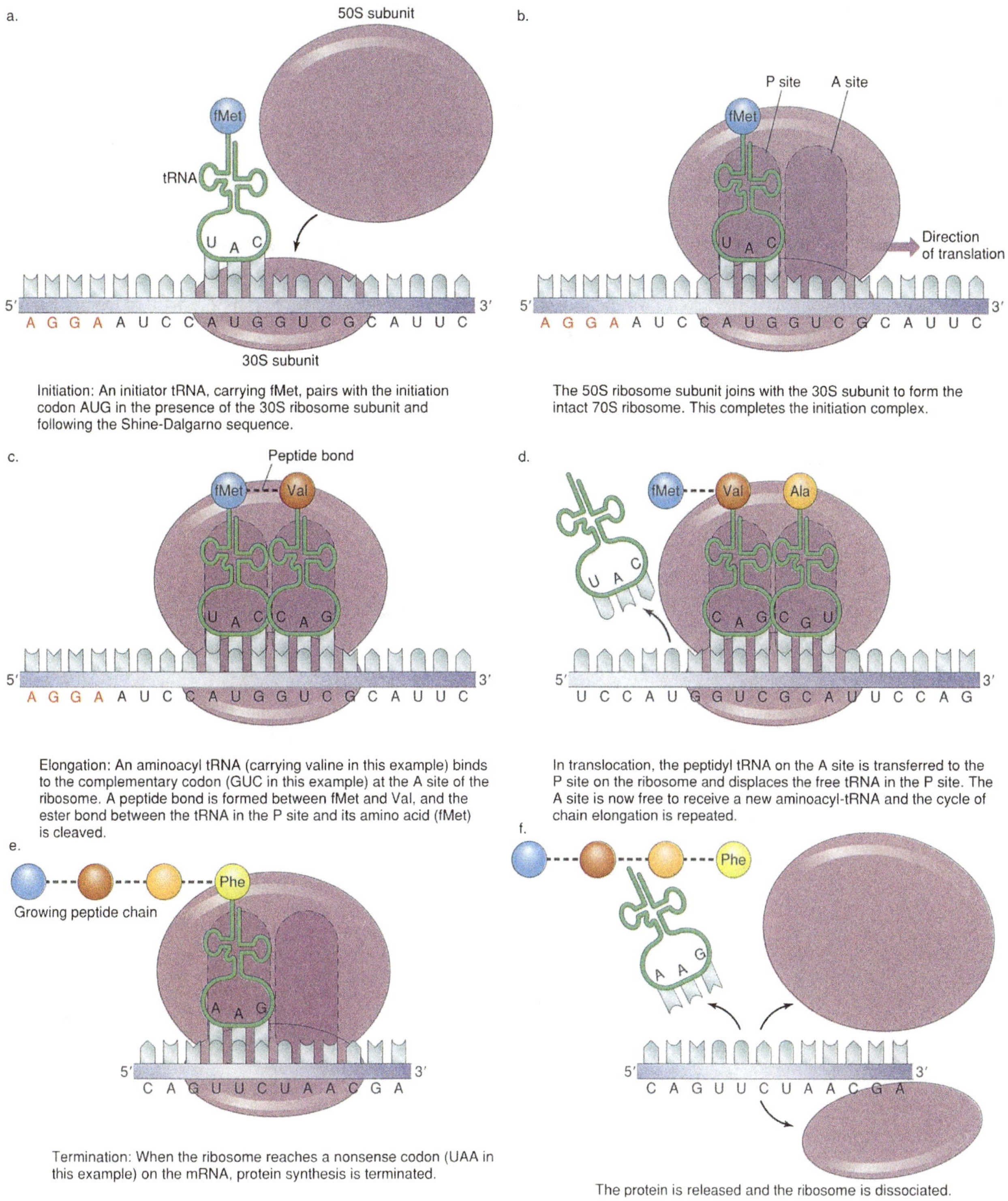
a.
50S subunit
fMet
tRNA
U A C
5′
3′
A G G A A U C C A U G G U C G C A U U C
30S subunit
Initiation: An initiator tRNA, carrying fMet, pairs with the initiation codon AUG in the presence of the 30S ribosome subunit and following the Shine-Dalgarno sequence.
b.
P site
A site
fMet
U A C
Direction of translation
5′
3′
A G G A A U C C A U G G U C G C A U U C
The 50S ribosome subunit joins with the 30S subunit to form the intact 70S ribosome. This completes the initiation complex.
c.
Peptide bond
fMet
Val
U A C
C A G
5′
3′
A G G A A U C C A U G G U C G C A U U C
Elongation: An aminoacyl tRNA (carrying valine in this example) binds to the complementary codon (GUC in this example) at the A site of the ribosome. A peptide bond is formed between fMet and Val, and the ester bond between the tRNA in the P site and its amino acid (fMet) is cleaved.
d.
fMet
Val
Ala
U A C
C A G
C G U
5′
3′
U C C A U G G U C G C A U U C C A G
In translocation, the peptidyl tRNA on the A site is transferred to the P site on the ribosome and displaces the free tRNA in the P site. The A site is now free to receive a new aminoacyl-tRNA and the cycle of chain elongation is repeated.
e.
Phe
Growing peptide chain
A A G
5′
3′
C A G U U C U A A C G A
Termination: When the ribosome reaches a nonsense codon (UAA in this example) on the mRNA, protein synthesis is terminated.
f.
Phe
A A G
5′
3′
C A G U U C U A A C G A
The protein is released and the ribosome is dissociated.

Fig. 9.10 Translation

Two common examples for studying gene expression are the *lac* operon and the *trp* operon. An **operon** is a group of genes that are transcribed together as a unit. There are more than one coding region between the promoter and the terminator. Operons may be **inducible**, which means that they are usually "off", but the genes are expressed when needed, or they may be **repressible**, meaning that they are usually expressed, but can be repressed.

The (Lactose) lac Operon: An Inducible Operon

Many bacteria prefer glucose before they utilize other compounds such as lactose as a growth substrate, when both are present in the medium. The lac operon in *E. coli* has a coding region for three functional proteins or enzymes necessary for the use of lactose as a food source. These enzymes are only needed when lactose is available and the normal food source, glucose, is in short supply. Therefore, the lac operon is not activated until glucose is in short supply, since it would be a waste of energy to produce these enzymes when glucose is plentiful. Transcription of the lac operon is controlled by two regulatory proteins: **activator** and **repressor**. Therefore, the regulatory region of the lac operon has two switches. The activator protein can bind to the activator binding site of DNA and facilitate transcription. The repressor protein can bind to the operator site of DNA and block transcription.

When glucose is present, the activator protein cannot bind to the activator binding site of DNA. Thus, RNA polymerase cannot bind to the promoter, and transcription of the lac operon does not occur. In addition, if there is no lactose in the medium, the lactose repressor binds to

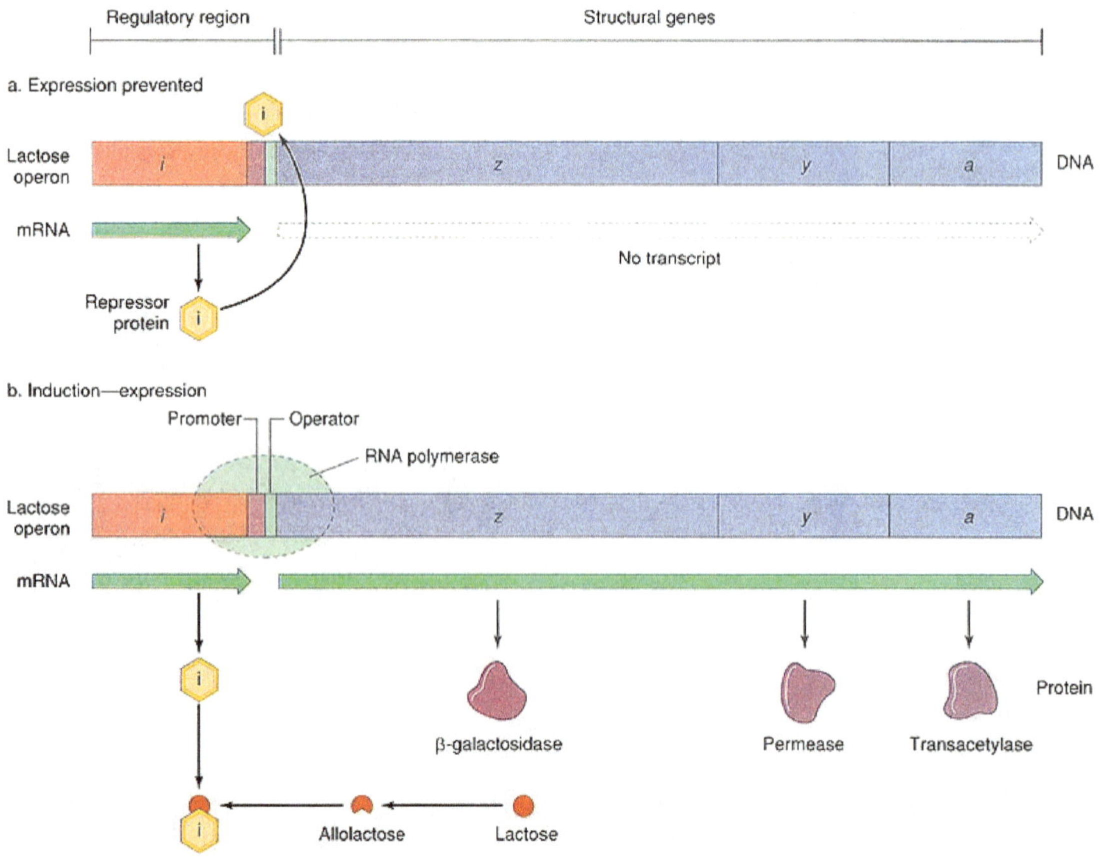

Fig. 9.11 The *lac* operon
© Kendall Hunt Publishing

the operator site of DNA and blocks transcription. When glucose is absent and lactose is present in the medium, the activator protein binds to the activator binding site of DNA and facilitates transcription. Lactose inhibits the repressor from binding to the operator site. RNA polymerase can now bind to the promoter and carry out transcription.

The (Tryptophan) trp Operon: A Repressible Operon

The *trp* operon is responsible for making enzymes involved in the synthesis of tryptophan, an amino acid used to make proteins. The *trp* operon is usually on, and the repressor protein remains inactive and does not bind to the DNA. However, if the concentration of tryptophan in the cell becomes too great, tryptophan will bind to the repressor protein and activate it. This will allow the repressor to bind to the DNA and prevent RNA polymerase from initiating transcription.

Phenotypic variation occurs when environmental factors regulate gene expression and influence the phenotype of the microorganism. Phenotypic variation is influenced by several variables, including temperature, pH, nutrient media, osmotic pressure, and exposure to ultraviolet (UV) light. Dramatic changes, affecting the great majority of the cells of a pure bacterial culture, frequently occur when the cells are transferred from one environment to another. These changes persist during subculture in the new environment. However, when the previous environmental conditions are restored, the cells regain their original properties. These variations are not to be confused with mutation. Phenotypic variations vary according to the environmental condition and are reversible. A particular gene which already exists is turned on or turned off according to the environmental needs. A mutation, on the other hand, is an alteration in the genetic make-up of the bacterium.

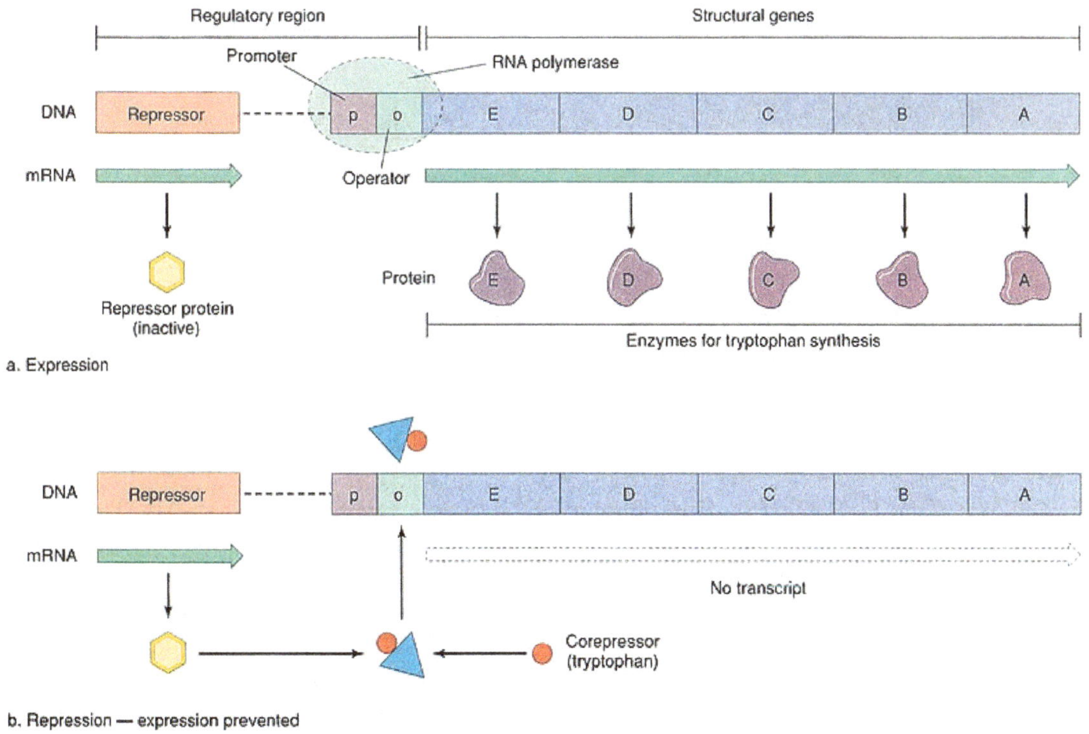

Fig. 9.12 The *tryp* operon

Phenotypic variation can be observed in *Serratia marcescens*, one of the few bacteria that produce bright pigments, most commonly red, white, or pink with colonies that are 1.5–2.0 mm with a convex elevation. *Serratia marcescens* is commonly found in the intestine and in soil. *Serratia* causes about 2% of nosocomial infections of the bloodstream, lower respiratory tract, urinary tract, surgical wounds, and skin and soft tissues of adult patients. Outbreaks of *S. marcescens* have occurred in pediatric wards. *S. marcescens* has a fondness for bread and communion wafers, where the pigmented, aged colonies have been mistaken for drops of blood. *S. marcescens* grows well in damp basements, on food stored in damp places, and in bathrooms. For example, the pink discoloration and slime on shower corners, toilet basins, and tile grout is *S. marcescens* bacteria that derive their nutrition from dead skin cells, skin oils, and soap and shampoo residue.

The pigment produced by *Serratia*, **prodigiosin,** offers protection against excessive UV in sunlight, serves as an antibiotic, enhances flagella production, and has cytotoxic qualities. This protection is very helpful in establishing the bacteria's domain in soil as it competes with fungi.

Biosynthesis of prodigiosin is a comprehensive pathway that has at least 10 different enzymes involved in its synthesis. In *Serratia marcescens*, pigment production is a temperature-regulated gene. If a culture of *Serratia* is incubated at normal body temperature (37°C), the colonies are creamish-white. However, if a culture of *Serratia* is incubated at room temperature (25°C), the colonies are dark red. This change in colony phenotype should be considered during the 'unknown' lab as a way to differentiate between *Serratia marcescens* and *Enterobacter aerogenes*, since even though *Serratia marcescens* is negative for lactose fermentation, it appears positive due to the production of this pigment. Gene regulation of the red phenotype is influenced by several variables, including temperature, nutrient media, and age of the colonies. Therefore, we will expect variability in our results. Most strains of *S. marcescens* are red under 27°C and white above 28°C. Pigment and flagella production stops at ~ 28° C.

Genetic Diversity and Evolution

Until now, we have been working under the assumption that all bacterial cells in a colony are identical. But that is an over-simplification; the cells in a colony are very similar, as they all come from a single progenitor cell, but they are not all identical.

Because bacteria reproduce asexually by binary fission, they have limited means for generating genetic diversity and evolution. Here, evolution simply means change over time. In order to evolve bacteria rely on **mutation** and **gene transfer.**

Everything that a cell can do is because of the directions in the cell's DNA. But sometimes, during DNA replication errors occur which change the sequence of bases on the DNA molecule. As you can see from studying protein synthesis, a change in the sequence of DNA leads to a change in the sequence of mRNA, which may lead to a change in the amino acid sequence of the polypeptide. The term "mutation" is usually associated with something bad, and most mutations are either harmful or deadly to the bacterial cell. However, some mutations are beneficial and some are silent, having no effect on the cell whatsoever.

The most common type of mutation is the **point mutation**, where only one base pair is affected. Point mutations can be:

- **Substitutions:** substituting one base for another
- **Insertions:** adding an extra base pair
- **Deletions:** leaving out a base

The effect of genetic mutation on protein synthesis:

Substitutions can cause:

- **Silent mutations:** One base is substituted for another, but the resulting amino acid remains the same.
- **Missense mutations:** A new amino acid is substituted in place of the old one.
- **Nonsense mutations:** A STOP codon is generated in place of an amino acid.
- **Frameshift mutations:** By inserting or deleting a base pair, the entire three-base reading frame shifts either to the left or right, causing all of the amino acids downstream of the shift to be changed.

Causes of Mutation

Some mutations are **spontaneous**, that is, they are just accidents that happen during DNA replication. Other mutations are **induced**, or caused by outside agents known as **mutagens.** Mutagens can be chemicals that **interfere with DNA replication, or they can be radiation, such as UV light that causes damage to DNA.**

Benefits of Mutation

Mutations are not deliberate acts by the cell! Mutations are *accidents* that occur during DNA replication. Most mutations are harmful or deadly and therefore said to be **deleterious.** They are deleted from the population because the mutated cell does not live long enough to pass the deleterious genes on to progeny. Some (very few) mutations are beneficial and give the cell an

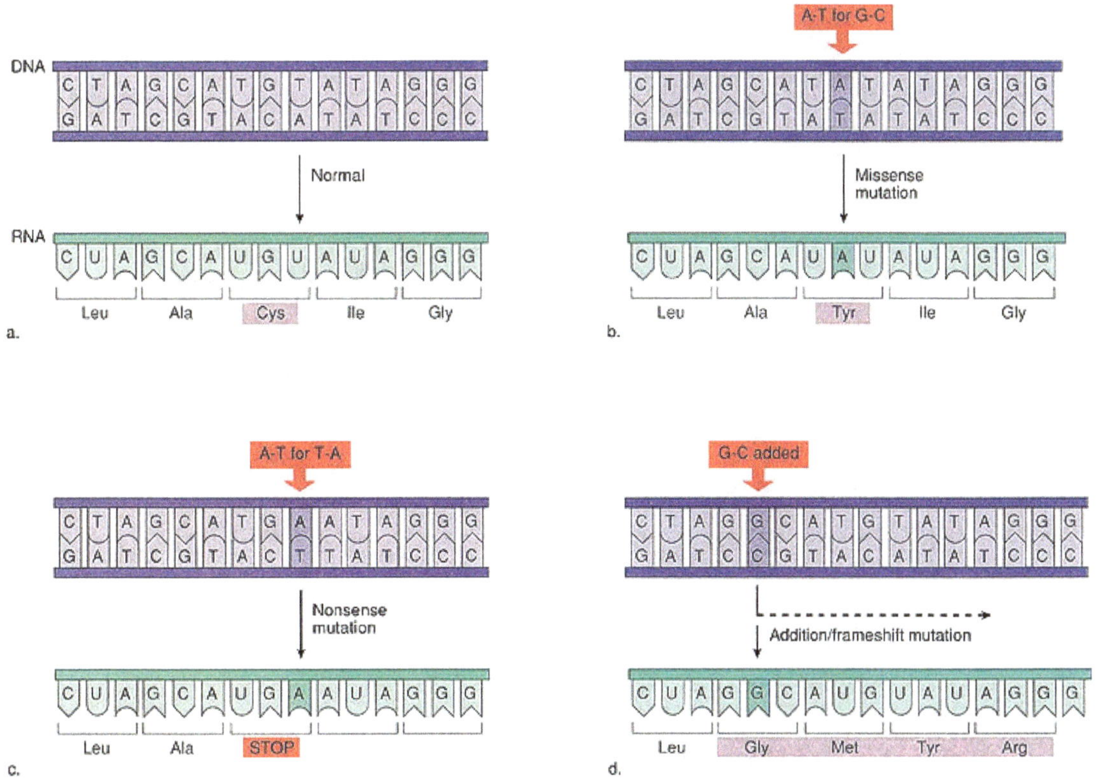

Fig. 9.13 Types of Mutations

© Kendall Hunt Publishing

advantage over other cells in the population. In the case of bacteria, these beneficial mutations often come in the form of some type of antibiotic resistance.

DNA Repair

Just because an error occurs during DNA replication does not mean that the result will be a mutation. Cells have a variety of repair mechanisms to fix damaged DNA.

Two mechanisms for repairing DNA in which the incorrect base has been added to the DNA strand are **base-excision repair**, in which just the single, incorrect base is removed and replaced, and **mismatch repair**, in which an entire section of DNA surrounding the incorrect DNA is removed and replaced. (Figure 9.14)

Two mechanisms also exist for repairing thymine-dimers that are the result of UV exposure. **Light repair** is when the enzyme photolyase is activated by visible light and cleaves the bond between adjacent thymine molecules, allowing the DNA to return to its original shape.

Another mechanism is **dark-repair**, which is when the cell does not have photolyase, and a repair enzyme removes the entire section of DNA containing the thymine-dimer and replaces it with new DNA.

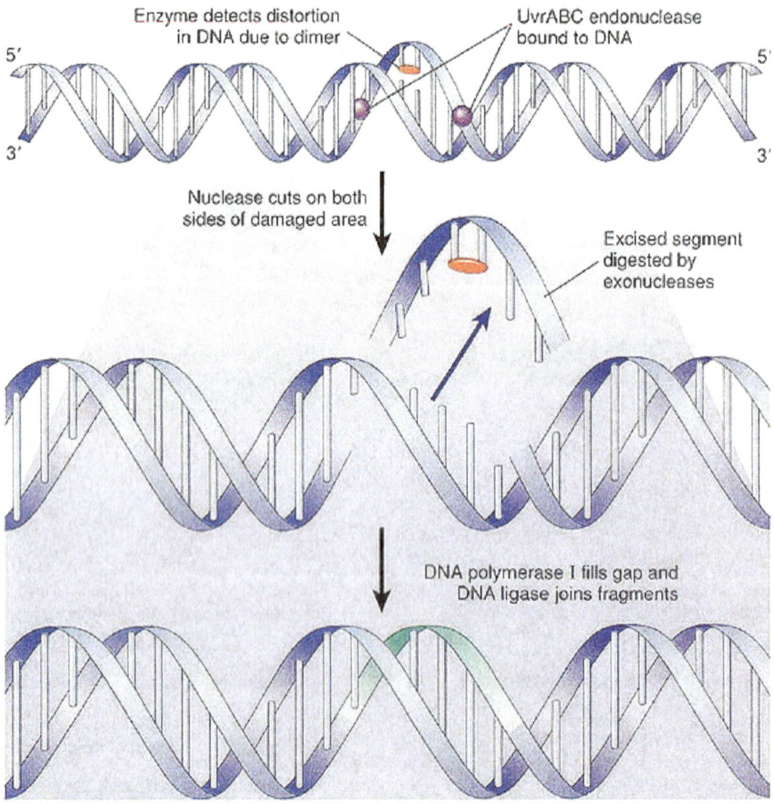

Fig. 9.14 Mutation Repair

Gene Transfer

The other mechanisms by which bacterial cells generate genetic diversity are through **gene transfer.** These mechanisms are particularly important because much of the spread of antibiotic resistance among bacteria is due to these mechanisms.

Horizontal gene transfer occurs when genes are transferred from one microbe to another microbe of the same generation (**vertical gene transfer** is the transfer of genes from parent cell to progeny).

There are three types of gene transfer: transformation, transduction and conjugation.

Transformation is the uptake of naked DNA by a competent cell. A **competent cell** is a cell that is capable of taking in naked DNA. The source of the free DNA may be by natural release from bacteria during death or it may be accomplished in the laboratory using plasmids. During natural transformation of a DNA fragment, double-stranded DNA binds to specific receptors on the surface of a competent cell. One strand of the donor DNA is degraded by nucleases while the other strand enters the cell. The single-stranded donor DNA pairs with a homologous region on the recipient DNA and is integrated into the recipient genome by the breakage and reunion mechanism called homologous recombination. (Figure 9.15) If there are any differences between the nucleotide sequences of the donor and recipients DNAs, the mismatch repair

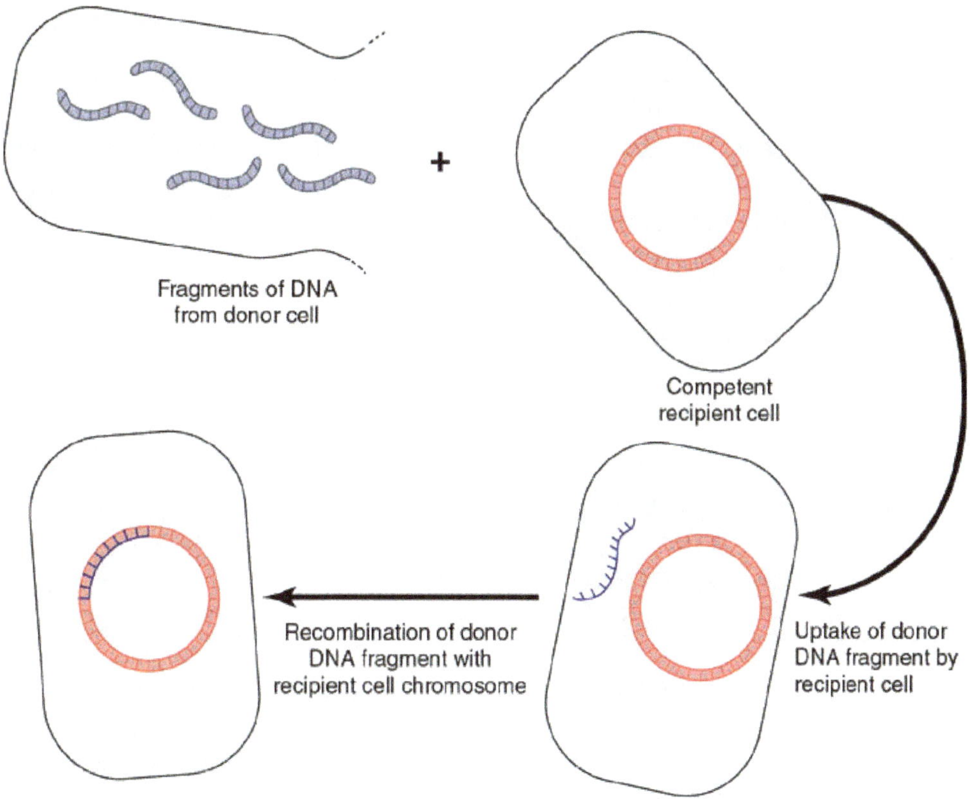

Fig. 9.15 Transformation IMAGE:
© Kendall Hunt Publishing

system comes into play. The repair system removes either the donor or the recipient strand, and replaces it with the complementary sequence. Since either strand may be repaired, some cells contain the new, donor DNA and others maintain the original DNA sequences. For genera such as *Streptococcus*, *Acinetobacter*, *Bacillus*, and *Haemophilus*, transformation appears to be a major means by which genetic transfer occurs. It is also known to occur in genera such as *Neisseria*, *Escherichia*, and *Pseudomonas*.

Some genera of bacteria are **naturally competent**, which means they are capable of undergoing transformation in the normal environment. They can take in free DNA from their environment and incorporate it into their chromosome. Other genera can be made **artificially competent** in a laboratory (i.e., altered in a laboratory to make them able to undergo transformation).

Transformation was discovered by Frederick Griffith in 1928 while doing experiments on mice with two strains of *Streptococcus pneumonia*.

- The S strain of the cells was smooth, because they had a capsule. The capsule allowed the cells to evade the immune system and cause pneumonia and death in the test mice.
- The R strain was rough, because it had no capsule, and therefore could not hide from the immune system and cause disease.

Griffith did several experiments, as listed below:

1. He injected test mice with the S strain of *S. pneumonia*. The mice developed pneumonia and died. Griffith found S strain bacteria in the lungs of the dead mice.
2. He injected test mice with the R strain of *S. pneumonia*. The mice did not develop any symptoms. The mice were sacrificed and their lungs examined for signs of bacteria. There were no bacteria in the lungs of the R strain-infected mice.
3. Griffith heat-killed the S strain bacteria and injected into the mice. The mice did not develop symptoms. No signs of bacteria were found in the lungs of these test mice.
4. In the next phase of his experiment, Griffith mixed heat-treated cells of strain S and live cells of strain R and the mice developed pneumonia and died. Upon examination, Griffith found S strain bacteria in the lungs of the dead mice.

Why did the mice die? Neither of the cells they were injected with caused pneumonia on their own!

It turns out that *S. pneumonia* is **naturally competent**, that is, it has the ability to take in naked DNA from the environment and incorporate that DNA into its own genome, thereby **transforming** itself into a different cell type. In Griffith's experiment, the R cells were able to take in the gene for capsule production, *transforming* themselves into the S strain and acquiring the ability to evade the immune system to cause disease.

Transduction occurs when bacterial DNA is transferred between cells by way of a bacteriophage. (Figure 9.16)

The goal of the phage is to take over the bacterial cell and use the cell to make more phage particles, but in some instances, the phage accidentally incorporate segments of bacterial DNA into the phage head. When the phage carrying the bacterial DNA infects another bacterial cell, that cell receives bacterial DNA rather than phage DNA.

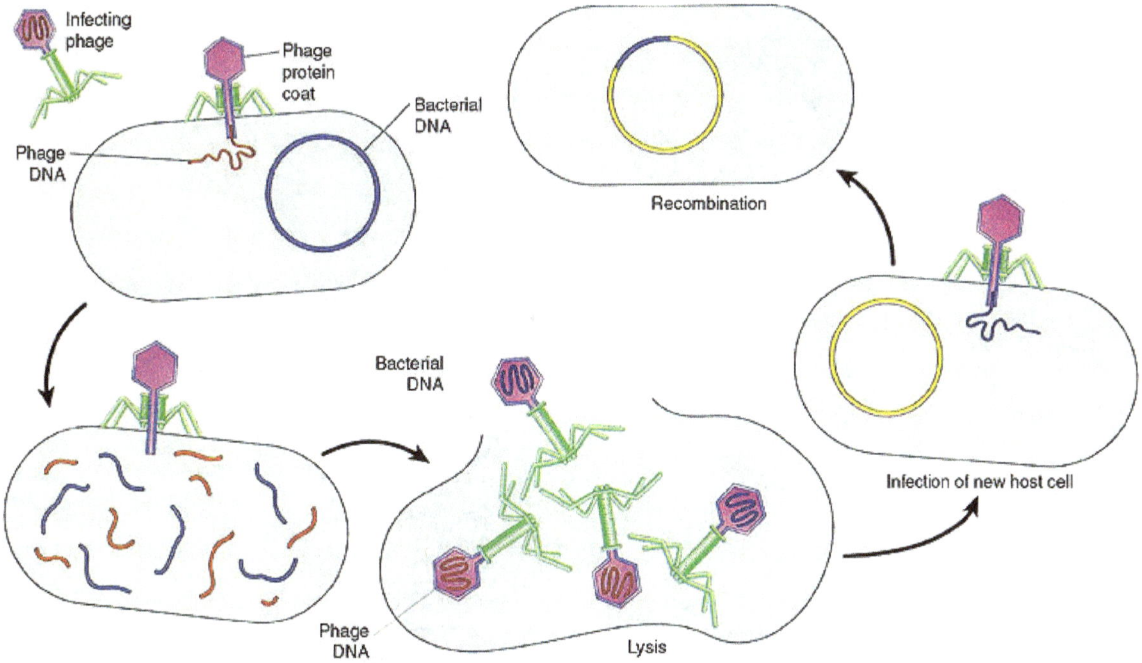

Fig. 9.16 Transduction IMAGE

Lysogenic conversion occurs when bacteria receive genes from bacteriophage that increase the pathogenicity of the bacterium. For example, the bacterium *Corynebacterium diphtheriae* causes illness by producing a toxin that disrupts protein synthesis in the cells of infected individuals. However, *C. diphtheriae* only produces the toxin after it has itself been infected by phage beta. The toxin gene is actually produced by the phage, not the bacterium.

Conjugation

Conjugation occurs when genes on plasmids are transferred from a live **donor** cell to a live **recipient** cell through a **sex pilus** (aka conjugation pilus). (Figure 9.17)

Plasmids are extrachromosomal segments of DNA. Plasmid size can vary from only a few genes to almost the size of a bacterial chromosome. Plasmids can carry genes for toxin production, antibiotic resistance and metabolic enzymes. Of great concern to humans is the ability of bacterial cells to transfer plasmids from one bacterium to another. This transfer can occur within bacterial genera, or across genera.

The gene that mediates the formation of the sex pilus is called the **fertility gene** and is found on the **F (fertility) plasmid.** Cells that have F plasmids (donors) are said to be F+, and cells lacking the plasmid (recipients) are F-.

A special type of conjugation, **Hfr (high-frequency recombination)** occurs when the fertility plasmid splices itself into the chromosome of the donor cell. Then, when the plasmid replicates and is transferred to the recipient cell, part of the donor's chromosomal DNA is transferred as well.

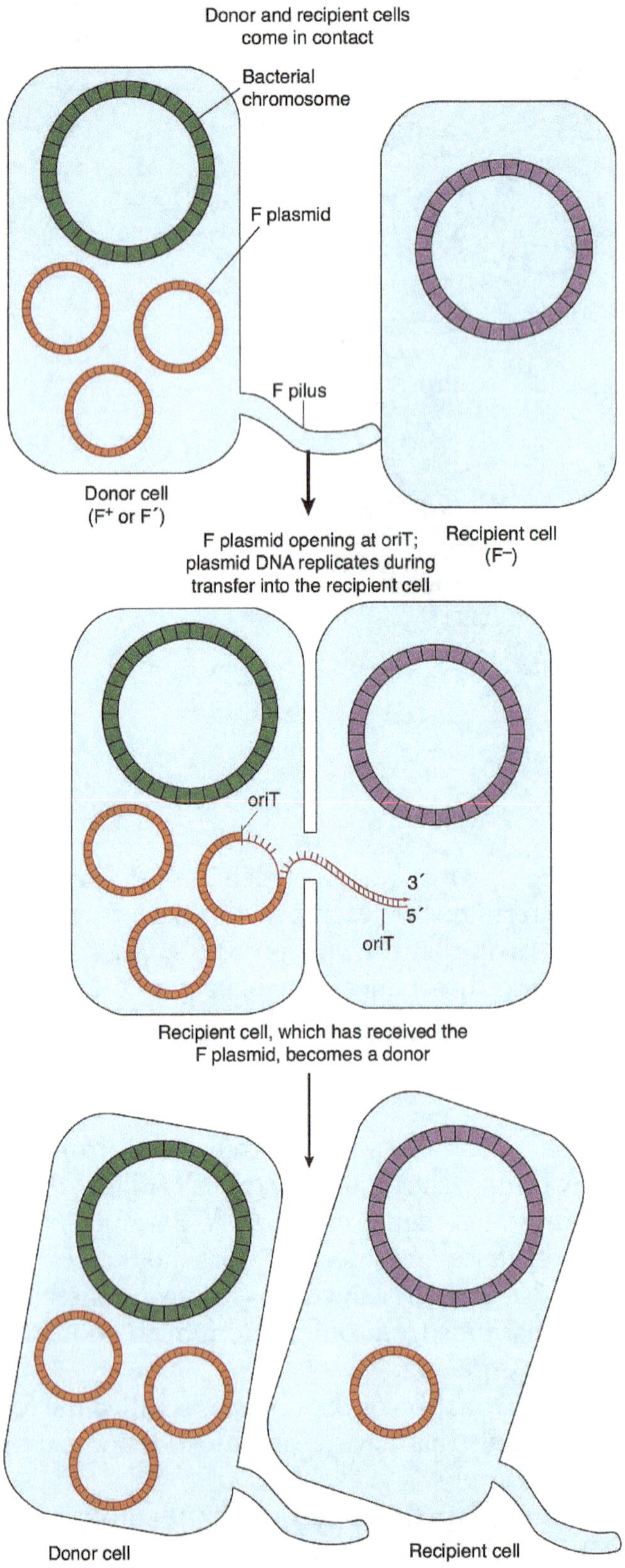

Fig. 9.17 Transfer of DNA During Conjugation, K-H Resources/Line-Art

Current Examples of Antibiotic Resistance Due to Plasmid Transfer

Some medically relevant examples of plasmid-mediated resistance are:

- VRE: vancomycin-resistant enterococci
- Penicillin-resistance in *S. pneumoniae*
- Resistance to ampicillin, tetracycline and kanamycin by *Pseudomonas* and *Enterobacteriae*

Antibiotic resistance can develop fairly quickly. The following table lists examples of drug resistance and when the resistance was first reported:

1941: Penicillin given to first patient	1942 Penicillin resistance reported
1956: Vancomycin introduced	1992: *S. aureus* gains vancomycin-resistance gene from *Enterococci* bacteria
	1997: Partial vancomycin resistance reported
	2002: Full vancomycin resistance reported
1960: Methicillin introduced	1961: Methicillin resistance reported
1999: Quinupristin/dalfopristin introduced	2000: Quinupristin/dalfopristin resistance reported
2000: Linezolid introduced	2001: Linezolid resistance reported
2003: Daptomycin introduced	2005: Daptomycin resistance reported
2005: Tigecycline introduced	???? Tigecycline resistance reported

Source: United States Center for Disease Control and Prevention, Atlanta, GA.

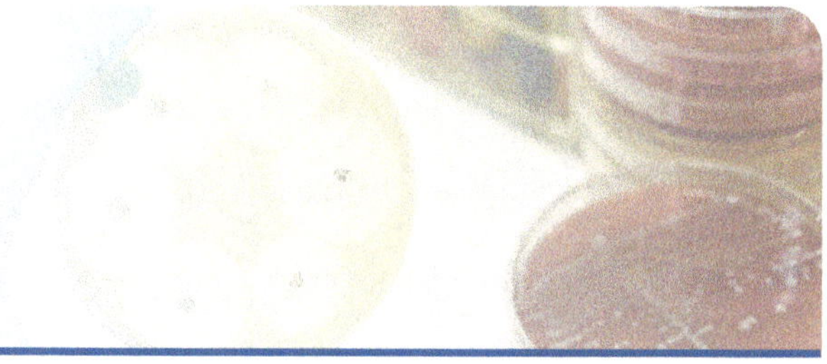

Lab Exercise 9.1

Genetics: Gene Expression and Phenotypic Variation

Student Objectives

1. Define the following terms: DNA, gene, RNA, transcription, translation, genotype, phenotype, operon, repressors, activators, gene expression, quorum sensing, phenotypic variation, and reversible variation.
2. Describe the action of an operon and the use of repressors and activators during gene expression.
3. Explain genetic expression and its effect on phenotype.
4. Evaluate how bacteria cope with changing conditions of the environment.
5. Evaluate environmental factors that regulate gene expression.
6. Perform a reversible variation experiment using a broth culture of *Serratia marcescens*.
7. Evaluate the effects of temperature on phenotypic variation.

Name ______________________________

Pre-Lab Exercise 9.1

1. What is phenotypic variation?

2. Define the following:

   Gene:

   Genotype:

   Phenotype:

   RNA polymerase:

   Promoter:

   mRNA:

   tRNA:

   Codon:

   Operon:

   Gene regulation:

3. Describe the chromosome in a prokaryote.

4. What is the purpose of transcription and translation?

5. Describe the major steps during transcription and translation.

6. Describe the significant differences in transcription between prokaryotes and eukaryotes.

7. Explain how a gene is regulated and how activators and repressors are involved.

8. Compare and contrast the lactose operon with the tryptophan operon.

9. Describe how quorum sensing and gene expression are related.

Week One—Phenotypic Variation Activity

Week One Materials

Per table: Two TSA plates and one broth culture of Serratia marcescens.

1. Streak two TSA plates for isolation using the quadrant streak method and the *Serratia marcescens* culture provided.
2. Using a china marker, mark one plate of the pair at 37°C (body temperature) and the other plate at 25°C (room temperature). Put your initials and table number on both plates.
3. Place the room temperature plate (inverted) on the appropriate tray. Place the 37°C plate (inverted) on the appropriate tray to go in the incubator.

Week Two—Phenotypic Variation Activity

Week Two Materials

Per student: Two TSA plates.

1. Retrieve your plates and record the growth and pigmentation on each in the report section at the back of the lab.
2. Using a china marker, label the new TSA plates as 25°C and 37°C. Put your table number on both plates.
3. Using the 25°C culture you just retrieved (the one with growth that you inoculated last week), carefully lift the lid with one hand, collect a sample of *Serratia marcescens* from the plate using a sterile inoculating loop with the other hand, and close the lid immediately. Use this sample to streak the newly labeled 37°C plate for isolation (using the quadrant streak method). Yes, you are changing temperatures! That's the point: reversible variation!
4. Using the 37°C culture you just retrieved (the one with growth that you inoculated last week), carefully lift the lid with one hand, collect a sample of *Serratia marcescens* from the plate using a sterile inoculating loop with the other hand, and close the lid immediately. Use this sample to streak the newly labeled 25°C plate for isolation (using the quadrant streak method). Yes, you are changing temperatures! That's the point: reversible variation!
5. Place the new plates (inverted) on the appropriate temperature trays to stay at room temperature or be placed in the incubator for next week. Last week's plates go on the autoclave cart.

Name ______________________

Exercise 9.1 Lab Report

1. **Week two**: Record the growth and pigmentation on each plate.

____

____

____

2. **Week three**: Record the growth and pigmentation on each plate.

____

____

____

3. Why is *Serratia marcescens* used in this laboratory experiment?

____

____

____

4. Describe the interactions between *Serratia marcescens* and society.

____

____

____

5. What are your conclusions with regard to phenotypic variation in *Serratia marcescens?*

____

____

____

6. Use the following terms to describe the process of phenotypic variation: environmental factors, gene, gene expression, activator, repressor, protein, genotype, and phenotype.

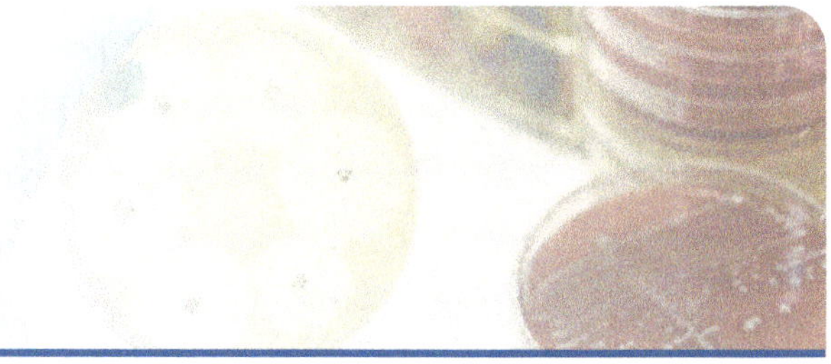

Lab Exercise 9.2

Genetics: Transformation

Student Objectives

1. Define the following terms: plasmid, transformation, natural competence, artificial competence, vector, recombinant, selectable marker, and recombinant DNA technology.
2. Perform a transformation experiment using *Escherichia coli* and pBLU plasmid.
3. Describe how transformation will create a new bacterial strain.
4. Evaluate the uses of transformation within recombinant DNA technology.
5. Understand new lab skills for spreading cultures on agar using glass beads and aseptically using a micropipette.
6. Explain why selectable markers are used in a transformation experiment.
7. Apply the uses of transformation within recombinant DNA technology.

Name ______________________

Pre-Lab Exercise 9.2

1. What is transformation?

2. Use the following terms to describe the process of transformation and explain how the enzymes are being made: plasmid, genes, DNA, genotype, transcription, translation, enzyme, and phenotype.

3. How is transformation used in medicine?

4. What does the term 'naturally competent' mean with regard to transformation?

5. What does the term 'artificially competent' mean with regard to transformation?

6. List several genera of bacteria that commonly undergo transformation.

7. During which step do you see potential for error, and how would those errors affect the results?

8. **Predict** your results in the chart below. Write "yes" or "no", depending on whether you think the plate will show growth. Give the reason(s) for your predictions.

LB –plasmid	Prediction:	Prediction:	LB +plasmid
	Reason:	Reason:	

LB+amp – plasmid	Prediction:	Prediction:	LB+amp + plasmid
	Reason:	Reason:	

LB+amp+Xgal – plasmid	Prediction:	Prediction:	LB+amp+Xgal + plasmid
	Reason:	Reason:	

INTRODUCTION

Transformation is the uptake of cell-free DNA from a competent bacterial cells environment. In this lab we will be using an ampicillin sensitive strain of *Escherichia coli* and genetically modify the organism using a transformation procedure to make our *E. coli* strain ampicillin-resistant. Yes, we are making antibiotic-resistant bacteria!

For this laboratory activity we will use a **plasmid** as the source of cell-free DNA. The plasmid we will use is referred to as a **cloning vector** because it can insert foreign DNA into a bacterium and as the host cell replicates many clones or copies of the plasmid DNA are generated. Plasmids are good cloning vectors because they are able to replicate independently from the chromosome within a cell. The plasmid we will use contains an ampicillin-resistant gene that allows us to use a selective media containing ampicillin. The use of the selective media allows us to see if our transformation was successful. The original *E. coli* before transformation is **ampicillin sensitive,** meaning the cells are inhibited (does not kill *E. coli*!) from growing when ampicillin is present. After transformation, the only cells that have taken up the plasmid with the ampicillin-resistant gene during transformation will grow in the presence of ampicillin, since **enzymes** (ampicillinase) that break down ampicillin are produced by the cell.

Plasmid DNA (ampicillin-resistant gene) → RNA → Protein (ampicillinase)

Experiments with cloning vectors and foreign DNA will have variable results. Due to the varied results and potential for contamination of ampicillin-resistant bacteria from the lab environment, a second gene is also on the plasmid. The second gene is the **lacZ gene**, which codes for the enzyme **beta galactosidase**. Most naturally-occurring *E. coli* strains contain the beta-galactosidase gene. However, we will start with a strain of *E. coli* that lacks the beta-galactosidase gene. Beta glactosidase can degrade a colorless chemical called x-gal, to form a blue compound. If you grow bacteria that produce beta-galactosidase on media containing x-gal, the colonies will be bright blue, therefore differentiating them from colonies that do not contain the plasmid. The formation of blue colonies is a second indicator of a successful transformation.

Plasmid DNA (lacZ gene) → RNA → Protein (Beta galactosidase)

In this laboratory exercise, a plasmid will be inserted into *E. coli* and plated on selective media so that only the transformed bacteria will grow. The plasmid we are using is called **pBLU** and contains **two genes**: ampicillin-resistance gene and the lacZ gene. The enzymes that are produced by the bacteria will allow the bacteria to grow in the presence of ampicillin and turn the colonies blue. Conversion of *E. coli* cells to a "**competent**" state can be accomplished by treating the cells with a solution of the chemical compound **calcium chloride** ($CaCl_2$) in a hypotonic buffer. Since water will move toward an area of high solute concentration, the cells will take up water and swell in response to the $CaCl_2$. The suspension is kept on ice to keep the cells from metabolizing and to preserve the cells in this state as well as maintain a temperature difference to increase the effectiveness of the heat shock. Exposure to $CaCl_2$ combined with heat shock (a brief period during which the cells are exposed to a temperature between 40–42°C) causes the cells to become competent. First a calcium phosphate-plasmid DNA complex is formed, which adheres to the cell's surface. Then the foreign DNA is internalized during the heat shock step. This treatment of calcium chloride and heat shock, while easy, is an inefficient procedure, and generally results in only 0.1–1% of the treated cells actually taking up and retaining the foreign DNA. Those cells that do take up and retain the DNA are transformed and the plasmid DNA will be passed on to subsequent generations.

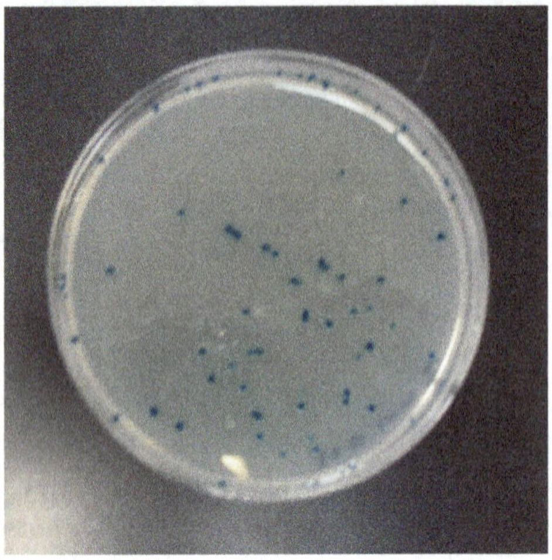

Courtesy of M. Pressler

Why Use Ampicillin in the Agar?

In a typical transformation, billions of bacteria are treated and exposed to plasmid DNA. Only a fraction (usually less than 1 in 1,000) will acquire the plasmid. Antibiotic-resistance genes provide a means of finding the bacteria, which acquired the plasmid DNA in the midst of all of those bacteria which did not. Adding an antibiotic to the media selects for bacteria that contain the antibiotic-resistance gene. So, by using a plasmid containing an antibiotic-resistance gene and growing on selective agar that contains that antibiotic, transformed cells containing the plasmid have the advantage.

Ampicillin is a member of the penicillin family of antibiotics. The ampicillin-resistance gene encodes for an enzyme (ampicillinase) that destroys the activity of ampicillin by breaking down specific chemical bonds in ampicillin. When a bacterium is transformed with a plasmid containing the amp-resistance gene, it expresses the gene and synthesizes the enzyme. The enzyme is secreted from the bacterium and destroys the ampicillin in the surrounding medium. As the ampicillin is broken down, the transformed bacterium regains its ability to form its cell wall and is able to replicate to form a colony. The colony continues to secrete the enzyme and forms a relatively ampicillin-free zone around it. After prolonged incubation, small **satellite colonies** of *non-transformed bacteria that are still sensitive to ampicillin* grow in these relatively ampicillin-free zones.

Why Use x-gal in the Agar?

The presence of X-gal in the media makes the media a differential media since the transformed colonies will be bright blue, therefore differentiating them from colonies that do not contain the plasmid. The formation of blue colonies is the second indicator of a successful transformation. If the bacteria do not produce b-galactosidase, the colonies will be the normal whitish color and are formed from contamination or in rare instances mutated non-transformed *E.coli.*

Application

Technology

Transformation is used for certain types of **recombinant DNA technology.** Foreign genes, from a totally unrelated organism, can be inserted into a plasmid. The plasmid can then be inserted into a competent bacterial cell via transformation. Human genes can and are being experimented upon outside the human body using this method. This same process in eukaryotic cells is termed **transfection**. It is an essential part of genetic engineering used with yeast, plants, and mice. It has been proposed as method for developing cures for genetic disorders.

Here's one example of how transformation is used today to produce a drug used in treatment. Alpha-2a interferon (sold as Roferon-A) is used to treat Kaposi's sarcoma and hairy-cell leukemia in AIDS patients. The human alpha-interferon gene is prepared for splicing into an *E. coli* plasmid. Once spliced, the resultant gene and plasmid combination is called a recombinant. Next, the recombinant is placed into a special laboratory strain of *E. coli* by using transformation. After a few hours of incubation, there can be billions of cells, each containing the interferon gene. Clones are selected by using a plasmid containing a specific drug-resistant gene. The *E. coli* cells are then induced to transcribe and translate the interferon gene, producing the product (alpha-2a interferon) which is then removed and purified.

The scale of the procedure can range from a test tube full of product to gigantic industrial vats that can manufacture thousands of gallons of product. The scale used is determined by the need for the product. Variations of the process can be used to mass-produce products as varied as human and animal hormones; plant, animal, and human enzymes; and agricultural products such as pesticides.

Examples of Current Protein Products Obtained by Recombinant DNA Technology

Immune Treatments:
Interferons (alpha, beta, and gamma): Used to treat cancers, MS (multiple sclerosis), hepatitis infections and genital warts
Tumor necrosis factor (TNF): Used to treat cancers
Interleukins (ILs): Used to treat cancers
Macrophage factors: Used with bone marrow grafts

Enzymes:
SOD: A superoxide used to treat brain traumas
Antitrypsin: Used to treat emphysema
rH DNase: Used to treat cystic fibrosis

Hormones:
Insulin: Used in diabetes
Human growth hormone (HGH): Used in dwarfism
Erythropoietin (EPO): Used to treat anemia
Plasminogen activating factor (tPA): Used in blood clots

Vaccines:
Hib (*Haemophilus influenzae* Type b) vaccine
Hepatitis B vaccine

Miscellaneous:
Bovine growth hormone (BGH): Used to increase milk production in cows
Spider silk: Used in parachutes and bulletproof vests

Week One—Transformation Activity

Week One Materials

Per Table: Supplies: 2 empty sterile test tubes, 1 container of approximately 20 sterile glass beads, 1 package sterile plastic 10 μL inoculating loops, 1 micropipette, 1 box sterile micropipette tips, 1 vial of sterile $CaCl_2$ solution (on ice), 1 thermos of ice.

Per table: Media and cultures: 1 *E. coli* (Beta-*galactosidase*-deficient strain) starter plate, 2 luria agar (LB) plates, 2 luria agar + ampicillin (LB+amp) plates, 2 luria agar + ampicillin + X-gal (LB+amp+X-gal) plates, 1 luria broth (LB) tube.

Access to: 42°C H_2O bath and class supply of pBLU plasmid on ice.

Micropipette Instructions

Turn the knob on top until the needed number of microliters (μL) shows in the "window". Open the sterilized tip box just long enough to push the pipettor into the tip so that it is securely attached. Close the box to keep the other tips sterile.

To draw up liquid: Push the top button/knob down to the first "stop". Place the tip into the liquid and release the button slowly so that there are no bubbles drawn up.

To expel liquid: Push the top button/knob down to the first and then the second "stop". Pull out of the container (still holding button down). Release the button.

To release tip: Depress the side button to shoot off the tip. Tips that have touched a culture must be placed in the biohazard bag.

1. Mark one sterile test tube as "+plasmid." Mark the other as "–plasmid." (Plasmid DNA will be added to the "+plasmid" tube; none will be added to the "–plasmid" tube.)
2. Refer to the micropipette instructions. Use a sterile tip on the micropipette to add 250 µL of ice-cold $CaCl_2$ solution to each tube. Have one person hold the micropipette to reuse this tip in Step 5.
3. Place both test tubes on ice.
4. Flame and cool a metal loop. Use it to transfer isolated colonies from the *E. coli* starter plate to the "+plasmid" tube. The total area of the colonies picked should be equal in size to the top of a pencil eraser.
   a. Be careful not to transfer any agar from the plate. Impurities in the agar will interfere and yield poor results later.
   b. Immerse the cells on the loop in the $CaCl_2$ solution in the "+plasmid" tube and vigorously spin the loop in the solution to dislodge the cell mass. Hold the tube to the light to observe that the cell mass has fallen off the loop.
5. IMMEDIATELY suspend the cells by repeatedly pipetting in and out of the micropipettor tip. Examine the tube against light to confirm that no visible clumps of cells remain in the tube or are stuck in the tip. The suspension should appear milky white. Discard this tip in the biohazard bag.

6. Return the "+plasmid" tube to ice. Repeat *Steps 4 and 5* for the "–plasmid" tube.
7. Return the "–plasmid" tube to ice. Both test tubes should now be on ice and contain 250 µL of ice-cold $CaCl_2$ solution plus suspended *E. coli* cells.
8. Use a sterile PLASTIC inoculating loop to add one loopful of plasmid DNA (on instructor's table) to the "+plasmid" tube. (The plastic loops are calibrated so that when the DNA solution forms a bubble across the loop opening, its volume is 10 µL. The metal loops that we usually use are not calibrated.) Immerse the loopful of plasmid DNA *directly into* the cell suspension and spin the loop to mix the DNA with the cells. Discard the plastic loop in the biohazard bag.
9. Return the "+plasmid" tube to ice and allow both tubes to incubate on ice for 15 minutes.
10. While the tubes are on ice, label your media plates as follows (include your table number):
    a. Label one LB plate, LB+amp plate, and LB+amp+Xgal plate "+plasmid" or just "+". These are your experimental plates.
    b. Label the one LB plate, LB+amp, and LB+amp+Xgal plates as "–plasmid" or just "–". These are negative control plates.
11. Following the 15-minute incubation on ice, "heat shock" the cells. Bring your thermos to the 42°C H_2O bath. Remove both tubes *directly from the ice and immediately immerse* them in the water bath for 90 seconds. Gently agitate the tubes while immersed. Return both tubes to the ice for one or more minutes.
12. Use a new sterile tip on the micropipette to add 250 µL of luria broth (LB) to each tube. Gently tap the tubes with your finger to mix the LB with the cell suspension. Place the tubes in a test tube rack at room temperature for a 5-15 minute recovery. Discard the tip in the biohazard bag.
13. The transformation process is complete in the "+plasmid tube". Now you will transfer cells from each tube to the appropriate plates ("+plasmid" tube cells go on "+plasmid" plates, "–plasmid" tube cells go on "–plasmid" plates) and spread them using a new procedure described below.
14. Use a sterile tip on the micropipette to add 100 µL of cells from the "–plasmid" transformation tube to each "–plasmid" plate (reuse the same tip for all "–plasmid" plates, then discard in the Biohazard bag). IMMEDIATELY spread the cells over the surface of the plate using the procedure below:
    a. "Clam shell" (slightly open) the lids and carefully pour four sterile glass beads onto each plate.
    b. Use a back and forth shaking motion (not swirling around and around) to move the beads across the *entire* surface of the plate. This should gently and evenly spread the cell suspension all over the agar surface.
    c. When you finish spreading, allow the plates to rest for several minutes to allow the cell suspensions to become absorbed into the agar.
    d. To remove the glass beads, hold each plate vertically over the "used bead" container, clam shell the lower part of the plate, and tap out the glass beads into the container.
15. Use a new sterile tip on the micropipette and repeat Step 14 for the "+plasmid tube" and plates.
16. Tape the plates and invert on the tray to be incubated for 24 hours at 37°C.
17. All test tubes, $CaCl_2$ vials, and used beads go on the autoclave cart. All used micropipette tips and the one plastic loop should already be in the biohazard bag.

Week Two—Transformation Activity

1. Retrieve your table's plates from the last session.
2. Record your observations in the lab report using the data and analysis questions as a guide.
3. Place the plates on the autoclave cart when finished.

Name ______________________________

Exercise 9.2 Lab Report

Transformation Activity

1. **Observe** the colonies through the Petri plate lids. Do not open the plates. Fill in the remainder of the chart below.

LB −plasmid	Observed result:	Observed result:	LB +plasmid
LB+amp	Observed result:	Observed result:	LB+amp
LB+amp+Xgal	Observed result:	Observed result:	LB+amp+Xgal

2. If your observed results differed from your predictions, explain what you think may have occurred.

____

____

____

3. A) Look at your LB+amp+X-gal+plasmid plate: What are you *selecting* for on this plate?
   B) Look at your LB+amp+X-gal+plasmid plate: What is this plate differential for?

   A) _______________

   ____

   B) _______________

   ____

4. On the LB+amp+X-gal+plasmid plate, what does the phenotype of the transformed colonies tell you?

____

____

____

5. If you could only look at one plate to determine if transformation successfully occurred, which one plate would you choose? Why?

____

____

____

6. There are two different types of colonies on the LB+amp+Xgal+plasmid plate. Notice how there seems to be individual colonies of one color with a bunch of little satellite colonies clustered close by. Explain the difference between the two different types of colonies. Explain why they grow so close together in groupings.

____

____

____

7. Suppose you carefully picked a **satellite colony** surrounding the blue transformed colonies and transferred it to a new LB+AMP+ X-gal plate. Would you predict growth? Would the colony turn blue? Explain.

8. Consider the transformed *E. coli* on the LB+AMP+ X-gal plate. List the different enzymes that are produced when the plasmid is successfully incorporated into *E. coli* and the colony turns blue. Briefly describe each enzyme.

Biotechnology

Biotechnology is the use of organisms or cell components to produce useful products, or any technological process that uses organisms. Biotech applications impact our daily lives. Products produced as a result of biotechnology include medicines, vitamins, and food. Technological processes include the use of stem cells for potential use in humans. You may have heard of genetically modified organisms (GMO); this term GMO is commonly associated with food products. These food products have been modified genetically through the use of recombinant DNA technology, so that the plant produces an enzyme that benefits the plant. Such modifications have caused controversy and questions about food safety. **Recombinant DNA technology** involves the insertion or modification of a gene (DNA) to produce a desired protein. This technology requires the use of a **vector,** which refers to the DNA that is used to carry the gene of interest into the cell; this DNA will replicate independently from the chromosome once inserted into the organism. Vectors include plasmids, which can be taken in by bacterial cells through transformation, or viruses, which will infect cells with the desired gene. There are numerous biotechnological processes and applications. In this chapter, we will highlight some processes that are significant for healthcare.

DNA Technology

The first step in many biotech applications involves the isolation or extraction of DNA from a cell in order for scientists to manipulate or insert genes. DNA is extracted by disrupting cell walls and cell membranes of bacteria and plants and by disrupting cell membranes of animal cells. The extraction of DNA can be through a chemical or a physical process. Once DNA is isolated from other cell materials, it can then be manipulated. One way in which scientists manipulate DNA is amplification, through a process called **polymerase chain reaction** (PCR). The process of PCR is basically that of DNA replication in a test tube. In a matter of a couple of hours, a scientist can replicate DNA more than a billion times. This amplification process is extremely important for downstream applications such as DNA sequencing, DNA fingerprinting, and detection of pathogens. PCR is made possible due to the heat stable enzyme ***Taq* polymerase,** which is purified from *Thermus aquaticus*, a bacterium found in hot springs. The PCR process involves repeated cycles of raising and lowering the temperature of the samples in an instrument called a thermocycler. DNA with the gene of interest is added to a small test tube along with DNA primers, *Taq* polymerase, and nucleotides. The samples are then loaded into the thermocyler. The first step of PCR involves heating the samples to 94° C, which denatures DNA (changes it from double-stranded to single-stranded). Once the DNA is denatured, the sample is cooled to 60° C so that DNA primers can anneal or bond to their complementary base pairs. **Primers** are small DNA molecules that target the region you are interested in amplifying and provide a 3` end for the replication process. The samples are heated to 72° C, which is the optimal temperature for *Taq* polymerase. *Taq* can now bind to the primer on the DNA strand; it then adds nucleotides to

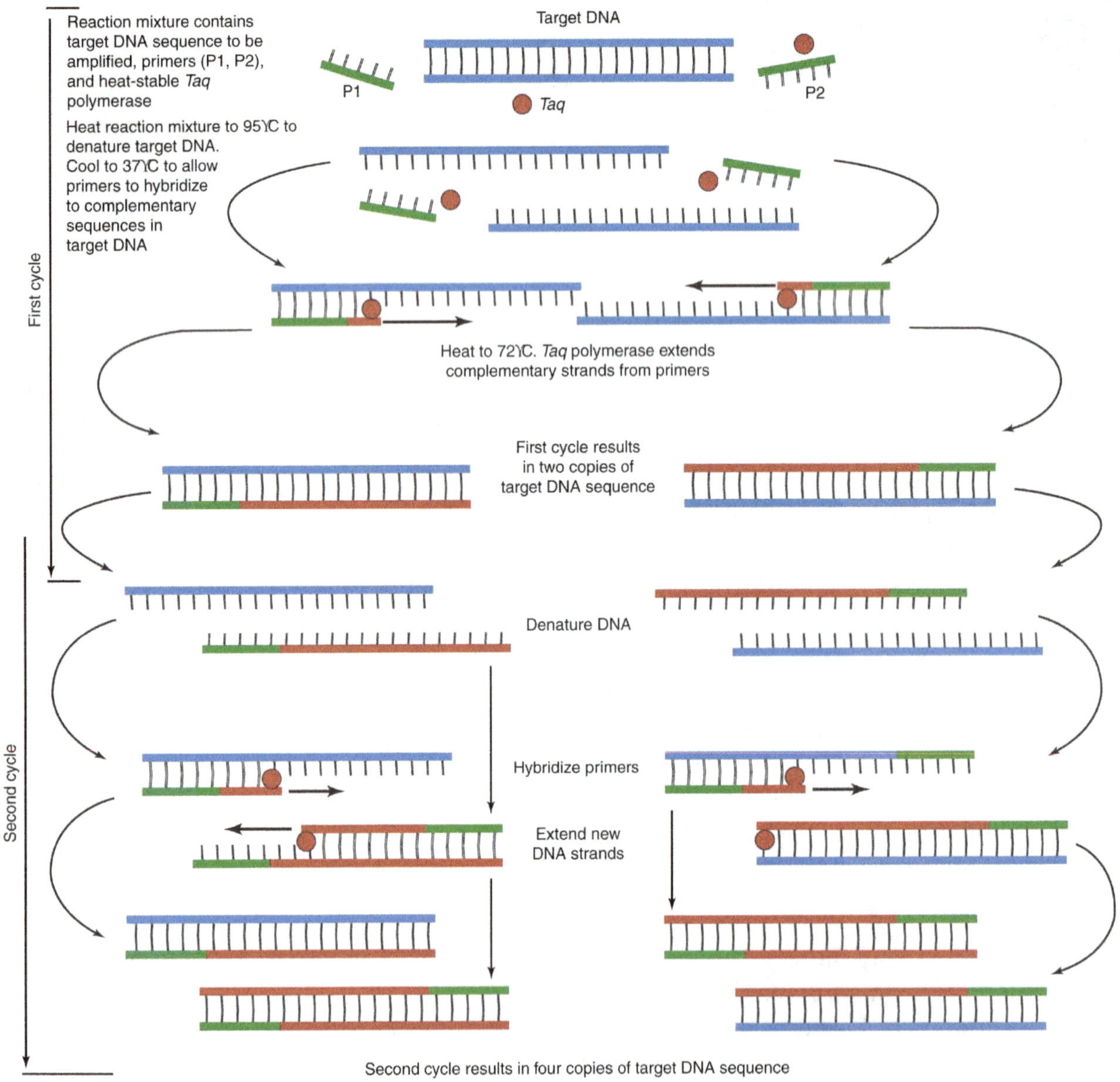

Fig. 10.1 DNA Amplification Using the Polymerase Chain Reaction

the 3` end of the primer and continues to read and add nucleotides to the template strand of the DNA. The cycle of heating and cooling is repeated 30–40 times, turning a few copies of your target gene into millions or even billions of copies of DNA for further analysis.

PCR and Medical Microbiology

DNA extractions and PCR is often used in the medical field to confirm diagnosis or to identify organisms causing disease. An example of one such disease is pertussis, commonly called whooping cough. Pertussis is caused by an aerobic, Gram-negative, bacillus, *Bordetella pertussis*. *Bordetella pertussis* is very difficult to isolate due to its extreme sensitivity to desiccation. Specimens are taken from nasopharyngeal aspirates and nasopharyngeal swabs and inoculated on a special media called *Bordet-Genou medium* (after the scientists who discovered the bacterium). Due to the difficulty in culturing the bacterium, hospitals in Michigan usually send patient specimens to the Michigan Department of Community health for PCR analysis. DNA is first extracted from the patient specimen and specific primers targeting regions specific to *B. pertussis* are used in a PCR. The presence of the genes specific for *B. pertussis* confirms that the patient does indeed have pertussis.

Pertussis was once almost completely eradicated from the United States due to the DTaP vaccine which protects against pertussis along with two other diseases. However, due to parental refusal to vaccinate their children for fear of side effects, there is an increase in the number of cases of this disease each year. Another reason that pertussis is on the rise is due to the vaccine's inability to give long-lasting immunity. Booster shots are required periodically, which most adults are unaware of. To find out more about the vaccination schedule for pertussis, visit www.cdc.gov.

PCR, Environmental Microbiology, and Bioterrorism

With the increase in homeland security following the September 11th attacks in New York and Washington D.C. and the 2001 anthrax attacks in the United States, government agencies developed protocols to screen certain government buildings against biological weapon attacks. One such system deploys air filters within existing air filters that monitor air quality. These air filters are taken to a lab where DNA is extracted from them. PCR is used to screen the samples for potential 'weaponizable' pathogens such as *Bacillus anthracis*, *Yersinia pestis*, and *Brucella* sp. Biological terrorism is an ongoing threat due to the ease of deploying such an agent (Table 10.1).

Restriction Enzymes

Polymerase Chain Reaction is an important technique, often used in combination with restriction digests in labs around the world. **Restriction digests** employ the use of **restriction enzymes**, which are enzymes that recognize 4–6 base-pair sequences on double-stranded DNA. The enzymes cut the DNA, which generates fragments of different lengths. These fragments are called **restriction fragment length polymorphisms** (RFLPs). Exposing DNA to restriction enzymes is called a restriction digest. Restriction enzymes such as *BamH1* (Table 10.1) cut double-stranded DNA and create what are called "sticky ends". The "sticky ends" can be used in recombinant DNA technology to insert a gene into a vector (Figure 10.2). *HaeIII*, on the other hand, creates blunt ends. Restriction digests that use enzymes such as *HaeIII* can be used in other downstream applications such as DNA fingerprinting.

Table 10.1 Example of 2 Restriction Enzymes. *BamH1* Creates "sticky ends" and *HaeIII* Creates "blunt ends"

Enzyme	Recognition Sequence
BamH1	↓ GGATCC CCTAGG ↑
HaeIII	↓ GGCC CCGG ↑

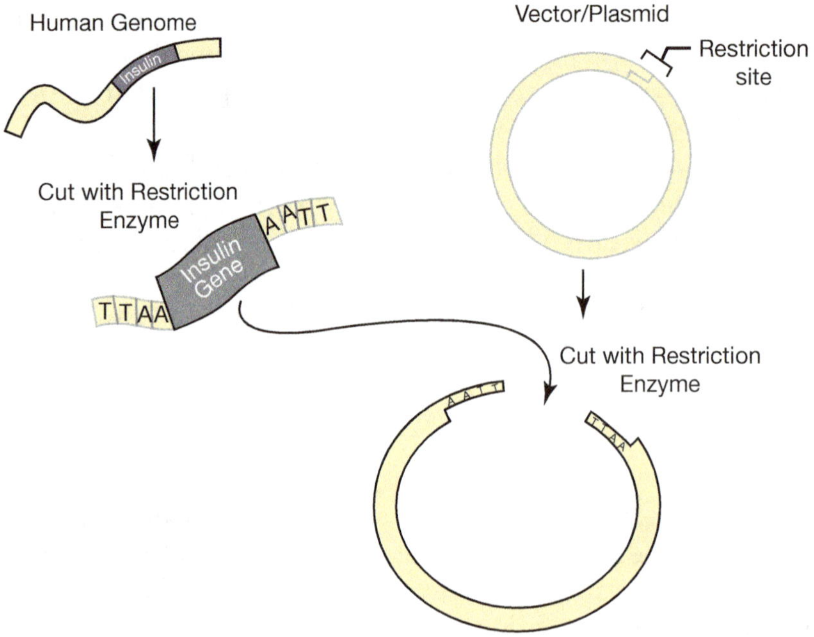

Fig. 10.2 Basic schematic of how recombinant DNA technology is used to create human insulin. Restriction enzymes cut the gene for human insulin out of the genome and the gene is inserted into a plasmid vector. Bacteria such as *E. coli* can then undergo transformation and produce the insulin protein.

DNA Fingerprinting

DNA fingerprinting is a DNA technology application involving PCR, restriction digests, and gel electrophoresis. DNA fingerprinting identifies individuals (human, bacteria, etc.) based on DNA fragment lengths. DNA fragments are generated using restriction enzymes. Each individual's genome varies, and therefore each individual would have their own unique "barcode" or fingerprint that identifies the individual. DNA RFLPs can be separated using Agarose gel electrophoresis.

Gel Electrophoresis

We use Tris-Acetate-EDTA (TAE) buffer for electrophoresis. The phosphate groups on the DNA fragments generated by the restriction digest will remain negatively charged. Therefore, the RFLPs will move toward the positive electrode when an electrical current is applied. The agarose gel forms a lattice or meshwork through which the DNA will travel. The fragments will migrate through this lattice according to their sizes. The smallest pieces of DNA will be able to move through the agarose matrix very fast, but the larger pieces will take longer because they will weave their way through the lattice. The DNA is not visible to the eye during electrophoresis, so a tracking dye is usually added to each sample in order to see the migration occurring. The tracking dye usually contains a sugar such as sucrose which is denser than the buffer, and this allows the DNA to sink to the bottom of the well. The tracking dye separates during the process into colored bands; some migrate rapidly and the others move very slowly. **Ethidium**

bromide is used in the electrophoresis buffer to allow the DNA fragments to be seen after the electrophoresis. Ethidium bromide intercalates between the bases in the DNA strand and fluoresces under UV light producing a visual image of the DNA band (Figure 10.3). When running a gel electrophoresis, scientists also run a **molecular base pair standard** (ladder) for fragment size comparison (Figure 10.3, shown as 100 bp ladder). Sample fragments can be compared to the molecular standard to estimate the sizes of each fragment.

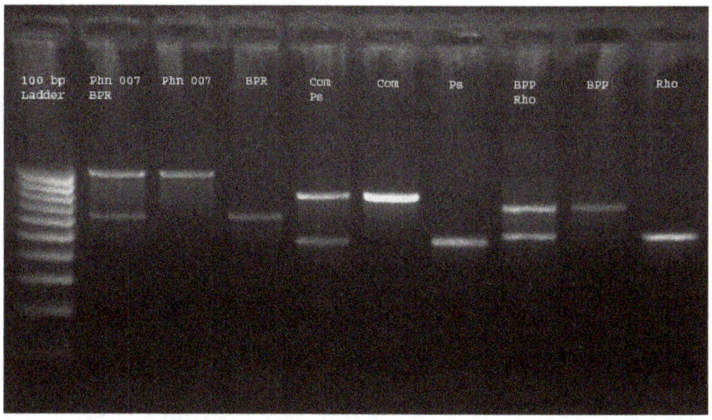

Fig. 10.3 Example picture of RFLPs after UV exposure. Lane one on the left is the molecular base pair standard; the band at the top is 1000 base pairs and each band below the top band will decrease by 100 base pairs

Courtesy of Author

When DNA from two different microorganisms are treated with the same restriction enzyme, the RFLP pattern may be compared creating a DNA "fingerprint". Because the two organisms will have a different sequence of bases in their DNA, the restriction enzyme will produce a different pattern of fragments. A comparison of the number and the sizes of the fragments can then be made. The more similar the patterns, the more closely related the organisms are expected to be. The same fingerprint means that the DNA is from the same source.

Restriction digests are useful in other applications such as crime scene analysis and paternity cases. In crime scene analysis, samples would be acquired from the victim as evidence from the victim (hair, skin cells, or semen, for example) and samples would also be taken from the suspects. DNA would then be extracted and restriction enzymes would be used to create fingerprints for each sample. Matching fingerprints would indicate the guilty party.

Probe Technologies

Probe technologies involve the use of fluorescence microscopy and single-stranded DNA or RNA probes (similar to a primer) with a fluorescent dye attached to the probe. The application of this technology is referred to as FISH (fluorescence in situ hybridization). FISH allows us to rapidly identify organisms by bypassing the need to cultivate organisms in the lab. An example as to how this is can be applied is with patients suspected to have tuberculosis. As you know, *Mycobacterium tuberculosis* has a very slow generation time of around 24 hours. Conventional biochemical testing can take a long time to generate results. However, using FISH, a sputum sample can be combined with probes specific for *Mycobacterium tuberculosis*. When samples are viewed with a fluorescent microscope, illumination of the probe would indicate a positive sample, and further testing for tuberculosis could then proceed (Figure 10.4).

Genetically Engineered Eukaryotes

The process used to genetically engineer eukaryotes is called **transfection**, which is a process very similar to transformation, except with the use of eukaryotic cells. Eukaryotic cells can also obtain new genes through transduction. **Transduction** is the process of infecting cells with a

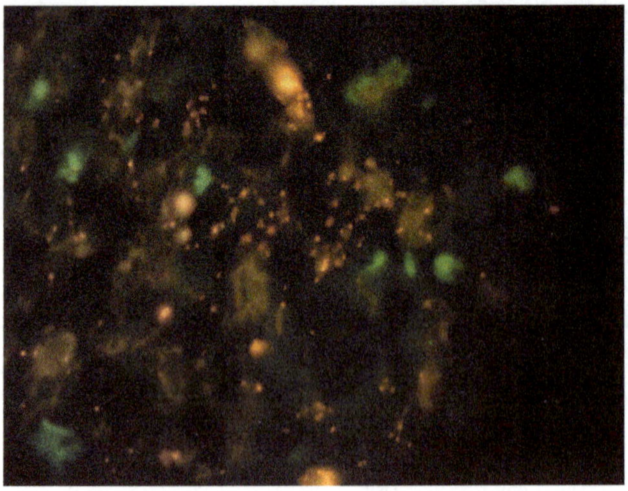

Fig. 10.4 Use of Two Different Probes to Stain Cells

virus. Plants such as corn, cotton, and potatoes have been genetically modified so that they produce BT toxin. BT toxin is produced by the bacterium *Bacillus thuringiensis*. The gene for BT toxin is inserted into the plant, so that the plant produces BT toxin; since BT toxin is toxic to insects, crop loss is avoided. This toxin is not toxic to humans, however, and therefore we can consume the toxin without coming to any harm. This same technology is responsible for crops such as Roundup TM ready crops, making the crop herbicide-resistant. Genetic modification technology is also being put to use to make crops more nutritious and also, potentially, to make edible vaccines.

Stem Cell Technology

Most of you have likely heard of stem cells, either in another class or in the media. There are different types of stem cells—Embryonic, Umbilical, Adult, and Induced Pluripotent Stem (iPS) cells, to name a few. These cells have great potential to advance medicine to a point where scientists can make many medical conditions and diseases a thing of the past (Figure 10.5). Embryonic stem cells have the greatest potential and can differentiate into any cell type. Adult stem cells have lesser potential, but they can be harvested from anyone. Induced pluripotent stem cells are adult cells that have been reprogramed genetically to "go back in time" and become a stem cell; however, these may pose a risk for use in humans. By learning biotechnological techniques such as those discussed in this chapter, you can apply the methods to all fields in biology. PCR is one of the most common tools that biologists use in laboratories today.

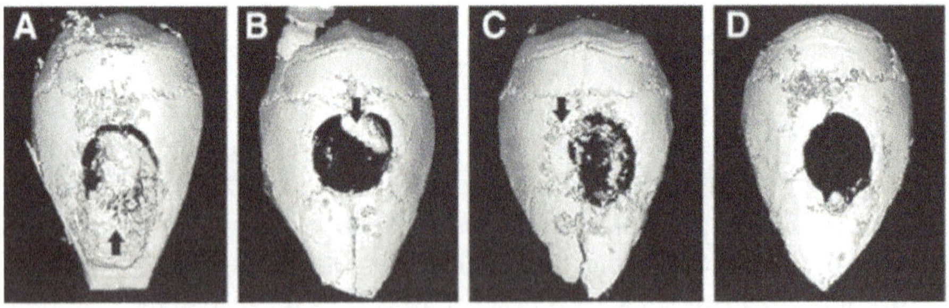

Fig. 10.5 Figures show rat skulls with a man-made defect. Slides A–C had human embryonic stem cells differentiated into osteoblast cells seeded on a cell scaffold and transplanted into the skull. Slide D is a control with no stem cells, only the scaffold. Slides A–C were shown to have human bone developing and repairing the defect. How do we know this is human bone and not rat bone? DNA is extracted from the bone tissue and human specific primers are used in PCR; the presence of human genes confirms that human bone tissue was formed

Courtesy of author

Lab Exercise 10.1

Genetics: DNA Isolation and DNA Technology

Student Objectives

1. Observe the physical appearance of DNA strands by extracting DNA from bacterial cells.
2. Describe the process of DNA extraction from bacterial cells.
3. Understand the importance of DNA technology for both environmental and medical microbiologists.
4. Understand applications for DNA technology.

Name ____________________

Pre-Lab Exercise 10.1

1. Define the purpose of the laboratory exercise.

2. What is the function of ethanol in the DNA isolation procedure?

3. Why are proteases added during certain DNA extractions?

4. Describe PCR.

5. How can DNA technologies benefit Homeland Security?

6. How can DNA technologies benefit the medical field?

7. Why is PCR needed to help diagnose whooping cough?

8. Why is pertussis on the rise in the United States?

INTRODUCTION

In many genetic studies, the first step is the extraction of DNA from a cell or a group of cells. Since prokaryotic cells can be found almost everywhere on earth, DNA can be extracted from many different substrates including water, feces, soil, air (captured on a filter), and even rocks, to name a few. Yet only less than 1% of prokaryotes have been cultivated or grown in the lab. DNA studies can help microbiologists determine what organisms are in an environment, based on genetic profiles made from sequencing DNA. Once DNA is extracted from a cell, it can then be used in numerous applications such as Polymerase chain reaction, Gel electrophoresis, or recombinant DNA technologies. To extract DNA from prokaryotes, we disrupt the cells by chemical or mechanical means and then collect the cell-free DNA. It can be difficult to extract DNA from some Gram-positive organisms due to their thick peptidoglycan layer, and hence a pumice-like substance is often added to the extraction vesicle to help mechanically disrupt the cell wall. Once the cell wall is disturbed, chemical additives can act on the cell membrane. In some applications, proteases are added during the process to break down proteins that may inhibit other applications such as PCR. In this lab exercise we will use a chemical disruption technique to isolate DNA from the Gram-negative *E. coli*.

It is important to understand what is happening at each step of the DNA isolation process. In the first step, a detergent—Sodium lauryl sulfate—is added to break down the phospholipid bilayer and proteins that form the plasma membranes by interacting with the phospholipids, and disrupting the hydrophobic attraction that holds the bilayer together. Since *E. coli* is a prokaryotic cell, there is no membrane-bound nucleus; once the phospholipid membranes are disrupted, the cell's contents, including its DNA, is now in solution. How do we get DNA out of solution? The addition of alcohol is the final step in isolating DNA from the rest of the cell solution. DNA clumps together and separates from the mixture when cold ethanol is added. As DNA begins to precipitate, millions of DNA fibers will clump together and become visible as a stringy white precipitate.

DNA Isolation

> **Materials** (two per table): *E. coli* pellet, ice bucket with ice, inoculating needle, 4 ml of 95% ethanol tubes, tube of detergent, vortex

Procedure:

1. Obtain a tube with a bacterial pellet.
2. Pour off the supernatant fluid into the waste beaker with disinfectant.
3. Slowly add 3 ml of detergent to the tube containing the bacteria.
4. Mix gently (for 20–30 seconds) using a vortex to re-suspend the bacterial pellet.
5. GENTLY pour the content of a 4 ml tube of cold 95% ethanol by letting it slowly run down the side of the tube. The liquids must be kept cold or else ethanol will dissolve in the water and not form layers. Do NOT mix the alcohol and the turbid aqueous layer. You should see two distinct layers in your tube. Ethanol precipitates DNA at the interface of the two layers.
6. DNA should immediately begin to precipitate out of the solution over the next 5–10 minutes; watch for the precipitation of DNA at the alcohol interface.
7. Collect DNA strands that appear at the interface by winding them gently onto a hook made with an inoculating needle. Wind in one direction only and slowly stir the ethanol and the water in a wide circle.
8. Lift the DNA strands out of the tube and observe them. Record your observations in the lab report.

Name _____

Exercise 10.1 Lab Report

1. Describe the appearance of the *E. coli* cell pellet.
   _____
   _____

2. Record your observations of the DNA.
   _____
   _____

3. Is *E. coli* Gram-positive or Gram-negative? What cell morphology does *E. coli* have?
   _____
   _____

4. What should be done to extract DNA from Gram-positive bacteria? Why is this necessary?
   _____
   _____

You will have to do some research to answer the following questions.

5. Why is *Bacillus anthracis* a suitable organism for biological warfare?
   _____
   _____

Chapter 10 Biotechnology

8. Describe the signs and symptoms of the disease caused by *Yersinia pestis*.

9. Describe Brucellosis. What is the natural source of *Brucella sp.?*

Lab Exercise 10.2

Agarose Gel Electrophoresis of DNA

Student Objectives

1. Learn basic principles and techniques of agarose gel electrophoresis and restriction digestion.
2. Learn why DNA fragment size and migration is important for DNA fingerprinting.
3. Understand the basics of polymerase chain reaction.

Name ______________________

Pre-Lab Exercise 10.2

1. Define the purpose of the laboratory experiment.

2. Differentiate between a gene and an RFLP.

3. Define the term 'restriction endonuclease' and explain why these enzymes are useful in molecular biology.

4. Explain why a DNA fragment that is 500 base pairs long will move through a gel faster than a 1000 base pair DNA fragment.

5. Why is a molecular weight standard used in Gel electrophoresis?

6. Why is a positive control often used in Gel electrophoresis?

7. What would happen if you forgot to use Ethidium Bromide in your running buffer? Why would this occur?

____

____

____

INTRODUCTION

In this exercise, you will perform gel electrophoresis on DNA from bacteria isolated from patients suspected of being infected with *E. coli* O157:H7. DNA collected from different bacteria has been cut with the same restriction enzyme; therefore the RFLPs may be compared by gel electrophoresis. A molecular weight standard will be run in Lane 1. The standard provides a reference for comparing the number of base pairs that each fragment contains. We have also included an important sample called a **positive control**. Sample #2 contains DNA from a pure culture of *E. coli* O157:H7 and shows the pattern of bands you would expect to see from *E. coli* O157:H7 digested with this restriction enzyme.

Pouring an Agarose Gel Activity

Materials (work in groups of 4 students): Agarose, gel casting tray/comb, and pre-measured molten agarose

Procedure: (Note: Depending on your instructor your gel may already be prepared)

1. Place black rubber ends on the casting tray. Place the comb in the slot on one end.
2. Pour warm agarose solution slowly into the tray to prevent bubbles from forming and immediately rinse the flasks. Allow the gel to cool undisturbed.
3. Let the gel harden (for about 20 minutes), carefully remove comb, and place the gel in the electrophoretic chamber.

Loading the Agarose Gel Activity

Materials (per table): Agarose gels, TAE buffer, electrophoresis apparatus, micropipetters and micropipette, pre-digested DNA samples

Procedure:

1. Place the tray + gel into the chamber with the wells at the BLACK (negative electrode) end of the chamber.
2. Be sure that both ends of the electrophoresis chamber are filled with TAE buffer. Add additional TAE buffer so that the buffer just covers the wells of your gel.
3. Carefully rub your finger over the wells to remove trapped air bubbles.

Proceed as described for loading the gel. Each group will have one gel and each student should load at least one lane. Make sure you record what samples were loaded into which lanes.

1. Set the pipette to 10 µl. Load a pipette tip onto the pipette. Press the pipette plunger and insert it into the first sample. Slowly draw liquid into the pipette tip by releasing the pipette plunger.
2. Locate the well you are about to load.
3. Insert the pipette tip into the buffer above the well you are loading. Insert the tip into the well and slowly press the plunger to expel the sample. ***Your instructor will likely demonstrate this technique before you load your gel.***
4. Remove the tip and place in the biohazard bag. Load a new tip and repeat steps for the remaining samples. Make sure you load each sample into its own well.

Running the Gel

1. Once your gel has been loaded, carefully replace the lid of the electrophoresis chamber without disturbing the chamber. (Red to Red; Black to Black).
2. **NOTE: Be sure the power supply is off!**
3. Let your instructor know that your group has finished loading the gel. Turn the power supply on (75V for 40 minutes) and start the current flow through the chamber.
4. After the gel has finished running, remove the cover and observe your gel within the chamber. (You will notice different colors that represent DNA in your sample).
5. Let the instructor know that your gel is ready for imaging.
6. Sketch the band patterns on the report page.

Name ______________________________

Exercise 10.2: Lab Report

Results:

Make a drawing of the results of the electrophoresis run. Draw your bands. **Label** the sample number of each lane/well. Lane 1 is the 100 base pair standard. Lane 2 is a positive control for *E. coli* O157:H7.

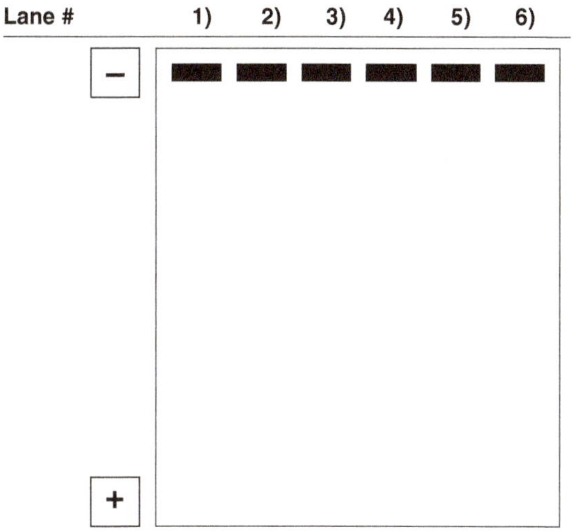

1. Which **patient or patients** (do not list a lane) were infected with *E. coli* O157:H7? Explain how you determined this.

   ____

   ____

   ____

2. How many base pairs long are the *E. coli* O157:H7 DNA fragments that were separated out on the gel? (There are more than one.)

   ____

3. How many base pairs long was the original *E. coli* O157:H7 gene before it was cut using a restriction enzyme? Explain how you determined your answer.

   ____

4. Consider a random mutation in a restriction site for one of the patient's samples that you ran today. Would the banding pattern match the positive control sample? **Explain** why or why not.

____

____

____

____

5. **Describe** the set-up of the electrophoresis chamber in terms of the orientation between the wells and the electrodes. Why does the DNA migrate toward the positive electrode? Explain the relationship between the size of a fragment and how far they migrate on the gel.

____

____

____

Antimicrobial Drugs

Chemotherapy is the use of chemicals for treatment of disease. **Chemotherapeutic agents** are chemicals used within the body for therapeutic purposes. Chemicals used against some kind of microorganism are commonly referred to as antimicrobials. An **antibiotic** is a naturally made chemical produced by one organism used to inhibit another organism. Antibiotics are used as a chemotherapeutic agent to control bacteria growth in animals.

There are three basic types of antimicrobials, classified according to their development: antibiotics (drugs produced by bacteria and fungi, like penicillin), semi-synthetics (naturally produced antibiotics which are modified in the laboratory to give the drug more action), and synthetics (drugs produced entirely in the laboratory, like sulfonamides).

Antimicrobials can be classified according to their activity: antibacterials, which work on bacteria; antivirals, which work on viruses; antifungals, which work on fungi; antiprotozoals, which work on parasitic protozoa; antimalarials, which work on malaria; and antihelminths, which work on parasitic worm infections.

Some antimicrobials do not actually kill the agent causing disease. The suffix "*-statis*" means to inhibit. Bacteriostatic means to inhibit the growth of bacteria; thereby, lowering their numbers within the host. The host's own defenses (i.e., non-specific host defenses and adaptive immune system) must help to destroy the pathogen. Antimicrobials that kill the agent are differentiated by the use of the suffix "*-cidal*" meaning to kill. Bactericidal for example would mean to kill bacteria directly.

A major host problem associated with the use of antibiotics is the loss of normal microorganisms from the urogenital and gastrointestinal tracts. That loss may occur because an antibiotic is not selective in its action or because the physician may have chosen a broad-spectrum drug to bring an infection under control. A **broad-spectrum** antibiotic is one that is able to control a wide range of microbes. An example of this would be tetracycline, which can be used in the treatment of infections caused by Gram-positive and Gram-negative bacteria, aerobic and anaerobic bacteria, spirochetes, rickettsias, mycoplasmas, chlamydia, and some protozoa. **Narrow-spectrum** antibiotics are those that are able to control a narrow range of microorganisms. An example would be polymyxin B, which is effective against certain Gram-negative bacteria. **Extended-spectrum** antibiotics are those that act on either Gram-positive or Gram-negative bacteria but have a limited effect on the other group. When nonspecific, broad-spectrum drugs are given orally, they may "accidentally" kill *E. coli* and other beneficial bacteria that normally live in the body. If they are lost, the normal microbiota becomes imbalanced and secondary infections such as antibiotic associated diarrhea caused by *Clostridium difficile* or yeast infections caused by the loss of *Lactobacillus* sp. in the vagina.

Six Characteristics of an Ideal Antimicrobial Drug

When research is conducted on a new antibiotic, investigators set their standards as high as possible. They look for the optimum antibiotic that will be most effective for the longest period. The ideal antibiotic:

- **Must demonstrate *selective toxicity*.** This means it must be toxic to the pathogen (i.e., destroy it) without harming the body tissues of the host.
- **Should be *microcidal*.** This means it should be capable of killing the targeted microbe.
- **Should be *stable*.** This means it should be soluble in body fluids, but it should not be broken down too rapidly in the body or eliminated too quickly from body (i.e., it should have a long half-life). Drug resistance is an ever-increasing problem. Antimicrobials should be structurally stable enough that it is difficult for the microbe to develop resistance to the agent.
- **Must be *complementary to host defenses*.** This means the antimicrobial should stimulate or enhance the immune response of the body. It should not suppress immunity in any way. It should not stimulate hypersensitivity reactions in most patients.
- **Should have *extensive tissue distribution*.** This means that depending on how the antimicrobial is administered to the patient (oral, intramuscular, intravenous, etc.) that the drug should be able to become well distributed throughout the body tissues, including entry across the blood-brain barrier as necessary.
- **Should *remain active in the presence of organic compounds*.** This means the antimicrobial must not become inactive due to the presence of organic compounds such as proteins or serum. If the antimicrobial is going to become bound to proteins, the active site on the agent must remain available for binding to the microbe. (If the active site were hidden, the antimicrobial would not be able to bind with the microbe, and would be ineffective.)

These are ideal characteristic that researchers strive for, with the understanding that they may never find or produce all of them in a single antibiotic.

Mechanisms of Action of Antibacterial Agents

This diagram illustrates the basic events in the life of a bacterial cell. The DNA controls transcription, which controls translation. The proteins created during translation are often enzymes, which in turn control the metabolism of the cell. The cell makes energy during catabolism, and makes other cell structures such cell membrane, cell wall, nucleic acids and proteins as a result of anabolism.

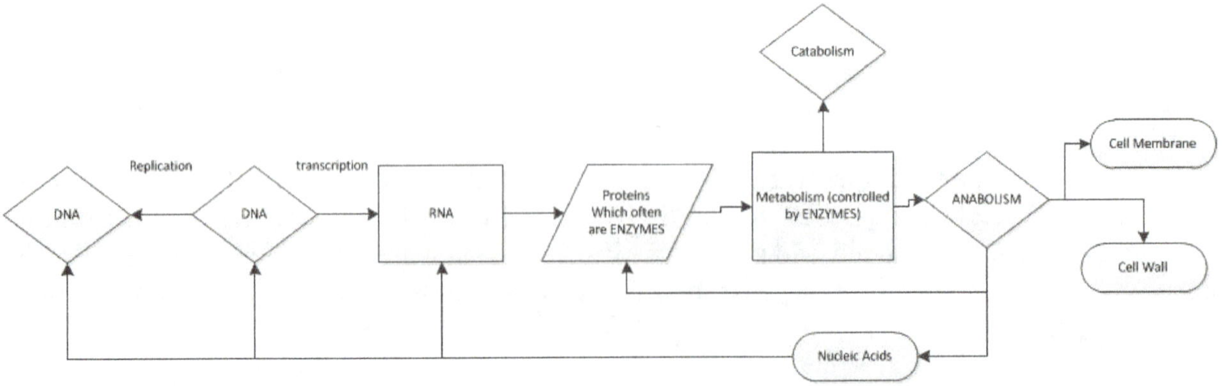

Fig. 11.1 Cycle of Events in the Life of a Bacterial Cell
Courtesy of A. Swarthout

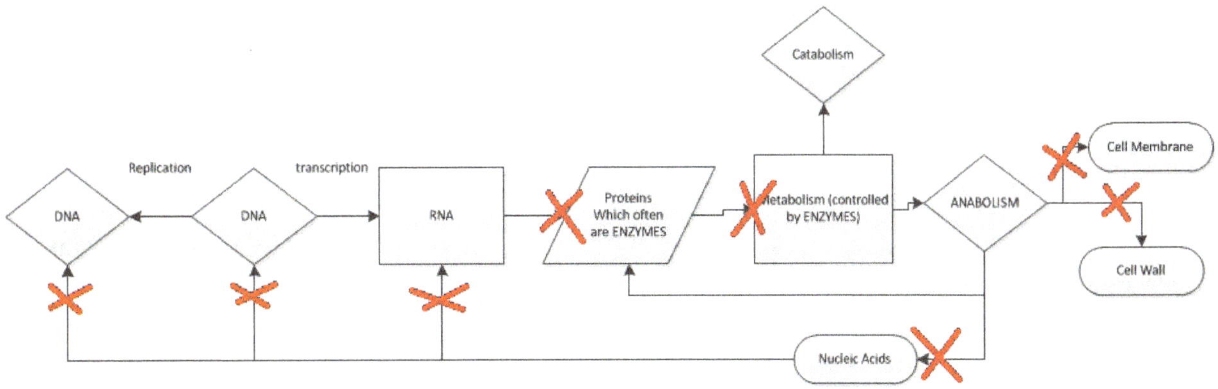

Fig. 11.2 Potential Targets for Disruption of the Cell Cycle
Courtesy of A. Swarthout

In order for the bacterial cell to live and reproduce, it must carry out all of the functions in the diagram above. But, if that bacterial cell is living inside of a human and making that human sick, we don't want the cell to be able to continue the cycle. So, we need to find points in the cycle where we can "screw up" the bacterial cell without harming the human host in the process. The diagram below indicates points in the cycle that we target with antibiotics.

1. **Interfere with nucleic acid synthesis**. Depending on the drug and the microbe, this can be interfering with RNA synthesis or DNA synthesis.

   Fluoroquinolones and quinolones are drugs that bind to gyrase, and enzyme that prevents DNA from being damaged by over-winding during DNA replication. With the drugs bound to gyrase, the enzyme cannot bind to the DNA and protect from over-winding. The DNA strands then tangle and break and the cell cannot complete DNA replication.

   Some drugs work by binding to RNA polymerase, which you should remember, is the enzyme that carries out transcription. Without transcription, there is no translation, and without translation no new proteins are made in the cell. That means that old, worn out proteins are not replaced and different metabolic pathways cannot be turned on when necessary. Rifampin is one drug that works in this way.
2. **Interfere with protein synthesis.** Most drugs that work by inhibiting protein synthesis work by binding to the ribosome. They may bind to either the 30S (small) subunit or the 50S (large) subunit. When tetracycline or aminoglycoside drugs bind to the ribosome, their presence prevents the proper binding of mRNA and tRNA to the ribosome, blocking translation.

   Most drugs that interfere with protein synthesis maintain selective toxicity by taking advantage of the differences between eukaryotic and prokaryotic ribosomes. Our ribosomes are 80S ribosomes and consist of a 60S and 40S subunit, while the 30S and 50S subunit of the bacterial ribosomes make up 70S ribosomes.

Endosymbiotic Theory is a theory that human mitochondria are actually descended from bacterial cells! According to the theory, an ancient eukaryote endocytosed an ancient bacterium and they established a symbiotic relationship that evolved into our current cell structure.

In fact the ribosomes found inside of mitochondria are 70S, like those of bacteria, and drugs that affect bacterial ribosomes will have an adverse effect on mitochondrial ribosomes as well.

3. **Interfere with metabolic pathways by interfering with enzymes involved in these metabolic pathways.** Sulfonamides are drugs that are also competetive inhibitors. They bind to the first enzyme in the folic acid synthesis pathway, stopping the conversion of PABA to dihydrofolic acid, and thereby stopping the production of folic acid. Folic acid is necessary to make DNA and RNA.

   Trimethoprim is another anti-metabolic agent, and it binds to the second enzyme in the folic acid synthesis pathway and prevents the conversion of dihydrofolate into tetrahydrofolate, again stopping the production of folic acid.

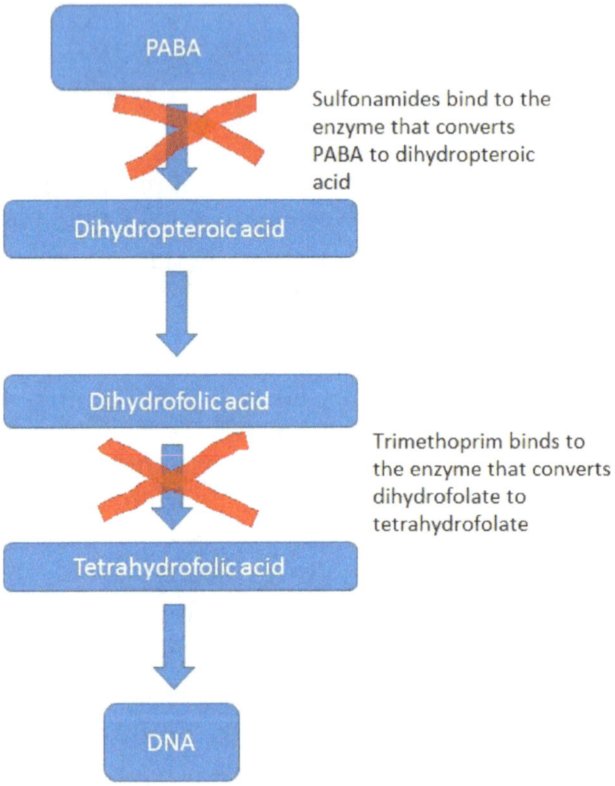

Fig. 11.3 Mode of Action of Sulfonamides and Trimethoprim
Courtesy of A. Swarthout

4. **Disrupt the integrity of the cell membrane.** There aren't many antibacterial drugs that work this way, in large part because the bacterial cell membrane is similar enough to ours that selective toxicity becomes an issue. Polymyxin antibiotics work this way, but are toxic to human kidneys, and are therefore used as topical agents.
5. **Interfere with cell wall synthesis.** As far as selective toxicity goes, this is a great way to target bacterial cells because we do not have peptidoglycan in our cells.

Penicillins and cephalosporins are both beta-lactam drugs. They both have a chemical structure, the **beta-lactam ring**, which is similar to the substrate that forms the pentaglycine crossbridges of peptioglycan. They bind competitively to the enzyme that builds the crossbridges and prevent crossbridge formation. This makes the peptidoglycan weak and the cell unable to resist osmotic pressure, ultimately causing lysis of the cell.

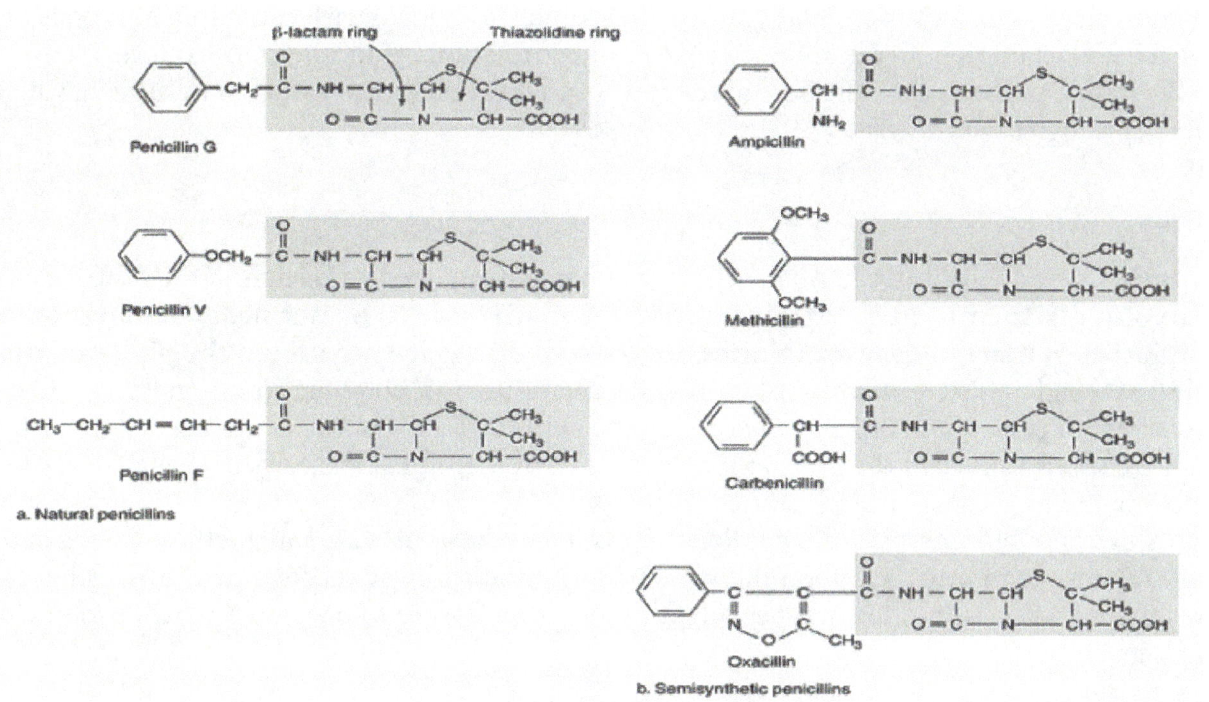

Fig. 11.4 Different Types of Penicillins
© Kendall Hunt Publishing

Penicillins

The side chains on penicillins influence the spectrum of activity for the antibiotics. Penicillins are divided into the following groups according to their modification and activity:

- Natural penicillins—such as penicillin G & penicillin V
- Penicillinase-resistant penicillins (PRPs)—such as methicillin, cloxacillin, nafcillin, oxacillin and dicloxacillin
- Broad-Spectrum penicillins—such as ampicillin, amoxicillin, ampillicin + sulbactam in combination, and amoxicillin + clavulanate potassium in combination
- Extended-Spectrum penicillins—such as mezlocillin, ticarcillin and piperacillin

Cephalosporins

First, second, and third generations of cephalosporins have been created by modifying the side groups of the beta lactam ring. These changes influence the spectrum of activity for the antibiotics. Cephalosporins are divided into the following groups according to their modification and activity:

- First generation cephalosporins—such as cephalexin, cefadroxil, cefazolin, cephalothin, cephapirin, and cephradine
- Second generation cephalosporins—such as cefaclor, cefamandole, cefmetazole, cefonicid, cefoteten, cefoxitin, cefprozil, cefurozime, and loracarbef
  Third generation cephalosporins—such as cefdinir, cefepime, cefixime, Cefoperazone, cefotaxime, cefpodoxime, ceftazidime, cefibutin, ceftizoxime, ceftriazone, and moxalactam

Vancomycin also disrupts peptidoglycan synthesis, but it is not a beta-lactam drug as it does not have a beta lactam ring, but it inhibits cross-bridge formation by another mechanism.

Bacitracin prevents the transport of the NAG and NAM subunits from out of the cell, which halts peptidoglycan synthesis.

Synergism

Sometimes one antibiotic is not enough to successfully treat a patient. With the rise in antibiotic resistant bacterial strains physicians may result to prescribe a combination of antibiotics. When multiple antibiotics are prescribed we are relying on the phenomenon called synergism. Synergism is when two drugs are used in combination to increase the effectiveness of the treatment while decreasing the concentrations of the drugs needed. By using smaller concentrations of the drugs, there is less risk of harmful side effects to the patient. Also, the overall effectiveness of the treatment is increased!

An example of synergy is the combination of trimethoprim with sulfamethoxazole, a sulfa drug. Because both drugs act upon the same pathway, the pathway receives two "hits." This ensures the pathway is shut down, especially if the bacterium were to develop resistance to one of the drugs.

A Special Case: Treatment of Infections Caused by *Mycobacterium Tuberculosis*

Mycobacterium tuberculosis is a Gram-positive bacillus that causes the disease tuberculosis. This organism is referred to as an acid fast bacterium and has several characteristics that make it intrinsically resistant to antibacterial drugs:

1. Mycolic acids in the cell wall make it waxy, enabling it repel drugs, disinfectants and antiseptics
2. It has a very long generation time
3. It mutates

Because of these factors, treatment of TB usually takes at least 6 months to treat and requires the patient to take a combination of two drugs at a time. TB drugs are divided into 1st line and 2nd line drugs. The 1st line drugs are the more effective, less toxic drugs. Examples include rifampin, isonizid, ethambutol and pyrazinamide. If the microbe develops resistance to the first line drugs, the less effective, more toxic 2nd line drugs must be used. Many TB patients are required to undergo **directly observed therapy**, whereby they must be observed taking their treatment by a qualified professional. This is due in part to the length of time antibiotic therapy takes. Many people stop taking their antibiotic before the infection is cleared from the system, leaving the more resistant microbes alive in the body to continue to reproduce.

Drug Resistance, Superinfections, and Drug Side Effects

Antibiotics act as agents of natural selection. The bacterial cells that have the genetic ability to withstand the effects of the drug will survive and become the "grandparents" of a new population of drug-resistant bacteria. Since one essential life characteristic is genetic change, the prevention of drug resistance is almost impossible. Worldwide, the incidence of antibiotic drug resistance is increasing at an alarming rate. Unchecked, some pathogenic bacteria will ultimately become uncontrollable by antibiotics.

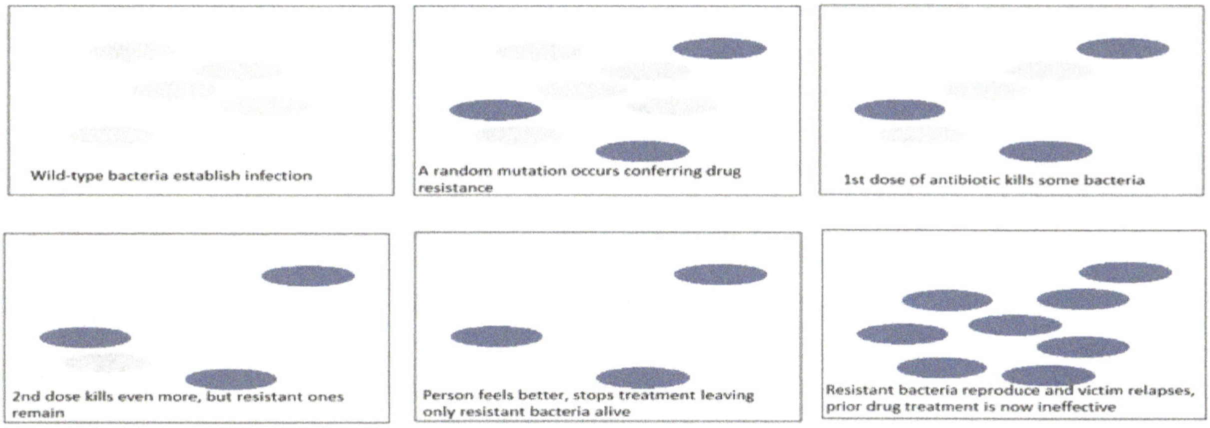

Fig. 11.5 Emergence of Drug Resistance
Courtesy of A. Swarthout

Bacteria can develop drug resistance through mutation or through gene transfer. Remember that mutations are random events that are statistically more likely to be harmful than helpful. However, because bacteria replicate and mutate so quickly, it is likely that a few beneficial mutations will pop up in the population. These spontaneous mutations "just happen"; *the bacteria are not mutating on purpose in order to become resistant.* In most cases, the new, mutated bacteria would not fare any better or worse than the rest of their population. But when pressure in the form of antibiotics is applied to the population, and some cells have developed resistance, the non-mutated cells are killed off leaving more space and nutrients for the resistant bacteria to grow. As these new bacteria grow, they can accumulate more mutations until they become fully resistant to an antibiotic, resistant to multiple antibiotics or both.

Bacteria can also share antibiotic resistance genes via transduction, transformation and conjugation. Plasmids that code for antibiotic resistance are termed **R (resistance) plasmids**. Gene transfer does not have to occur between bacteria of the same type, it can occur between two different genera of bacteria.

If the normal microbiota is wiped out (due to such actions as use of antibiotics) the pH rises, which allows the yeast to become dominant. This overgrowth of yeast is known as a **superinfection.** A superinfection occurs when opportunistic microbiota are allowed to increase in numbers uncontrollably, leading to an infection. There are individuals who do experience harm when taking antibiotics to control an infection. This kind of harm is referred to as a side effect and may include anemia, kidney toxicity, allergic reactions, and tooth discoloration.

The antigenic properties of antibiotics are also important. These drugs should not be able to stimulate the immune system of the host. If that should occur, the patient could: (a) show an allergic response to the presence of the drug by producing antibodies against the antibiotic; or (b) produce antibodies that would destroy the drug.

Mechanisms of Drug Resistance

Resistance to antibiotics is due to the added capabilities of the resistant microbes. In order to resist the effects of antibiotics, bacteria can:

- produce enzymes that actively degrade antibiotics. Beta-lactamase degrades beta-lactam drugs. Penicillinase is beta-lactamase that specifically degrades penicillin. The production of beta-lactamase occurs in many penicillin and methicillin resistant strains of *Staphylococcus.*

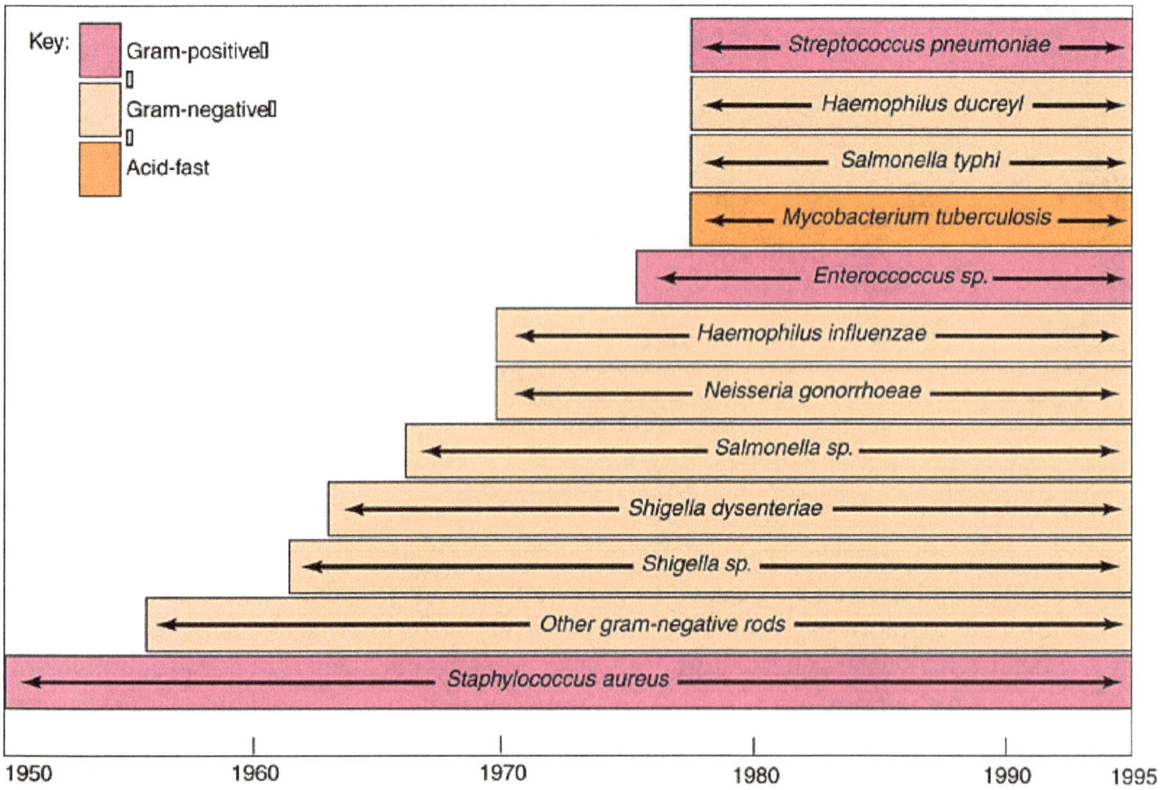

Fig. 11.6 Drug Resistant Microbes

- develop efflux pumps that specifically bind to the antibiotic once it has entered the cell and then pump it out
- alter the target of drug, such as changing the shape of the ribosome slightly, causing protein-synthesis inhibitors to stop working
- altering their metabolic pathways, for example, some bacteria will stop producing their own folic acid when exposed to sulfa drugs, they will absorb it from the environment instead
- more tightly regulate what comes into the cell via porins in the cell wall
- form biofilms to protect themselves from coming in contact with the antibiotic

Multiple drug resistance (MDR) occurs when a microbe is resistant to 2 or more drugs. Cross-resistance occurs when resistance to one drug results in resistance to other, similar drugs. For example, bacteria that are resistant to penicillin may also be resistant to cephalosporin because they both have the same mode of action.

Slowing the Emergence of Antibiotic Resistance

In the fight against bacteria and antibiotic resistance, it often seems as if we simply can't win. But there are things that we can and should do to help prevent the emergence of further antibiotic resistance:

- always finish all antibacterial drugs as prescribed
- don't take antibacterial drugs for viral infections (cold & flu)

- limit the use of antibiotics in cattle, chicken, pigs and other animals sold for human consumption. This is a major source of antibiotic resistance.
- keep patients with infections caused by antibiotic resistant microbes from coming in contact with other patients with other resistant microbes and use proper PPE, Standard and Universal Precautions when dealing with these patients
- Educate ourselves and others about the dangers of the over-use of antibiotics and antimicrobial products!

Mechanisms of Action of Antiviral Drugs

Unfortunately, the treatment choice for viral infections is much more limited than for bacterial infections, because viruses have fewer targets than bacteria. A major complication is the fact that they don't carry out any metabolism on their own, and most drugs work by targeting and interfering with some type of metabolic process. A virus that is not active has virtually no targets, and one that is active is taking advantage of our cellular metabolism, meaning we have to essentially target ourselves, and we must be very careful in how we do this.

A few viral targets:

1. **Prevent viral uncoating:** think back to learning viral replication. The virus must "uncoat" or separate the nucleic acid from the capsid so the nucleic acid has access to our cell's organelles. If the virus stays in the coat, it can't access our cellular machinery to take over. A few drugs, rimantidine and amantadine (used to treat Influenza A) work this way.
2. **Interfere with nucleic acid synthesis:** nucleoside and nucleotide analogs are substances that "look like" the nucleic acid bases adenine, guanine, cytosine, thymine and uracil, but are fakes. They look enough like the real thing that the enzymes involved DNA replication and transcription use them in place of the real thing, but the DNA and RNA created using these analogs are non-functional. These are selectively toxic because viruses do not "proofread" their genetic material as well as eukaryotic cells, which would catch these fakes and remove them.

   The HIV treatment AZT and the herpes treatments acyclovir and valcyclovir are examples of nucleoside drugs.
3. **Inhibit the action of Protease:** Viruses often make long, non-functional polypeptides during translation that are later cut or cleaved into the smaller, functional proteins by the enzyme protease. Interfering with the action of protease stops the production of functional viral proteins. Protease inhibitors are used to treat HIV.

Antifungal Drugs

Because fungi are eukaryotes, many of their cell structures and metabolic pathways are similar enough to ours to cause problems with selective toxicity. Most antifungal drugs work by interfering with the integrity of the fungal cell membrane. Fungal cell membranes contain ergosterol, a sterol that is similar to cholesterol in structure and function. Allylamines and azoles inhibit ergosterol synthesis which disrupts cell membrane integrity. Because humans do not make ergosterol allylamines and azoles are harmless to humans. Polyenes and Amphtericin B work by binding to the ergosterol in the membrane in order to disrupt membrane integrity. Due to ergosterol's similarity to cholesterol these drugs can bind to cholesterol (found in our cell membranes) and can be harmful to us in high concentrations.

Antiprotozoal and Antihelminthic Drugs

Antiprotozoal and Antihelminthic drugs are very limited categories of antimicrobial drugs. Treatment of protozoal and helminth infections is complicated by the fact that these organisms are also eukaryotes, and many of them have complex life cycles with different targets at different stages. A further issue is the fact that most of these types of infections occur in under-developed parts of the world where people do not have money to pay for the drugs or the development of drugs.

Testing the Efficacy of Antibacterial Drugs

One way to test whether an antibacterial drug will be effective against a certain of bacterium is to conduct a Kirby-Bauer Test. Disks impregnated with different antibiotics are placed on a Petri plate which has been inoculated with the suspected bacterium. After incubation, the size of clearing around the disk indicates the effectiveness of the antibacterial. If there is a large zone, this would indicate sensitivity, however the size of the zone must be compared to the standardized chart provided later in this lab exercise. If there is no zone (i.e., the bacterium was able to

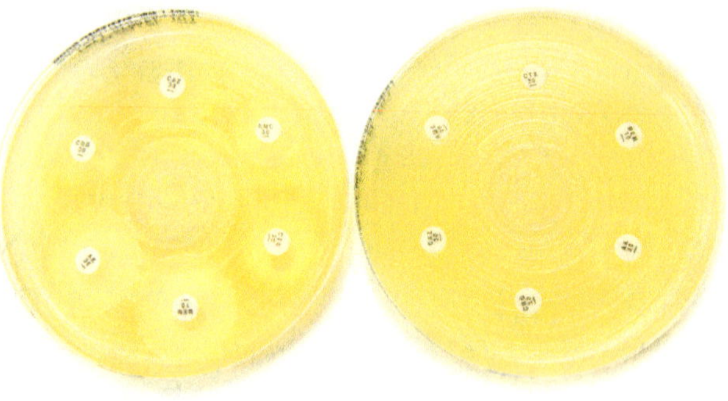

Fig. 11.7 The Kirby Bauer Test
Janeness/Shutterstock.com

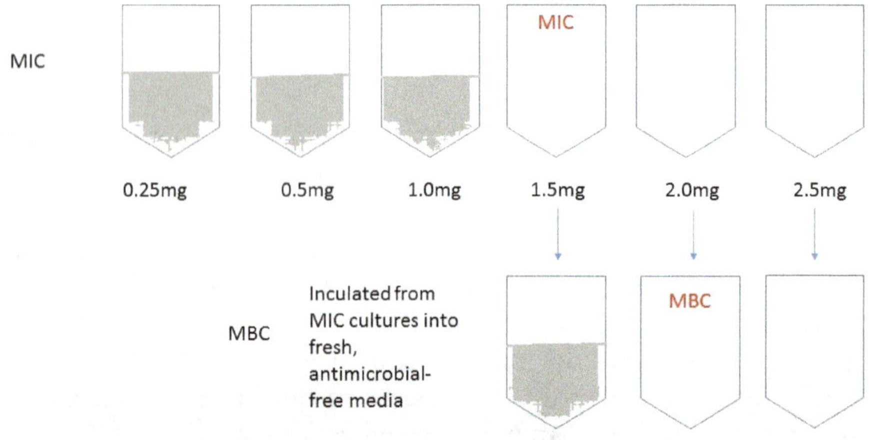

Fig. 11.8 Minimum Inhibitory Concentration and Minimum Bactericidal Concentration Tests
Courtesy of A. Swarthout

grow right up to the edge of the disk), this would indicate resistance. Evaluation tables have been published which standardize this process of determining susceptibility or resistance. This method is used to determine the types and concentrations of antibiotics that could be selected for treatment of a disease.

A similar test that provides drug concentration information is the **E test.** Rather than disks, a strip containing the antibiotic is used, and the concentration along the strip gradually increases from zero.This test shows not only that the antibiotic works, but the minimum inhibitory concentration as well.

The **minimum inhibitory concentration** (MIC) of a drug is the lowest dose that inhibits microbial growth. It can be found by creating a series of serial dilutions of drug, then inoculating them with bacteria. The lowest concentration that inhibits growth is the MIC.

The **minimum bactericidal concentration** (MBC) is the lowest dose that actually kills the microbes and is found by taking bacteria from the MIC tube and using them inoculate new media, and finding the lowest concentration at which the bacteria are killed. This allows the physician to more accurately prescribe the smallest concentration of an antibiotic known to kill or inhibit the infecting bacteria.

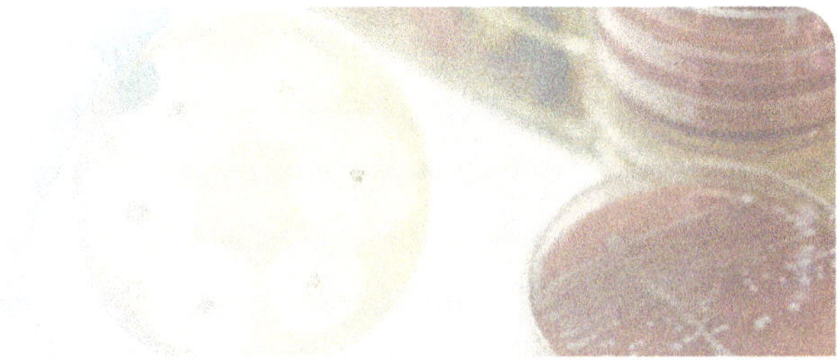

Lab Exercise 11.1

Control Methods: Antibiotic Testing

Student Objectives

1. Define the following terms: antibiotic, C & S, MIC, resistance, bacterial lawn, broad spectrum antibiotic, narrow spectrum antibiotic, and extended spectrum antibiotic.
2. Perform proper aseptic technique when preparing a bacterial lawn.
3. Interpret a Kirby-Bauer sensitivity plate.
4. Identify antibiotic resistant bacteria on a sensitivity plate.
5. Know the mechanisms of action of the major categories of antibacterial antibiotics.
6. Discuss how different categories of antibiotics work on Gram + and Gram – cell walls.
7. Interpret the antimicrobic zone of inhibition chart.
8. Describe various methods used to avoid the development of drug resistance.
9. Know the various factors that must be considered before prescribing an antibiotic.

Name ______________________________

Pre-Lab Exercise 11.1

1. Define the following terms:

   Chemotherapeutic agent:

   ____

   ____

   Antimicrobial:

   ____

   ____

   Antibiotic:

   ____

   ____

   Microcidal:

   ____

   ____

   Microstatic:

   ____

   ____

   Narrow-spectrum drug:

   ____

   ____

   Broad-spectrum drug:

   ____

   ____

MIC:

MBC:

2. Explain susceptibility, intermediate, and resistant used in the Kirby-Bauer test.

3. Name and describe the six characteristics of an ideal antibiotic.

4. Describe synergistic drug action.

5. Name some examples of beta-lactam drugs.

6. Describe the beta-lactam ring and explain how penicillins differ from each other.

7. Describe how cephalosporins are created.

Antibiotics Activity—Week One

Week One Materials

Laboratory Materials: Antibiotic disk dispensers (preloaded) and forceps.

Broth Cultures per Table: *Staphylococcus*, *Bacillus*, *E. coli*, and *Pseudomonas*.

Materials per Student: 1 Mueller-Hinton agar plate, 1 sterile hockey stick, 1 sterile 1 mL pipette, and mechanical pipetter.

Antibiotics Susceptibility Procedure

1. Each student per lab table will use one of the four different bacteria (so that all four bacteria are used at each table). **Record** the name of the bacterium you are using.
2. Using a sterile 1-ml pipette, transfer **0.1 mL** of your assigned bacterium onto the surface of the agar. Do NOT open the lid of your Petri plate until you are ready to transfer. **Dispose** of the used pipette in the **Nalgene container**.
3. Use a sterile hockey stick to spread the inoculum over the entire surface of the agar. The inoculum MUST evenly cover the entire agar surface to create a bacterial lawn. *Cover the Petri plate*. Dispose of the used hockey stick in a biohazard container.
4. Remove the cover of the Petri plate; place the pre-loaded antibiotic disk dispenser over the exposed bacterial lawn. Press carefully on the dispenser handle. This will drop the antibiotic-impregnated disks onto the lawn.
5. If any of the disks fail to lie flat on the agar surface, use a sterile pair of forceps to carefully adjust them. Sterilize the forceps both before and after each use.
6. Tape the plate, place in a storage can (do not invert), and put on the cart to be incubated at 37°C for 24 hours.
7. Identify the antibiotics used in your dispensers and go to the Lab Report to **record** the names of the antibiotics in your **lab report**.

Antibiotics Activity—Week Two

Lab Materials

Laboratory Materials: Metric rulers.

1. Examine your plates. Recall how you inoculated each by reviewing the procedure portion of the activity. Compare your results first with those at your lab table, then with everyone in the lab. Recall each table was testing different antibiotics.
2. Use a metric ruler to measure the diameter of the zones of clearing (i.e., zones of inhibition) surrounding each of the disks. Be certain to measure in **millimeters**, not centimeters.

Note: The diameter is equal to two times the radius. If you cannot measure all the way across the zone of clearing to obtain a diameter measurement, take a radius reading and multiply by two. Measure from the center of the disk to the outermost edge of the inhibition zone (i.e., radius) and multiply that number by two.

Courtesy of Authors

3. Refer to the *Antibiotic Zone of Inhibition Table*, in the next several pages, to determine whether your organism is sensitive (S), intermediately sensitive (I), or resistant (R) to the various antibiotics you tested. Use your measurements to make your determination.
4. **Record** your results in the *C & S (Culture & Sensitivity) Results Table* in the **lab report** section. Record whether the organism is resistant (R), intermediately sensitive (I), or sensitive (S) to each antibiotic tested.
5. Your instructor may draw a table on the board, so that all students can fill out the table. Once all the information is recorded discuss the implications of the results with the class.

Name of Antimicrobial Agent	Abbreviation	Class of Antimicrobial Agent	Resistant (R) mm	Intermediate (I) mm	Sensitive (S) mm
Amoxicillin/ Clavulanic Acid—*Staphylococci*	AMC	PRP	19	N/A	20
Amoxicillin/ Clavulanic Acid—other organisms	AMC	PRP	13	14–17	18
Ampicillin—Gram-negative bacteria	AM P	Penicillin	13	14–16	17

(*continued*)

Name of Antimicrobial Agent	Abbreviation	Class of Antimicrobial Agent	Resistant (R) mm	Interme-diate (I) mm	Sensitive (S) mm
Ampicillin—Gram-positive bacteria	AM P	Penicillin	28	N/A	29
Ampicillin/Sulbactam—Gram-negative & *Staphylococci*	SAM	PRP	11	12–14	15
Bacitracin	B	Penicillin	8	9–12	13
Cefazolin	CZ	1st G Cephalosporin	14	15–17	18
Ceftazidime	CAZ	3rd G Cephalosporin	14	15–17	18
Chloramphenicol (Chloromycetin)	C	Chloramphenicol	12	13–17	18
Chlortetracycline	A	Ointment	14	15–18	19
Ciprofloxacin	CIP	Fluoroquinolone	15	16–20	21
Clindamycin	CC	Lincosamide	14	15–20	21
Doxycycline	DO	Tetracycline	12	13–15	16
Erythromycin	E	Macrolide	13	14–22	23
Gentamicin	CN	Aminoglycoside	12	13–14	15
Imipenem	IPM	Carbepenem	13	14–15	16
Nalidixic acid	NA	Quinolone	13	14–18	19
Novobiocin	NV	Hydroxycoumarin	17	18–21	22
Polymyxin B	PB	Polymixin	8	9–11	12
Rifampin	RD	Antitubercular	16	17–19	20
Trimethoprim	TMP	Folate antagonist	10	11–15	16
Triple Sulfa (Sulfadiazine, Sulfamerazine and Sulfamethazine)	S3	Sulfonamide	12	13–16	17
Vancomycin	VA	Glycopeptide	9	10–11	12
Cefaclor	CEC	2nd Gen. Cephalosporin	14	15–17	18

Name ______________________________

Exercise 11.1 Lab Report

Antimicrobics Activity

1. Antibiotic results. Measure and record the diameter of the zone around the disc using a ruler, in millimeters. Record 'S' for sensitive, 'I' for intermediate, and 'R' for resistant.

ANTIMICROBIAL SENSITIVITY RESULTS				
Name of Antibiotic	*Staphylococcus*	*Bacillus*	*E. coli*	*Pseudomonas*

2. Which antibiotic works best for *Staphylococcus*? Explain how you came to your conclusion.

3. Which antibiotic was least effective against *Pseudomonas*? Explain how you came to your conclusion.

4. A lab report on a patient with a staphylococcus infection has come back showing that staph has a Kirby-Bauer result of 31 mm for ampicillin. Why must the physician take each of the following items into consideration when deciding whether or not to prescribe ampicillin?
   a. Kirby-Bauer results

   b. Patient history

   c. Staph resistance data (use the *Antibiotic Zone of Inhibition Table*)

   d. MIC

5. A patient arrives at his physician's office with a severe skin infection. The physician does a quick diagnosis and speculates it to be streptococcal infection. The patient is immediately placed on high concentrations of a broad spectrum antibiotic.
   a. Why was it essential that the physician handle the patient in this manner?

   ____

   ____

   ____

   b. What is the disadvantage of this prescriptive therapy?

   ____

   ____

   ____

Acellular Agents: Viruses, Viroids and Prions

Viruses

Viruses are acellular infectious agents that are much smaller than bacteria and are usually measured in nanometers (Figure 12.1). They have either DNA or RNA genomes and never have both types of nucleic acids within the virus capsid. Viruses cause many infections of humans, animals, plants, and bacteria and **cause most of the diseases that plague the industrialized world (HIV, SARS, herpes, polio).** Most viruses infect only certain host cells and can be said to be species specific. Some viruses may be so specific they only infect a certain cell type (ex. HIV infects human T lymphocytes). There are plenty of viruses however that cross species barriers

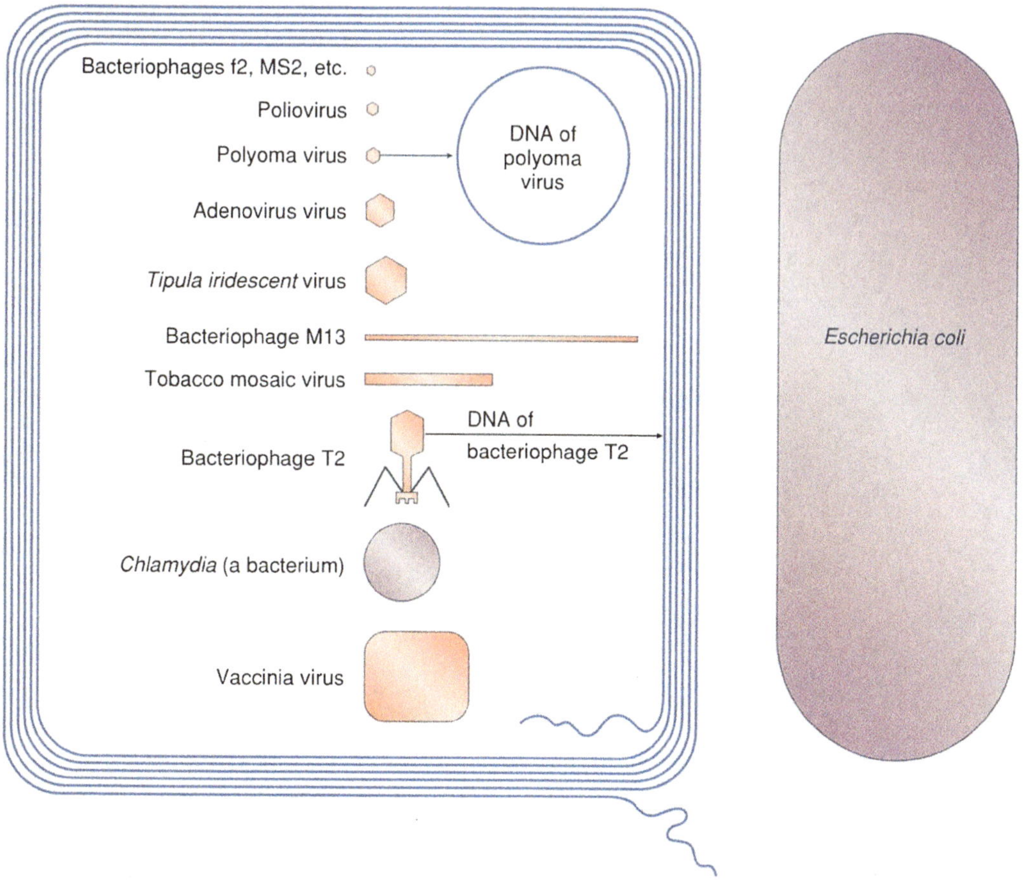

Fig. 12.1 Comparative Sizes of Viruses to E. Coli, Which is Approximately 2 Microns Long

and can infect multiple hosts, these are referred to as generalists (ex. Influenza virus and Rabies virus).

There is some debate on whether viruses are living organisms or not. Some believe they are the least complex living entity, others that they are complex pathogenic chemicals, which is supported by the numerous non-living characteristics. The arguments for both sides are outlined below.

Living Characteristics	Non-Living Characteristics
Can Reproduce (only in living cells)	Acellular (no cytoplasm or organelles)
Can Mutate	Do not grow or respond to the environment
Have ability to take control of host cell	Have DNA or RNA never both
	Cannot reproduce independently
	Do not metabolize chemicals for energy
	Have an intracellular and an extracellular state
	Recruit the cells metabolic pathways to increase numbers

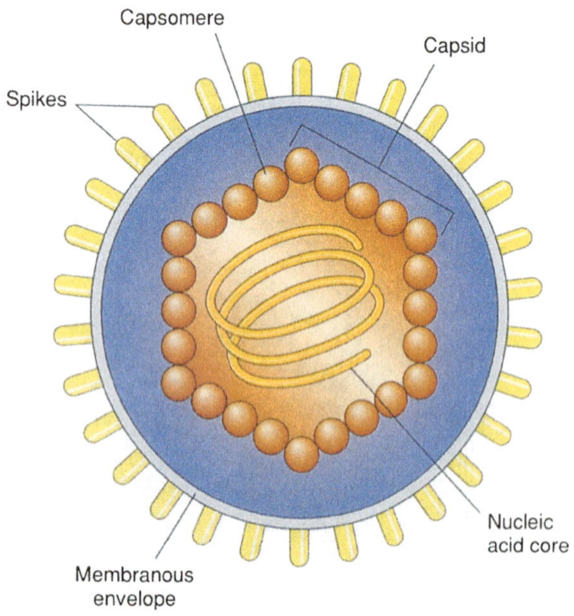

Fig. 12.2 Enveloped Virus Structure. A Naked Virus is Similar but Lacks the Envelope

Animal Viruses

Virus Structure

Viruses have two states: the extracellular state and the intracellular state. In the **extracellular state** the virus is called a **virion.** The virion can either be characterized as "Naked" or "Enveloped". A **Naked** virion is made up of a nucleic acid genome surrounded by a protein "coat" called the **capsid.** The genome + the capsid together make up the **nucleocapsid.** The viral genome may be made of single-stranded DNA (ssDNA), double-stranded DNA (dsDNA), ssRNA or dsRNA.

The capsid is composed of smaller protein subunits, the **capsomeres** and can vary in structure: helical (tube/rod shaped), polyhedral, or complex, meaning the shape varies or has multiple parts. **Enveloped** virions have a phospholipid bilayer that surrounds the capsid. The envelope is often derived from the host cell it is infecting and provides an extra layer of protection (Figure 12.2).

The outermost layer of the virus (the envelope if it enveloped, or the capsid if it is naked) protects the viral genome and contains **attachment proteins,** also known as "spikes". These proteins bind to receptors on the cells of the virus host and are responsible for **virus-host specificity**. Viruses are classified based on the shape of their capsid, the type of nucleic acid they have, the presence/absence of an envelope and their size.

In the **intracellular state** of the virus exists solely as nucleic acid. Once the virus infects the cell the capsid is degraded and the virus will begin the process to take over the internal cell components.

Replication/Life Cycle of Animal Viruses

Viral replication is dependent on host cell organelles and enzymes therefore they must invade the cell in order to replicate. The steps for viral replication are outlined below (Figure 12.3):

1. Recognition and Attachment: capsid or envelope proteins recognize host cell receptors and the virus attaches to the cell
2. Entry: The animal virus can enter by fusion with cell membrane (enveloped only) or endocytosis (naked and enveloped)
3. Synthesis: DNA viruses replicate in the nucleus and viral proteins are translated in the cytoplasm.
4. Assembly: Viral proteins are assembled along with replicated viral DNA in the cytoplasm.
5. Release: The release mechanisms vary depending on the virus the 3 basic release mechanisms are budding, exocytosis, or lysis of the host cell.

Because of the way virus particles are made, one cell can make hundreds to thousands of copies of a single virus.

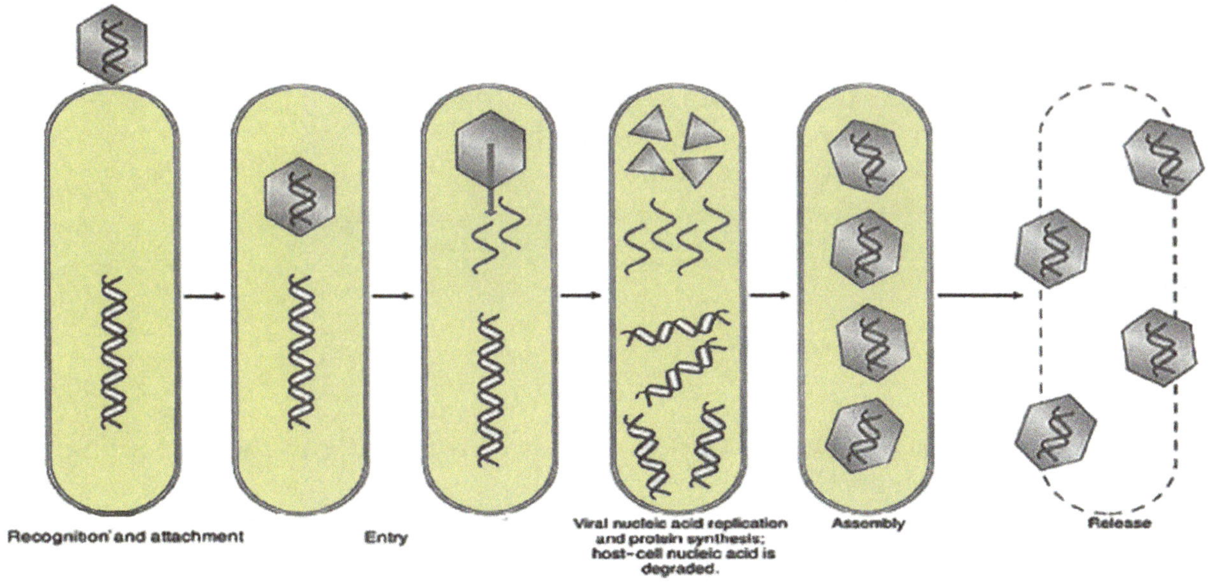

Fig. 12.3 Outline of Animal Virus Replication

Influenza Virus Replication Animantion : http://www.npr.org/blogs/krulwich/2011/06/01/114075029/flu-attack-how-a-virus-invades-your-body

Latency and Latent Infections vs. Acute Infections

The capability of enveloped viruses to leave the host cell by budding and not killing the host cell allows some viruses to cause **latent infections**. Latency occurs when the virus is present in the cell but dormant for long periods and then becomes active, going through a cycle of viral replication and producing symptoms in the host. Some latent viruses incorporate their viral DNA into the host cell DNA, causing the entire host cell progeny to become infected, while other latent viruses keep their genome separate from that of the host. Examples of latent infections include herpes (both genital and oral), in which the virus can be dormant for months or years and become active for a short time, producing symptoms, and chickenpox, in which a person contracts the virus (usually as a child) and is symptom free for decades before the virus becomes active and symptoms appear as shingles. People who develop latent infections do not develop immunity and can experience symptoms again and again.

Viruses that do not become latent cause acute infections. Acute infections are characterized by sudden onset symptoms that are short in duration with active replication of the virus by the host cell. Infections by these types of viruses usually result in long lasting future immunity for the host. Examples of acute infections include, the cold, influenza and norovirus.

The Role of Viruses in Cancer

Animal genes contain sequences of DNA called protooncogenes (literally pre-cancer genes). These protooncogenes are bound by repressor proteins, which effectively keep the genes off (in the pre-cancer state). Viruses contribute to the development of cancer in a two-step fashion: they insert promoters in front of the protooncogene (remember, in order for a gene to be transcribed, mRNA binds to the promoter), then, they disrupt or eliminate the gene for the repressor protein. Without the repressor, the protooncogene can be turned "on", thereby becoming an oncogene.

Viruses cause 20–25% of human cancers

- Some viruses carry copies of oncogenes as part of their genomes
- Some viruses promote oncogenes already present in host
- Some viruses interfere with tumor repression

Specific viruses are known to cause ~15% of human cancers

- Burkitt's lymphoma
- Hodgkin's disease
- Kaposi's sarcoma
- Cervical cancer

Environmental factors often combine with viruses to contribute to the activation of oncogenes. Examples include:

- Ultraviolet light
- Radiation
- Carcinogens

Retroviruses

Retroviruses are viruses that have RNA as their genome. This RNA can be used directly as mRNA to initiate protein synthesis and make viral capsids and other viral components. The

RNA of retroviruses is transcribed into DNA by the viral enzyme, **reverse transcriptase**. This viral DNA can then be incorporated into the host cell genome. This makes the virus a permanent part of the host cell and any host cell progeny. This splicing may also disrupt repressor genes and contribute to the development of cancer within the host. An example of a prominent retrovirus is Human Immunodeficiency Virus (HIV), the causative agent of AIDS.

Evolution of Animal Viruses

Animal viruses evolve via two mechanisms: **antigenic drift** and **antigenic shift**. Before we discuss these in detail, let's take a closer look at the influenza virus, which we will use as a model for studying drift and shift:

The two major types of influenza are Type A and Type B. Both types have envelopes and 8 segments of RNA as their genome. Both types also have 2 major types of spikes on their envelopes:

- Hemagglutinin: allows the virus to bind to receptors on pulmonary epithelial cells and facilitates endocytosis by those cells, there are 16 varients of hemagglutinin
- Neuraminidase: helps the virus penetrate the lung by hydrolyzing mucus in the lungs, and also helps facilitate the release of new virus particles from infected host cells, there are 9 varients of neuraminidase

Influenza viral strains are named based on the varients of hemagglutinin and neuraminidase they have. For example, the 2009 flu epidemic was caused by H1N1. The so-called "Avian Flu" is H5N1.

Antigenic drift occurs when point mutations occur in the RNA coding for the neuraminidase and hemagglutinin. These point mututions change the virus just enough so that the immune system of a person infected with it would not recognize and remember the virus. The virus would be seen as something new upon re-exposure, and would cause illness while the immune system goes through a primary immune response.

Antigenic Shift is a larger change in the genome, usually due to the mixing of genetic information from two strains of virus that are infecting a host simultaneously. In the case of the influenza virus, this may involve the infection of an intermediate host, such as a pig or a bird. Antigenic shift leads to the creation of entirely new strain of virus.

Replication/Life Cycle of Bacteriophages

A **bacteriophage** (or "phage" for short) is a virus that infects bacteria. Felix d'Herelle coined the term bacteriophage, meaning "bacteria eater" in 1917. There are many kinds of bacteriophage, each infecting a specific type of bacteria. Bacteriophages such as the T-even phage used in the lab exercise at the end of this chapter have a complex structure comprised of a polyhedral head called the capsid that contains DNA. These phages first attach to the bacterial cell and then inject their DNA into the bacterium (much like a hypodermic syringe injection). There are two outcomes of this injection, depending on whether the phage is lytic or lysogenic. The cycles are outlined below:

- **Lytic replication**
  Replication cycle results in lysis and death of host cell about 30 minutes after infection. Lysis releases approximately 100–200 progeny.

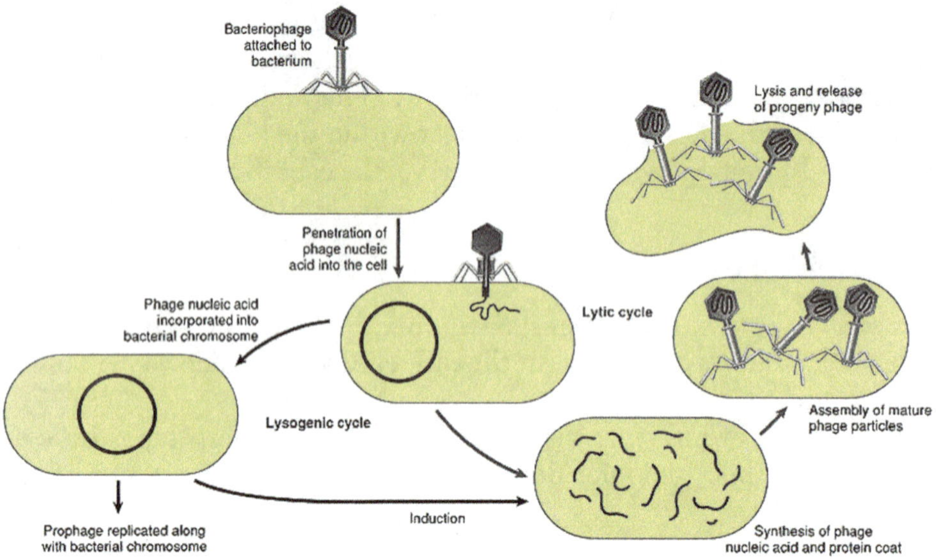

Fig. 12.4 Image showing lysogenic cycle and how induction causes the lytic cycle. Note: the color of the viral genome is red and the host chromosome is in blue. Notice how during assembly some of the phage particles have "blue" bacterial DNA in the capsid. This is how the gene transfer mechanism of transduction works

- **Lysogenic replication** (Figure 12.4)
  Modified lytic cycle where the viral genome is incorporated into the host cell genome. The bacteria will reproduce for many generations until the virus is induced. **Induction** is the process in which the viral genome is "cut" out of the host cell genome and enters into the lytic cycle. Induction can be caused by numerous environmental factors such as DNA damaging chemicals, UV light, and other sub optimal conditions. Bacteria that contain the DNA of the phage usually cannot be re-infected or lysed by the same type of phage.
- **Basic stages of lytic replication cycle**
  1. Recognition and Attachment: Bacteriophage spikes recognize complementary receptors on bacteria surface.
  2. Entry: Injection of viral genome
  3. Synthesis: Replication of viral genome and translation of viral proteins
  4. Chromosome degraded: virally coded enzymes degrade the host chromosome.
  5. Assembly: **Note**: the gene transfer mechanism of transduction relies on mistakes when the virus is assembling. Short fragments of bacterial chromosome or plasmids may be packaged inside the viral capsid. Once released the bacteriophage will infect another bacteria with the bacterial DNA.
  6. Release: is by lysis of the cell resulting in cell death.

It is important to study phage and learn to manipulate them for the following reasons.

1. Animal viruses, including human pathogens, are grown in tissue culture cells in the same fashion as phage on bacteria. Tissue cultures are animal cells grown in bottles or on plates. The animal virus can form plaques by causing cells to degenerate or die. It is convenient to

learn viral techniques using bacteriophage, because human pathogenic viral manipulation in lab could be disastrous if you used poor aseptic technique.

2. Phages are used in recombinant DNA experiments and are also useful in studying the genetics of bacteria. They are useful tools for transfer of genes and pieces of DNA from cell to cell.
3. Phages are used to identify certain strains of bacteria, because one type of phage will only infect one or a few specific species or strains of bacteria.
4. Lysogeny has served as a model for viruses that insert their DNA into animal cells. The life cycle of lysogenic phage resembles that of retroviruses (e.g. HIV).

Lysogenic conversion occurs when the genes carried by the phage alter the phenotype of the bacterial cell in some way. The toxins produced by *C. diphtheriae* and *C. botulinum* are the result of lysogenic conversion.

Culturing Viruses

Culturing of viruses is problematic because they are obligate intracellular parasites: they must be grown in their host cells.

In the case of bacteriophage, a **plaque assay** can be done (Lab 12.1). The phage particles are mixed with bacterial cells then mixed into warm, molten agar which is then poured into Petri plates and allowed to solidify. Mixing the bacteria into the agar this way should create a lawn. Wherever the phage have infected and killed bacterial cells, holes called **plaques** occur in the lawn. Plaques contain high numbers of bacteriophage which can be cultured again in this manner.

Human and animal viruses can be cultured in cell or tissue culture. First, the host cells must be grown in tissue culture media. Sometimes, cell culture lines are available commercially, but some types of cells are not and they must be harvested fresh from the host organisms. For example hepatocytes had to be cultured fresh each time. This meant harvesting liver cells from mice and rats each time I needed to perform an experiment. Cell culture is also prone to contamination because the media and much of the equipment is heat-sensitive and cannot be autoclaved. After establishing a culture of host cells, the virus is introduced to the culture and allowed to infect it. Animal viruses can also be cultured in embryonated chicken eggs (Figure 12.5). Chicken eggs are free from microbes and contain nourishing yolk to provide energy for viral replication.

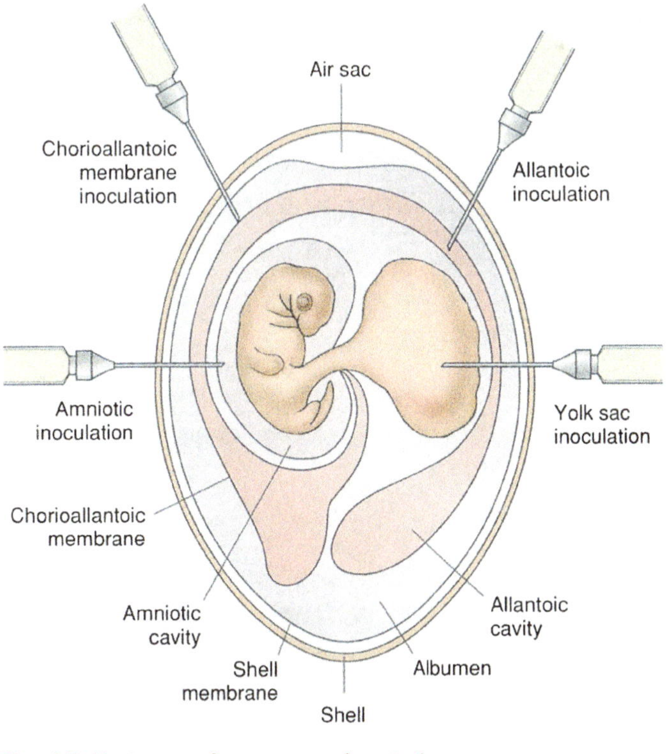

Fig. 12.5 Sites of Injection for Culturing Viruses
© Kendall Hunt Publishing

Other Parasitic Particles: Viroids & Prions

Viroids

Viroids are extremely small, circular pieces of RNA that are infectious and pathogenic in plants, although little is known about how they infect and cause disease in plants. They are similar to RNA viruses, but lack capsids. It is not known if viroids infect animals or humans.

Prions

The term "prion" is short for **pro**teinaceous**in**fectious agent (if you rearrange the letters a bit). Prions are misfolded PrP proteins that can induce other proteins to change their shape to become prion proteins. PrP proteins are made by all mammals, prion PrP proteins convert normal PrP into prion PrP. Prions in a sense sort of act like an enzyme slowly converting good proteins into bad proteins.

Prions affect the brain matter causing spongiform encephalopathies**.** These result in always fatal neurological degeneration, fibril deposits and loss of brain matter. On examination, the brain appears to have large vacuoles in it, giving it the characteristic spongy appearance.

There is still a lot that is unknown regarding prion diseases and their transmission. As of now, there are no diagnostic tests for them beyond autopsy, and there are no treatments. It is known that prions can be transmitted by ingesting contaminated neural tissue or by receiving transplanted tissue from a donor who had a prion disease, but it is not known if prions can be transmitted in any other way.

Prion disease include mad cow (cattle), scrapie (sheep), varient Creutzfeld-Jakob syndrome (humans) and chronic wasting disease (deer and elk).

In order to destroy prions, they must be autoclaved in 1N NaOH or incinerated.

The mad cow outbreak in Great Britain in the mid-1980s was a result of two disturbing practices: 1. using the carcasses of dead cattle to boost protein levels in cattle feed and 2. not rendering the carcasses at high enough temperatures to destroy prions prior to adding them to the feed. The lower temperature was adopted as a way to save on energy costs. Because we do not fully understand the transmission of prions, people from England and Americans who visited/work in England during this time period are not allowed to donate blood in the U.S.

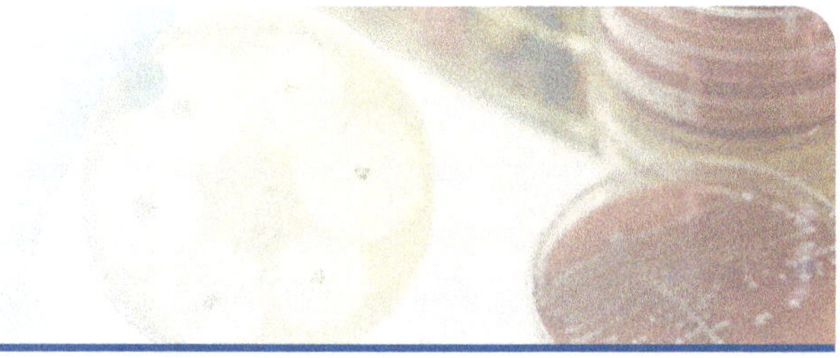

Lab Exercise 12.1

Bacteriophages

Objectives

Describe and perform cultivation of bacteriophages.
Determine the titer of a bacteriophage sample using the plaque assay.
Differentiate between a **lytic phage** and a **lysogenic phage**:
Describe how to recognize a plaque on the plate:

Name ______________________

Pre-Lab Exercise 12.1

1. What is the purpose of this laboratory exercise:

2. In which step of the procedure do you see the potential for error? What could go wrong if you were to make a mistake during the step you mention?

3. What is a plaque?

4. What does titer refer to?

5. How will you determine the number of phage per milliliter?

6. How do lysogenic plaques differ from lytic plaques?

7. What is the target range for the number of plaques on a petri dish?

INTRODUCTION

Bacteriophages are viruses that infect bacteria. Although structurally similar to some human viruses, they belong to separate virus families and do not cross species barriers, that is, each phage is specific to certain species of host bacterium. Because of their structural similarities to human viruses, observing the occurrence and survival of bacteriophages can give us some ideas about how human viruses behave under certain environmental conditions. Phages are also useful for mastering techniques for working with viruses, since they are simpler to work with and faster growing than human viruses.

Lytic Cycle Replication

In this lab exercise we will focus on the lytic cycle, after entry into the cell the phage DNA acts as a template for production of phage proteins. These proteins replicate the phage and take control of the cell, eventually causing lysis and death of the host cell. A bacteriophage particle is much smaller than a bacterial cell therefore, is beyond the limits of resolution of the light microscope and can be seen only with electron microscopes. Bacteriophages will actually pass through a bacterial filter that have pores small enough to remove bacterial cells but allow sterile liquids to pass through. Fortunately, we can use a technique similar to the colony-counting technique used to measure the number of bacteria to count phage particles, known as a **plaque assay**. **Lytic phages** are quantified using this method.

Phage and their matching host cells are mixed in a small tube of soft agar and then poured on top of an agar base plate. Soft agar contains about half the concentration of standard agar so that the phage can disperse more easily. The plates are then incubated overnight at 37°C. During incubation, the bacteria multiply and cover the entire surface of the agar creating a **lawn of bacteria**, except in those places where phage have infected the bacteria leaving clear zones called **plaques (Figure 12.6)**. Since each plaque originated with one phage, plaques can be counted to determine the number of phage present in the original sample. The number of phage per milliliter can be calculated and is referred to as the **titer**. Although the appearance of the plaques can be influenced by many factors, in general, lytic phage produce clear plaques. **Lysogenic phage**s produce cloudy plaques because many bacteria in the plaque were lysogenized instead of lysed and thus continue to grow and multiply. The plaques do not increase in size indefinitely because phage can only replicate in actively growing and dividing bacteria and each infected cell will usually release about the same number of phage.

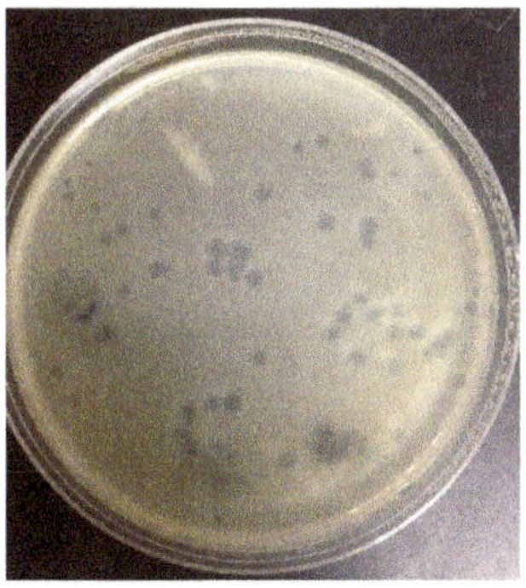

Fig. 12.6 Petri Dish Showing a Lawn of Bacteria with Bacteriophage Plaques
Courtesy of Author

Day One Procedure

Materials: 5 TSA plates, 5 Dilution blanks, 5 Soft agar tubes (in water bath), Micropipette, pipette tips, bacteriophage.

It is essential to use a "young" bacterial cultures since the virus particles are replicated only in cells that are actively growing.

1. Label the BOTTOM of the nutrient agar plates 10^{-1}, 10^{-2}, 10^{-3}, 10^{-4}, and control.
2. Label the dilution blank tubes (on the glass part) 10^{-1}, 10^{-2}, 10^{-3}, 10^{-4}, and control.

Diluting the Phage:

3. Aseptically add 0.1 ml of phage suspension to tube 10^{-1}. Mix the tube carefully by pipetting the liquid into and out of the tube (Figure 12.7).
4. Using a sterile second pipette tip, transfer 0.1 ml from tube 10^{-1} to the tube labeled 10^{-2}, mix carefully.
5. Using a sterile third pipette tip, transfer 0.1 ml from the 10^{-2} tube to the 10^{-3} tube. Continue to dilute the phage through tube 10^{-4}. Save all your diluted broth tubes for the next steps.
6. **Do not inoculate the control tube with virus.**

Plating the Phage:

Work carefully, but quickly because the agar will harden in the soft agar overlay and will not spread evenly on your base plate. The agar will **not** re-melt if you put it back into the water bath. **Therefore, only work with one soft agar tube at a time.**

7. Remove one molten soft agar tube from the water bath, wipe the water off with a paper towel so it doesn't drip onto your plate and cause a contamination problem.

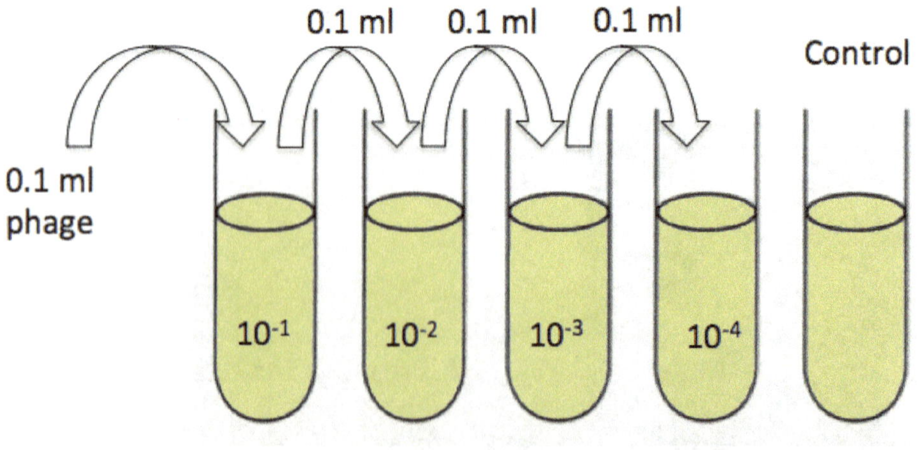

Fig. 12.7 Serial dilution of the bacteriophage. Make sure to mix well before transferring phage to the other dilution blanks. Image made by author.

Courtesy of Author

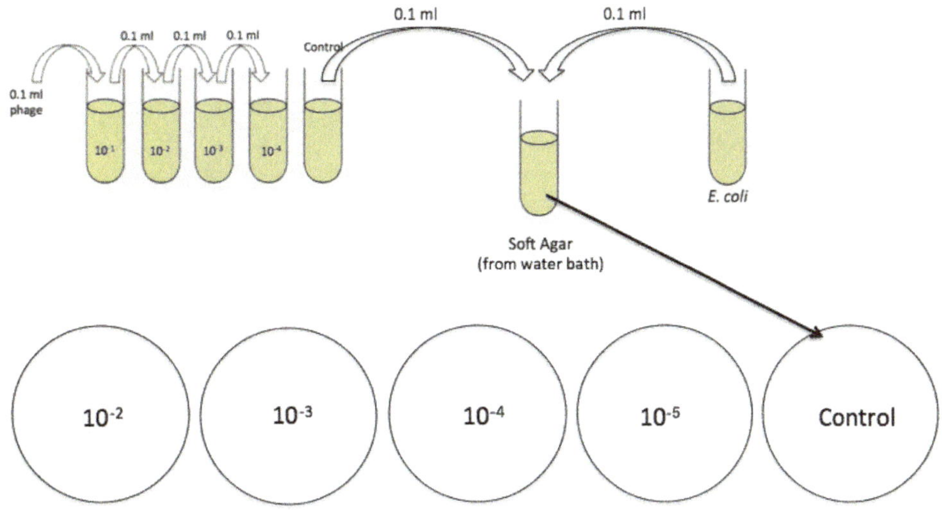

Fig. 12.8 After diluting the phage you will add the dilution blank media to a soft agar tube along with 0.1 mls of E. coli. Pour the soft agar onto the appropriate plate and swirl in a figure 8 motion to spread soft agar evenly. Repeat this procedure for all dilutions and control

Courtesy of Author

8. Aseptically transfer 0.1 ml from the control dilution blank tube into the first molten soft agar tube (Figure 12.8).
9. With a second pipette tip add 0.1 ml of your *E. coli* culture to the same control tube. Mix well and pour onto the agar in the plate labeled control. Distribute the agar over the agar already in the plate by moving the plate in a figure 8 pattern.
10. Using the a different pipette tip aseptically transfer 0.1 ml from the dilution blank tube from the tube labeled 10^{-2} to a second molten soft agar tube, with a second pipette tip add 0.1 ml of your *E. coli* culture to the same molten soft agar tube, mix well and pour into the plate labeled 10^{-2}. Distribute the overlay onto the agar in the plate.
11. Repeat this procedure for the remaining dilutions.
12. Let the plates sit for 10–15 minutes so the agar has time to solidify. Then invert the plate and place on the tray for incubation. The plates will be incubated at 37°C.

Day Two Procedure

Materials:

Marking pencil or pen   calculator

Procedure:

1. Examine your set of plates. Select a plate containing between 30 and 300 plaques if possible. A count of fewer than 30 colonies is inaccurate and more than 300 is too difficult to count. As you count the plaques, place a dot with a marking pencil or pen on each plaque on the bottom of the Petri dish.
2. Estimate the number on a second plate. They should vary by a factor of 10 as the dilution increases or decreases if you did your dilutions carefully.
3. To determine the titer, use this formula: # of plaques × 1/dilution factor = PFU/ml
4. Record your results and answer the questions on the report page.

Name ______________________________

Exercise 12.1 Report

Sketch the plate that was used in the pfu/ml calculation in the circle below.

Dilutions and Plate Counts

What was the dilution of the plate that was counted for E. coli? ______________________________

How many plaques were on the countable plate for E. coli? ______________________________

Determine the Titer

1. How many phage per ml (PFU/ml) were in the original suspension? Show how you set up your calculation for full credit.

____

____

2. In theory should you get the same titer from all of your plates? Why?

____

____

3. Are all the plaques the same size? Briefly explain why or why not.

4. Briefly describe the lytic and lysogenic cycles of bacteriophages.

5. Assume your goal is to develop a pure solution of bacteriophage in a liquid medium.
   a. Starting with your countable plaque assay plate, where on the plate would you find the bacteriophage to transfer into a nutrient broth culture?

   b. After incubation of your broth culture, what sterilization process could you use to separate the bacterial host cells from the phage particles to obtain a liquid with only phage?

6. Suppose you attempted this same laboratory exercise but you used *Staphylococcus aureus* instead of *E. coli.* However after incubation you observed no viral plaques on any of the dilution plates. Give a reasonable explanation as to what might have happened.

7. Why is the study of bacteriophages important?

Disease and Microbial Mechanisms of Pathogenesis

Microbes and humans exist together in one of three types of symbiotic relationships, mutualism, commensalism, or parasitism. These relationships are studied in the field of Medical Microbiology. There are many examples of the interactions that take place between humans and microbes (Table 13.1). Most of the examples involve normal microbiota. Normal microbiota may switch categories if introduced into part of the body they are not accustomed to and become an opportunistic pathogen. Normal microbiota may also become opportunistic if you are immune-compromised, or if there are changes in the normal microbiota from antibiotic use.

Disease

The terms infection and disease are often used interchangeably, however they are different. **Infection** is when an infectious agent enters the body and begins to colonize. **Disease** is a result of infection and occurs when normal body function is altered and signs and symptoms are present. **Symptoms** of disease are felt by the patient and are not measureable by a physician, such as a headache. **Signs** are objective manifestations that are observed and measured by others, such as fever. There are scenarios where the diseased individual is **asymptomatic**, meaning the patient lacks symptoms but may still have signs of the disease. Asymptomatic patients are considered **carriers**, which means they can transmit the disease to others. The majority of infections are actually eliminated by the immune system before disease progresses. Once infected with a pathogen and the body fails to fight off the infection disease will follow a predictable pattern as the disease manifests. There are 5 stages that typically occur:

1. Incubation period: the time between infection and first symptoms and signs of disease occur.
2. Prodromal period: a short period of time, mild symptoms and signs (may not occur at all)
3. Illness: signs and symptoms are most severe and the microbial load in the body is the highest.

Table 13.1 Examples of symbiosis

	Organism 1	Organism 2	Example
Mutualism	Benefits	Benefits	*E. coli* in human intestinal tract
Commensalism	Benefits	Neither benefit or are harmed	Micrococcus on skin
Parasitism	Benefits	Harmed	*Streptococcus pyogenes* causing strep throat

4. Decline: gradual decline of symptoms and signs as the body returns to normal due to antibiotic treatment or immune response.
5. Convalescence: patient returns to normal no signs or symptoms present.

**Depending on the disease the patient may transmit the disease to others at any stage.*

Classification of Infectious Disease

Infectious diseases can be classified in different ways based on the amount of time it takes for signs and symptoms to appear and whether or not the disease is recurrent. There are 3 main categories an infectious disease may fall under:

1. Acute disease: Rapid onset of signs and symptoms, which last for a relatively short period of time. Example: common cold.
2. Chronic disease: Disease manifests slowly, long incubation time, signs and symptoms develop slowly. Example: Tuberculosis.
3. Latent disease: Pathogen remains inactive for a long period of time before becoming active. After initial exposure mild signs and symptoms may have occurred or different signs and symptoms occurred. Example: Varicella Zoster virus, which causes chicken pox initially, then later in life may cause shingles.

Acquiring Disease Causing Organisms

In order for a human to develop an infection, which leads to disease, we must come in contact with a reservoir of infection. A **reservoir of infection** is the site where a pathogen is maintained as a source of infection. There are 3 main types of reservoirs:

1. **Human Reservoir:** Considered the most common reservoir. Individuals with signs and symptoms transmit disease to healthy individuals. Carriers, those who are asymptomatic, can transmit disease to healthy individuals. Carriers with HIV, syphilis, or Tuberculosis, may remain asymptomatic for long periods of time infecting many people unknowingly.
2. **Animal Reservoirs:** Animals may transmit disease causing what are called Zoonosis or Zoonotic diseases. Zoonotic diseases are transmitted naturally from animal host to humans through a variety of routes. Direct contact with animals or animal waste (*E. coli* O157:H7), animal (rabies) or insect bites (Malaria), or through the air (Influenza).
3. **Nonliving Reservoirs:** Soil, water, and food can serve as reservoirs of infection. The presence of microorganisms in nonliving reservoirs is often due to contamination from animal urine or feces, yet the organism is maintained in the nonliving reservoir (Parasitic Eggs: Giardia). The organism can also be found naturally in the environment (*Clostridium tetani:* Tetanus)

Portals of Entry

The movement of a pathogen into a host is through a portal of entry. There are four general pathways a microbe may enter into the human body.

1. Skin: the outer layer of skin is composed of dead cells that act as a barrier to pathogens. Most pathogens need to enter the skin through a cut or abrasion. There are very few that can burrow into the skin and cause infection.
2. Mucous membranes: Respiratory tract is the most common site of entry, eyes, nose, and mouth. Urinary tract and reproductive tracts are also included here.

3. Placenta: The placenta typically forms an effective barrier to pathogens. There are few pathogens that can cross the placenta and infect the developing fetus causing spontaneous abortion, birth defects, or premature birth. Organisms such as *Listeria monocytogenes* are capable of doing crossing the placenta, pregnant women are cautioned by their physician or midwife to not eat certain foods like lunch meat and unpasteurized foods such as certain cheeses where Listeria is commonly found in low numbers.
4. Parenteral route: Not a true portal of entry but is a way a portal of entry can be bypassed by mechanically depositing pathogens directly into tissues beneath the skin or mucous membranes. Hypodermic needles, thorn punctures, nail punctures are examples of ways to bypass the portals of entry.

Portals of Exit

The movement of a pathogen out of a human reservoir is through a portal of exit. Many portals of exit are the same as the portals of entry. Pathogens often leave the host in materials the body secretes or excretes such as urine, semen, respiratory droplets, and feces.

Modes of Disease Transmission

Transmission refers to the transfer of a disease causing microbe from a reservoir of infection to a susceptible hosts portal of entry. There are three modes of disease transmission:

1. **Contact Transmission:** which involves picking up the microbe through **direct** contact with a reservoir (hand shaking or fecal-oral route), **indirect** contact which is the transfer of pathogens through inanimate objects, called fomites (clothing or bedding), or through respiratory droplets in which you inhale small-large microbe saturated droplets (coughing or sneezing) that would fall to the ground within 5 feet of the source.
2. **Vehicle Transmission:** involves inhalation or ingestion of a pathogen through air, food, or water. **Airborne** transmission is very difficult to control since the pathogen can be suspended in the air for long periods of time and can travel by air currents to other parts of a building through ventilation systems. **Foodborne** transmission is the ingestion of pathogens that are naturally harbored by an animal such as *E. coli* or *Salmonella*, this type of transmission is easily avoided through proper cooking time and temperatures. Organisms can also be added to food from the food handler either by failing to wash hands (*S. aureus*) or cross-contamination while preparing the food (cutting boards for vegetables and meat). Foodborne illnesses can be categorized as either **foodborne infection**, which is ingestion and colonization of the microbe (*E. coli* or *Salmonella*) or **foodborne intoxication** which involves ingesting a toxin that was produced by a microbe growing on the food (*S. aureus* and Botulism). **Waterborne transmission** involves the ingestion of contaminated water.
3. **Vector Transmission:** can be either biological or mechanical. **Biological vector transmission** involves an organism transmitting the pathogen to the susceptible host. Usually the vector is involved in the life cycle of the pathogen, such is the case with malaria, where part of the life cycle of the protozoan, plasmodium, occurs inside the Anopheles mosquito. **Mechanical vector transmission** involves the direct contact of a vector to the susceptible host, where the vector carries the pathogen and lands on the host. This occurs during the transmission of *Chlamydia trachomatis*, which can cause blindness, the organism is carried on the feet of a fly from an infected persons eye to the eye of a new host.

Microbial Mechanisms of Pathogenesis

The mechanism a bacterium uses to cause harm to the body is referred to as **pathogenesis**, therefore a microbe must exhibit pathogenic properties or **pathogenicity.** In order for a pathogen to cause disease they must possess some sort of virulence factor, the pathogenic properties, in order cause harm to the body. **Virulence factors** are traits such as toxin production or capsule production, which not only allow a bacterium to evade the immune system but also allow them access to nutrients. Pathogens do not all share the same level of **virulence**, or relative ability to cause disease. Pathogenicity and virulence do not relate to the severity of disease, for example the pathogen that causes rabbit fever is highly virulent but causes only mild disease. On the other hand a less virulent pathogen such as *Pseudomonas aeruginosa* can cause life-threatening disease in immunocompromised individuals and is usually only an opportunistic pathogen. Another contributing factor to the onset of infection is the number of microbes in the inoculum. This is referred to as the **infectious dose (ID)**, organisms with high virulence tend to only need a small infectious dose whereas organisms with low virulence need a high infectious dose. The infectious dose for gonorrhea is only about 1,000 cells in contrast cholera's ID is around 1,000,000,000 cells! If the number of microbes exceeds the ID the onset of disease can be extremely rapid.

Pathogens in a sense have a "goal", bacteria are not conscience of this goal but they evolved interacting with human hosts for thousands of years. The goal is to enter a host through a portal of entry, attach to host cells, gain access to nutrients, evade the immune system in order to colonize, and leave the host through a portal of exit in order to infect other susceptible hosts (Figure 13.1). There are many examples of pathogens that can survive for extended periods of time outside of its primary reservoir. However, most human pathogens cannot survive very long outside of a human host. A successful pathogen only needs to overcome the host immune system long enough to acquire nutrients for replication and leave the host. Over time a pathogen tends to become less pathogenic by a phenomenon called **balanced pathogenicity**. It is a disadvantage for a pathogen to kill a host, since the opportunity to be transmitted may be limited and the pathogen loses the source of nutrients.

Virulence Factors

Adherence. In order to colonize a host and cause disease the pathogen must possess virulence factors. Attachment can be considered one of the most important virulence factors since in order

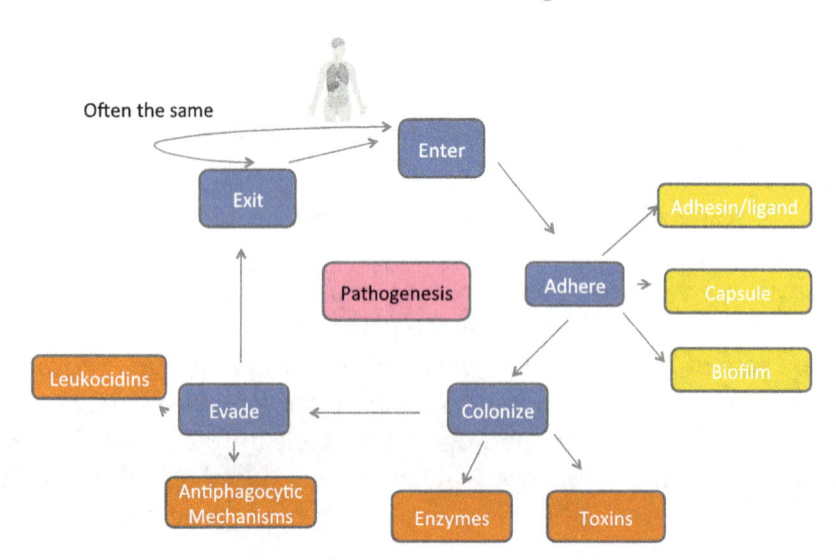

Fig. 13.1 Route a pathogen must take to successfully cause disease.
BluezAce/Shutterstock.com

What Do Bubonic Plague and HIV Have In Common?
If a host lacks receptors for a specific adhesin the microbe will not be able to attach to host, and therefore will not be able to successfully colonize the host. An example of this is the CXCR2 receptor on human T cells. The adhesins of the plague bacterium, *Yersina pestis*, are specific to CXCR2 allowing it to attach to and invade the T cell helper cells as part of its pathogenesis. Some humans have a mutated receptor, called Delta 32. In people who are heterozygous for Delta 32, binding of the microbe to the host cell is impaired, resulting in less severe cases of plague. If someone is homozygous for the Delta 32 mutation, binding and disease do not occur. Because people without the mutation were more likely to die of plague than people with the mutation, the mutation was selected in the population.

In the present day, HIV attaches to the CXCR2 receptor via its spikes. Descendants of plague survivors who have two copies of the Delta 32 mutation have genetic immunity to HIV.

for a bacterium to infect a cell or tissue they must attach to a cells surface. Bacteria possess **adhesion factors**, which are factors a bacteria possesses in order to attach to a surface. An example of an adhesion factor are proteins on their cell surface called **adhesins**, these adhesins allow the bacteria to bind to complementary receptors on host cells. Viruses have similar attachment proteins that are an absolute necessity for viral infections. **Capsule** and **biofilm** production is also an adhesion factor, capsules allow bacteria to attach to smooth surfaces. *Neisseria gonorrhea* for example has adhesins that are specific to human epithelial cells and adhere to cells that line the urethra and vagina, and can even attach to sperm cells providing a mode of entry into the female reproductive tract.

Colonization and Accessing Nutrients. Virulence factors that allow access to nutrients include the production of extra cellular enzymes (exoenzymes), which are produced to dissolve structural chemicals in the body (Table 13.2). The production of such enzymes allows the bacterium to further invade tissues and access to nutrients, which are not normally exposed.

Table 13.2 Microbial Enzymes Associated with Virulence

ENZYME	MODE OF ACTION	EXAMPLES OF BACTERIA PRODUCING THE ENZYME
COAGULASE	CLOTS PLASMA	STAPHYLOCOCCUS AUREUS
COLLAGENASE	DEGRADES COLLAGEN IN MUSCLE TISSUE	CLOSTRIDIUM PERFRINGENS, PSEUDOMONAS AERUGINOSA
DEOXYRIBONUCLEASE (DNASE)	DEGRADES DNA IN PUS	STAPHYLOCOCCUS AUREUS, STREPTOCOCCUS PYOGENES
HEMOLYSIN	LYSES RED BLOOD CELLS AND OTHER CELLS	STAPHYLOCOCCUS AUREUS, STREPTOCOCCUS PYOGENES
HYALURONIDASE	DEGRADES THE GROUND SUBSTANCE OF CONNECTIVE TISSUE	CLOSTRIDIUM PERFRINGENS, STAPHYLOCOCCUS AUREUS, STREPTOCOCCUS PYOGENES
LECITHINASE	SPLITS LECITHIN IN PLASMA MEMBRANES	CLOSTRIDIUM PERFRINGENS
LEUKOCIDIN	DESTROYS WHITE BLOOD CELLS	STAPHYLOCOCCUS AUREUS
STAPHYLOKINASE, STREPTOKINASE	CONVERTS PLASMINOGEN TO PLASMIN, DISSOLVING BLOOD CLOTS	STAPHYLOCOCCUS AUREUS, STREPTOCOCCUS PYOGENES

© Kendall Hunt Publishing

Another way cells may gain access to nutrients is by producing toxins, **Exotoxins** (Table 13.3) can be secreted by a bacterium in order to destroy host cells. There are three main categories of exotoxins:

1. Cytotoxins: kill host cells or damage host cells altering their function.
2. Neurotoxins: damage or kill nerve cells.
3. Enterotoxins: target cells lining the gastrointestinal tract.

Endotoxin is another type of toxin only associated with Gram-negative bacteria. Endotoxin is also called lipid-A, which is the lipid portion of lipopolysaccharide in the outer membrane of Gram-negative bacteria. Lipid-A is release when the cells die naturally, during binary fission, or when digested material from phagocytes is released from the cell. The effect of lipid-A on the body may include: fever, inflammation, diarrhea, or shock.

Evasion of the Immune System. Perhaps some of the most interesting virulence factors are the ways bacterial cells avoid the immune system. Many pathogens have developed ways to avoid phagocytic cells, these are referred to as **antiphagocytic factors**. **Capsule** production is one example as the capsule is composed of chemicals found naturally in the body, therefore phagocytic cells do not recognize the bacterium. Also, the capsule is a smooth and slippery surface, which makes attachment to the bacterium difficult. Some pathogens produce **antiphagocytic chemicals**, bacteria that are ingested by a phagocyte can secrete chemicals that prevent the fusion of the lysosome with the phagocytic vesicles. Therefore, the pathogen survives inside of the phagocyte and can hitch a ride to other areas in the body.

Another interesting mechanism bacteria may use to avoid the immune system is **avoiding antibodies**. Antibodies are proteins produced by activate B cells and secreted to bind to foreign material in the body for removal. Some pathogens can **secrete proteases**, which degrade a class of antibodies called IgA. Since IgA is associated with mucous membranes and can bind to bacterial adhesins on pili the production of protease degrades the antibody and allows the bacteria to attach to cells. Other ways that pathogens avoid IgA is to **rapidly produce new pili**, the pili with antibody bound can disassociate from the bacteria, which will allow the organism to attach. Pathogens can also **alter their pili genetically** therefore the pili will be composed of a different sequence of amino acids and the immune system will have to produce a different IgA antibody to recognize the new pili. The last way bacteria may avoid antibodies are by producing an Fc receptor. Antibodies are Y shaped proteins with variable binding sites at the top of the Y and an Fc stem. Antibodies bind to foreign materials with the variable sites so the Fc stem then acts as a "red flag" for phagocytosis. Bacteria that produce Fc stems reversibly binds to the antibody (Figure 13.2) therefore a phagocytic cell will not recognize them. Organisms such as *Neisseria gonorrhea* can produce chemicals that prevent fusion of lysosome with the phagocytic vessicle and secretes a protease, which breaks down IgA in mucus. You will learn more about antibody interactions and the outcomes of antibody binding in Chapter 15.

Table 13.3 Bacterial Exotoxins

Bacterium	Disease	Toxin	Mode of Action
Bordetella pertussis	Whooping cough	Pertussis toxin	Necrosis
Clostridium botulinum	Botulism	Neurotoxin	Blocking the release of acetylcholine, resulting in flaccid paralysis
Clostridium perfringens	Gas gangrene, food poisoning	α-toxin	Lecithinase
		β-toxin	Necrosis
		ε-toxin	Necrosis
		ι-toxin	Necrosis
		θ-toxin	Hemolysis
Clostridium tetani	Tetanus	Neurotoxin	Blocking the release of inhibitory transmitter, resulting in spastic paralysis
Corynebacterium diphtheriae	Diphtheria	Diphtheria toxin	Inhibition of the elongation step in eukaryotic protein synthesis
Escherichia coli (enterotoxigenic strains)	Gastroenteritis	Heat-stable toxin (ST)	Activation of guanylate cyclase, resulting in fluid and electrolyte loss from intestinal cells
		Heat-labile toxin (LT)	Activation of adenylate cyclase, resulting in fluid and electrolyte loss from intestinal cells
Shigella dysenteriae	Bacillary dysentery	Neurotoxin	Inhibition of protein synthesis
Staphylococcus aureus	Pyogenic infections (for example, impetigo)	α-toxin	Hemolysis
		β-toxin	Hemolysis
		γ-toxin	Hemolysis
		δ-toxin	Hemolysis
		Enterotoxin	Unknown
		Leukocidin	Degranulation of leukocytes
Streptococcus pyogenes	Pyogenic infections (for example, impetigo), scarlet fever	Streptolysin O	Hemolysis
		Streptolysin S	Hemolysis
		Erythrogenic toxin	Localized erythematous reactions (abnormal redness of the skin)
Vibrio cholerae	Cholera	Choleragen	Activation of adenylate cyclase, resulting in fluid and electrolyte loss from intestinal cells
Yersinia pestis	Bubonic plague, pneumonic plague, septicemic plague	Plague toxin	Necrosis (possible)

Table 13.4 Comparison of Endotoxins and Exotoxins

Characteristic	Endotoxin	Exotoxin
Bacterial source	Gram-negative bacteria	Primarily gram-positive bacteria, some gram-negative bacteria
Location in bacterium	Lipopolysaccharide	Product of bacterial cell, released extracellularly
Chemical structure	Toxic activity resides in lipid portion of lipopolysaccharide	Protein
Heat stability	Stable; can withstand 121°C for 1 hr	Unstable; usually destroyed by heating at 60°C to 80°C (except enterotoxins)
Toxicity	Low	High
Toxoid production	No	Yes (except enterotoxins)
Representative symptoms	Fever, inflammation, increased phagocytosis, rash, septic shock	Neurological complications, cell necrosis, loss of fluid and electrolytes from intestines
Representative diseases	Septic abortion	Diphtheria, botulism, tetanus, cholera

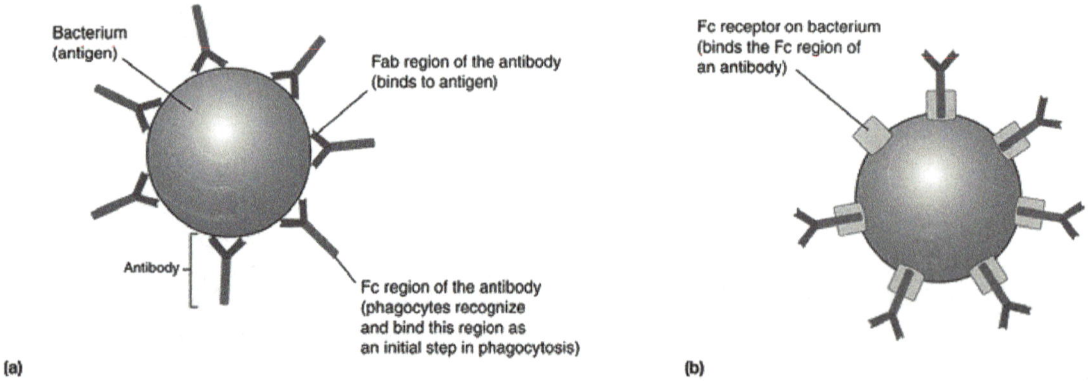

Fig. 13.2 **A)** Demonstrates how antibodies normally bind to a bacterium (antigen), with the Fc region exposed to the environment. **B)** Some bacteria produce Fc receptors so the antibodies bind to the bacteria the wrong way making them ineffective

Lab Exercise 13.1

Medical Microbiology: Urinalysis and Sexually Transmitted Diseases

Student Objectives

1. Understand and perform proper urine specimen collection techniques.
2. Perform and interpret a urinalysis test-strip to evaluate the physical characteristics of a urine specimen.
3. Operate a centrifuge as part of the sediment concentration procedure.
4. Perform a Gram stain and carry out a microscopic evaluation of urine sediment.
5. Perform a serial dilution and standard plate count to determine the number of colony-forming units per mL (CFUs/mL) present in a urine specimen.
6. Perform proper aseptic technique while inoculating special culture media.
7. Identify the types and concentrations of bacterial cells present in a urine specimen that indicate a urinary tract infection (UTI).

Name ____________________

Pre-Lab Exercise 13.1

1. Define the following UTI pathology terms:

   True bacteriuria: ____________________

   Dysuria: ____________________

   Pyuria: ____________________

   Cystitis: ____________________

   Acute pyelonephritis: ____________________

2. What are three types of abnormal crystals that may be found in urine?

3. What types of human cells may be found in normal urine?

4. What is a cast?

5. What are the three routes by which bacteria can enter the urinary tract?

6. List the bacteria responsible for causing most urinary tract infections (UTIs).

INTRODUCTION

In this lab you will be analyzing your own urine to look for signs of bacterial infection and signs of other physiological conditions. It is an absolute requirement that you bring your own urine sample and not someone else's! **Urine** is the liquid by-product of the body's metabolic processes. Urine is secreted by the kidneys, which extract soluble waste products from the bloodstream and regulate electrolyte balance. The composition of urine thus reflects dietary intake, physical activity, organ function, infection, and other measures of health. The urethra normally supports the growth of some microbiota, mainly avirulent species of *Staphylococcus*, *Streptococcus*, and *Lactobacillus.* In both males and females, microorganisms colonizing the urethra rarely move into the bladder, up the ureters, and infect the kidneys. Opportunistic pathogens and sexually transmitted microbes can not only infect the urinary tract but they can also infect the reproductive systems.

Bacterial infections of the urinary tract are commonly derived from the normal microbiota usually through fecal contamination. There are bacteria that are not a part of the normal microbiota that are transmitted sexually and can be identified by microscopic examination, urinalysis, and biochemical testing. Bacteria involved in sexually transmitted diseases do not survive on cool, dry surfaces such as toilet seats therefore their transmission is only through sexual contact.

A **urinalysis** is the analysis of urine composition, including its physical characteristics and presence of dissolved organic and inorganic substances. In this laboratory exercise you will perform a urinalysis using physical observations, test-strip method, microscopic evaluation, and special cultures of your urine specimen. Record your results in the Lab Report. You will also observe prepared slides of common sexually transmitted pathogens. You will draw and record your observations in your Lab Report.

Urinary tract infections (UTIs) are bacterial or fungal infections of the urinary tract. They are most common in females, due to the close proximity of the urethral opening to the anus and the moist, protective environment of the labia. In fact millions of females in the United States suffer from bacterial UTIs each year. This includes the approximately 600,000 patients who acquire nosocomial UTIs annually.

Routes of Invasion

Microorganisms enter the urinary tract through one of three routes: ascending, hematogenous, and lymphatic. The *ascending route* is the most common pathway of infection in females. Improper wiping (from back to front) following restroom use brings fecal bacteria forward to the vaginal and urethral openings. To avoid this route of infection, females should be taught to wipe from front to back. Sexual intercourse can also force bacteria into the bladder through the ascending route. The use of spermicides inhibits the growth of the normal vaginal microbiota *Lactobacillus*, leading to colonization by *E. coli* bacteria from the colon.

Catheterization can lead to UTIs via the ascending route. About 1% of patients undergoing a single catheterization develop a UTI, and virtually all patients with indwelling catheters develop a UTI within three to four days.

The *hematogenous route* refers to infection via the blood. It is rare cause of UTIs, and almost always involves Gram-positive bacteria such as *Staphylococcus.*

The *lymphatic route* refers to infection via lymphatic flow, and is quite rare.

Types of Bacteria

The most common bacteria associated with UTIs are members of the *Enterobacteriaceae* family, including *E. coli*, *Enterobacter*, *Klebsiella*, and *Proteus*. Over 80% of UTIs are caused by *E. coli*, which possess several factors that enhance their ability to cause infection and disease. First, the uropathogenic strains have a very short *generation time* (20–50 minutes), allowing them to accumulate between normal bladder emptying times. Next, these strains have *fimbriae* which allow them to adhere to the body's mucosal cells and resist the normal flushing action of the bladder. Furthermore, these strains often possess *flagella*, enhancing their motility and allowing them to travel up the urinary tract through the ascending route. Fourth, the pathogenic *E. coli* produce *hemolysin*, an exotoxin that lyses the body's red blood cells and thus causes tissue damage. Some *E. coli* may also produce an endotoxin which is released if their lipopolysaccharide layer is destroyed by the body's immune response. Last, many uropathogenic strains are surrounded by a capsule, enabling them to evade detection by phagocytic cells that would normally engulf and destroy foreign bacteria.

Enterobacter aerogenes and *E. cloacoa* can both cause UTIs. They are motile and produce the enzyme *urease*, which allows them to hydrolyze urea as a food source.

Although *Klebsiella oxytoca* and *K. pneumoniae* are nonmotile, they are associated with UTIs due to their fimbria, capsules, and ability to hydrolyze urea.

Proteus mirabilis and *P. vulgaris* possess fimbriae for adherence, flagella for motility, and urease for urea metabolism.

Types of UTIs

Some urinary tract infections do not produce symptoms and are referred to as *asymptomatic bacteriurias*. *True bacteriuria* is indicated when the bacteria are present in significant numbers (100,000 CFUs/mL) and the patient displays symptoms of infection. Symptoms range from *dysuria* (difficult or painful urination) to *pyuria* (pus in the urine). Should the bladder wall become infected, the condition progresses to *cystitis* in which the patient experiences a frequent and urgent need to urinate. If the infection moves to the kidney, it is referred to as *acute pyelonephritis*, with systemic symptoms including chills, fever, nausea, vomiting, and back pain.

Treatment

Patients with chronic UTIs are often treated with Bactrim® or Septra® (co-trimoxazole), a medication composed of sulfamethoxazole and trimethoprim. This compound reduces the production and functioning of fimbriae.

Sexually Transmitted Pathogens

Sexual contact between partners results in the transmission of microbes. When the microbes are potential pathogens, this type of transmission is known as a **sexually transmitted infection (STI)**. Diseases resulting from a STI are sexually **transmitted disease (STD)**, which are also known as venereal diseases. The World Health Organization (WHO) estimates approximately 333 million cases of STDs occur each year. The only way to prevent STDs is through mutual monogamy or abstinence. Latex and polyurethane condoms reduce the risk of infection but they are not 100% safe. Properly used condoms have a failure rate of 17%–25%. Circumcision has been shown to reduce a man's chances of acquiring several sexually transmitted diseases including genital warts, AIDs, syphilis, herpes, and gonorrhea. There are examples of viral, protozoan, and bacterial sexually transmitted diseases causing a wide variety of signs and symptoms. We will briefly survey the 3 organisms, and the diseases they cause, you observed during this laboratory activity.

Gonorrhea

Neisseria gonorrhoeae, as the name suggests this is the causative agent of the STD gonorrhea, which only causes disease in humans. Physicians have known about gonorrhea for centuries and were often confused with syphilis until the 19th century. Men will usually develop acute inflammation 2–5 days after infection in the urethra. Women are often asymptomatic 50–80% of the time.

Neisseria gonorrhoeae, also known as gonococcus, is a Gram-negative bacterium, which usually forms pairs of cells (diplococcus) that have fimbriae and polysaccharide capsules. The diplococcus bacteria will also secrete a protease enzyme that breaks down secretory IgA in mucus. Gonococci adhere to human cells using their capsules and fimbriae, and as few as 100 cells is all that is needed in order to establish and cause disease. Interestingly, *N. gonorrhoeae* can attach to sperm as they swim past the cervix and hitchhike to uterine tubes leading to pelvic inflammatory disease in about 25% of females. The gonococci can infect the urethra during sexual intercourse. Anal intercourse can lead to inflammation of the rectum (proctitis), while oral sex allows for infection of the gums or pharynx, resulting in gingivitis or pharyngitis. Phagocytized bacteria are able to survive and multiply in neutrophils, traveling inside the leukocytes to other areas of the body. In rare cases *N. gonorrhoeae* can travel to joints (causing arthritis), meninges (meningitis), or heart (endocarditis).

Treatment of gonorrhea in recent years has been complicated due to the emergence of multidrug resistant strains of *N. gonorrhoeae*. Strains of the diplococcus have shown resistance to penicillin, tetracycline, erythromycin, aminoglycosides, and fluoroquinolones. Currently the Center for Disease Control (CDC) recommends broad-spectrum drugs such as ceftriaxone or cefixime to treat gonorrhea. Individuals who acquire gonorrhea do not develop long-term immunity, which can be explained by the variable surface antigens this organism produces. This means immunity against one strain provides no protection against other strains. The highly variable surface antigens *N. gonorrhoeae* produce also prevents the development of an effective vaccine.

Syphilis

Syphilis was first recognized in 1495 and to this day afflicts millions of people worldwide. Syphilis is caused by the bacterium *Treponema pallidum*, this organism is a narrow spirochete difficult to see using a regular light microscope. This bacterium lives naturally only in humans and absolutely cannot survive in the environment. The spirochete's environmental and nutritional requirements are so specific that no scientist has ever cultured this organism in cell free media.

There are 4 phases of syphilis:

Primary signs: A small, painless, red, hard lesion called a chancre, forms at the site of infection 10–21 days post exposure.

Secondary signs: sore throat, headache, mild fever, malaise, myalgia, lymphadenopathy (diseased lymph nodes) and a widespread rash.

Latent: no signs or symptoms

Tertiary: Years later, if untreated some patients experience dementia, paralysis, heart failure, blindness, syphilitic gummas (rubbery lesions on bone, nervous tissue, or skin).

Congenital syphilis: Results when *Treponema* crosses the placenta from an infected mother. If the mother is experiencing primary or secondary syphilis this often results in death to the fetus. If transmission occurs during latent phase the fetus often suffers from mental retardation and malformation of organs.

Diagnosis of syphilis is done using specific antibody tests against antigens of *Treponema pallidum* such as the microhemagglutination assay against *T. pallidum* (MHA-TP). Spirochetes can also be observed in fresh discharge from lesions however, the microscopic examination must be conducted immediately since *Treponema pallidum* rarely survives the trip to the laboratory. Primary, secondary, and latent phases, and congenital syphilis are effectively treated with penicillin G, but it does not work for the tertiary phase.

Chlamydial infections

Chlamydia trachomatis is the most common sexually transmitted bacterium among humans. There were approximately 1,200,000 cases in 2009 while another 3–4 million cases go unreported. Chlamydia cells stain Gram-negative and have 2 membranes similar to Gram-negative bacteria however, they lack a peptidoglycan cell wall. This organism is known to cause congenital diseases and several sexually transmitted diseases. Approximately 85% of chlamydial genital tract infections in women are asymptomatic. In men 75% show signs and symptoms similar to gonorrhea which include painful urination, pus discharge, and urethritis. Babies, when infected during the birthing process, can develop an eye disease called trachoma, which is the leading cause of blindness in humans worldwide.

A severe form of chlamydial STD is lymphogranuloma venereum, which shows an obvious genital lesion at the site of infection on the penis, urethra, scrotum, cervix, vagina, or external female genitalia. The lesion is followed by a bubo, a painfully inflamed lymph node in the groin. Buboes may enlarge and rupture, producing a draining sore.

Physicians prescribe tetracycline, azithromycin, or erythromycin to treat chlamydial genital infections in adults. Newborns with eye infections caused by *Chlamydia trachomatis* are treated with erythromycin cream for 10–14 days. As with all STDs abstinence or monogamy is the only preventative measures. Latex condoms, even when used properly, are not 100% effective.

Specimen Collection

Materials

Laboratory materials: urine collecting container, antiseptic wipe

1. Proceed to the restroom with a urine-collecting container and antiseptic wipe.
2. Wash your hands.
3. Enter the stall. Use the antiseptic wipe to clean the urinary meatus. Females should clean from front to back; males should clean with a circular motion from the tip of the penis outward.
4. Void a small amount of urine (about 25 mL) before collecting your sample, as this first void generally contains an abnormally high microorganism count.
5. Collect a midstream sample of at least 50 mL of urine into your collection cup. Avoid contaminating the inside of the cup with fingers or clothing. Seal the lid.
6. Wash your hands before returning to the lab.

Physical Observations

Materials

Cultures: urine specimen

1. Describe the *color* of your urine sample. Normal urine varies from pale yellow to dark amber, depending on the presence of urochrome pigments produced as by-products of normal hemoglobin breakdown. Colorless urine often indicates over-hydration, while deep yellow urine may indicate dehydration. A bright fluorescent yellow-green color is often seen in the urine of individuals taking a vitamin B supplement. Other foods and medications may also alter the appearance of the urine specimen.
2. Describe the *turbidity* of your urine sample. Normal urine is usually transparent (clear enough to see through). Cloudy urine may indicate the presence of various substances such as vaginal secretions, seminal fluid, bacteria, epithelial cells, fat droplets, or inorganic salts such as calcium phosphate.
3. Describe the *odor* of your urine sample. Urine has a distinctly recognizable odor. A fruity odor is due to the presence of ketones, and is often associated with diabetes mellitus. A foul-smelling odor usually indicates a urinary tract infection (UTI). Some foods, such as asparagus, give urine an unusual odor due to the presence of asparagusic acid

Test-Strip Analysis

Materials

Laboratory materials: Pro Advantage urine test-strip with interpretive chart

1. Aseptically obtain a Pro Advantage urine test-strip and interpretive chart.
2. Quickly dip the test-strip in your urine specimen, ensuring that all reagent blocks on the test-strip come into contact with your urine sample.
3. Immediately remove the test-strip from your sample, as oversaturation produces erroneous results. Tap it against the rim of the specimen cup to remove any drippings. Hold the test-strip horizontally to prevent the reagent blocks from mixing.
4. Compare the colors of the reagent blocks with the corresponding colors on the interpretive chart. To prevent contamination, *avoid resting the test-strip on the chart! **Make sure that you read each reaction at the appropriate time interval. Color changes occurring after the specified time are misleading and should not be interpreted.***
   a. ***Glucose***—Glucose is usually not found in the urine of healthy individuals, because it is reabsorbed from the renal tubules back into the blood. Slightly elevated glucose readings may occur after the consumption of a high-sugar meal. A glucose concentration above 180 mg/dL is known as glycosuria, and is used to diagnose and monitor the treatment of diabetes mellitus.
   b. ***Bilirubin***—Bilirubin is a pigment molecule that results from the breakdown of hemoglobin from old red blood cells. It is normally secreted into the bile and passed through the small intestines. Bilirubinuria, the presence of bilirubin in the urine, usually indicates liver dysfunction or bile duct obstruction. *A bright pink color on the bilirubin reagent block often results from oversaturation. Perform the test again with a new test-strip.*
   c. ***Ketones*** *(ketone bodies)* – Ketones result from the metabolism of fats. Ketonuria, or the presence of ketones in the urine, occurs with insulin deficiency diabetes, fasting, or low-carbohydrate diets, in which the body breaks down fats rather than glucose for energy.
   d. ***Specific gravity***—Specific gravity is a measure of the concentration (density) of solid substances in the urine. The term is used to compare how heavy the urine is with respect to an equal amount of distilled water, which has a specific gravity of 1.0. The specific gravity of urine varies from 1.003 (dilute urine in a well-hydrated individual) to 1.035 (concentrated urine in an under-hydrated individual). Early morning urine samples generally have a higher specific gravity reading than samples taken later in the day.
   e. ***Blood***—Urine generally does not contain red blood cells, except in the case of females who are menstruating. Hematuria, the presence of red blood cells in the urine, indicates bleeding in the genitourinary tract. *The urine test-strip may give a false-positive blood result if pigments such as hemoglobin or myoglobin are present, so urine specimens with positive blood test-strip results should be further examined microscopically for the presence of red blood cells.*
   f. ***pH*** – Urine pH varies from 4.6 to 8.0. A diet high in fruits and vegetables tends to increase pH (basic/alkaline), while a meat-based diet tends to decrease pH (acidic). Acidity can contribute to the formation of uric acid stones in the kidneys.

g. ***Proteins***—Large proteins are not present in the urine, but smaller proteins such as albumin may appear in trace amounts, particularly following physical exertion or stress. Pathological causes of protein in the urine include glomerulonephritis (an infection of the kidneys) and hypertension (high blood pressure). Protein in the urine is known as proteinuria.
h. ***Urobilinogen***—Urobilinogen results from the intestinal breakdown of bilirubin by normal microbiota. Approximately half of this is excreted in the feces, while the rest is reabsorbed through the intestines. Urobilinogen is generally not present in the urine; high levels may indicate liver disease or a hemolytic disorder.
i. ***Nitrite***—Nitrituria, the presence of nitrite in the urine, indicates infection with coliform bacteria which reduce nitrate to nitrite in the process of cellular metabolism. This test is used as a diagnostic screen for urinary tract infections (UTIs).
j. ***Leukocytes***—White blood cells in the urine is known as leukocyturia, indicating an infection in which neutrophils and macrophages are actively responding.

5. When you are finished recording all of your observations, dispose of the contaminated test-strip in the biohazard bag.

Sediment Concentration

Materials

Laboratory materials: sterile 10-mL pipette, sterile centrifuge tube, centrifuge, inoculating loop, Bunsen burner and sparker

1. Sediment, or the microscopic solids dissolved in a urine sample including cells, crystals, and casts, can be separated from the solution and concentrated for further study. Begin by aseptically transferring 10.0 mL of your urine sample from the collection cup to a sterile centrifuge tube, using a sterile 10-mL pipette. Cap the centrifuge tube and label the side of it with your initials. Dispose of the pipette in the Nalgene container.
2. Load your tube into the centrifuge with a proper balance: tubes should be loaded in pairs across from one another. *Operating a centrifuge out-of-balance is dangerous and causes damage to the machine.* When the centrifuge is loaded and balanced, spin at 2,000 rotations per minute (RPMs) for 10 minutes.
3. Allow the centrifuge to stop completely. Remove your tube. Carefully decant all but a very small amount of the urine (about 0.5 mL) down the staining sink drain, being careful not to disturb the pellet at the bottom of the tube.
4. Sterilize an inoculating loop using a Bunsen burner flame. Resuspend your pellet by scraping the bottom of the centrifuge tube with the sterile loop, breaking up the sediment
5. particles into the remaining volume of urine.

Special Culture Inoculation

Materials

Laboratory materials: inoculating loop, Bunsen burner, igniter, 1 Eosin Methylene Blue (EMB) plate, 1 Tryptic Soy Agar plate.

Cultures: resuspended concentrated sediment

1. Obtain an EMB and TSA plate and label the bottom of it with your initials.
2. Perform a quadrant streak for isolation onto your EMB and TSA plate, using a sterile loop containing one loopful of your resuspended materials.
3. Tape the plate closed and place it upside down (inverted) on the incubation tray. It will be cultured for 24 hours at 37° C.
4. Dispose of your centrifuge tube in the biohazard bag.
5. Pour any remaining urine from your collection cup into the toilet, recap the lid, and dispose of the cup in the biohazard bag.

Microscopic Evaluation

Materials

Laboratory materials: (2) microscope slides, Bunsen burner and igniter, inoculating loop, Gram stain reagents, microscope, urine sediment diagram

Cultures: resuspended concentrated sediment

1. Prepare a smear-prep and perform a Gram stain using one loopful of your resuspended materials.
2. Observe the sediment under the microscope, working your way up to the 40× objective lens. Be sure to observe classmates' slides as well, since each sample will contain different sediments. Refer to the classroom urine sediment diagram to identify the crystals, cells, and casts present in the specimens.
   a. ***Crystals***—Crystals are found in trace amounts in a normal urinalysis. The types of crystals present depend on the pH of the urine and the diet of the individual. Abnormal crystals include cysteine (indicating an inherited autosomal-recessive) trait that prevents cysteine uptake), leucine (indicating hepatitis or liver cirrhosis), and tyrosine (indicating liver disease).
   b. ***Cells***—Epithelial cells may be found in normal urine sediment due to the sloughing off of older cells in the urinary tract. Red blood cells may be present due to menstruation or strenuous exercise. The presence of more than one or two red blood cells per high-powered field (HPF) is considered to be hematuria. White blood cells are frequently observed in small amounts (less than 5 per HPF). Large numbers of white blood cells (greater than 350 per HPF) indicate acute bacterial infection of the urinary tract.
   c. ***Casts***—Casts are cylindrical masses of cells that form in the distal tube of the nephron, and are flushed out by the flow of urine. Normal casts include hyaline and granular, both of which increase during exercise and stress. Abnormal casts include blood cell (indicating hemorrhage or inflammation), epithelial (indicating kidney disease or viral infection), and crystal (indicating renal obstruction). Bacteria and yeast caught up in the formation of a cast will be seen covering the cast surface. Increased amounts of mucus usually result from vaginal contamination.

Microscopic Examination of Sexually Transmitted Pathogens

1. Obtain a prepared slide of each of the following: *Neisseria gonorrhea.* Depending on your instructor you may be instructed to perform a Gram stain of this organism.
2. Using proper microscope procedures, view the prepared slides at the 10×, and 40× objective lenses.
3. Locate the bacterial cells and draw a few cells in the lab report at 40× magnification (or 100× if you prepared your own Gram stain).

Name ______________________________

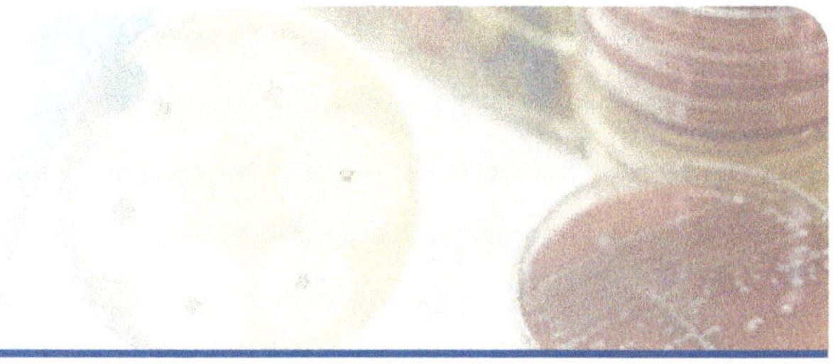

Exercise 13.1 Lab Report

1. Describe the color, turbidity, and odor of your urine specimen.

____

____

____

2. Record your test results in the report form below.

URINALYSIS

Urinalysis

Report Form

Pro ADVANTAGE® by NDC

Patient Name: ______________________________

Age: __________ ❑ Male ❑ Female

Physician's Name: ______________________________

Collection Date: __________ Test Date: __________ Tester's Initials: __________

Physical Characteristics:

Color: ❑ Colorless ❑ Yellow ❑ Amber ❑ Orange ❑ Green ❑ Red ❑ Other

Appearance: ❑ Clear ❑ Hazy ❑ Cloudy ❑ Turbid

Chemical Measurements: (circle one)

Urobilinogen (mg/dL)	Normal	2	4	8			
Glucose (mg/dL)	Negative	50	100	250	500	1000	
Ketone (mg/dL)	Negative	trace/5	+/15	++/40	+++/80	++++/160	
Bilirubin	Negative		+	++	+++		
Protein (mg/dL)	Negative	trace		+/30	++/100	+++/300	++++/2000
Nitrite	Negative	Positive (any pink color is considered positive)					
Leukocytes	Negative	trace	+	++	+++		
Blood	Negative	trace	mod	trace	+/small	++/mod	+++/large
		Non-Hemolyzed		Hemolyzed			
pH	5	6	6.5	7	8	9	
Specific Gravity	1.000	1.005	1.010	1.015	1.020	1.025	1.030

Microscope Examination:

WBC __________/HPF Crystals __________ Spermatozoa __________

RBC __________/HPF Bacteria __________ Artifacts __________

Casts __________/LPF Yeast __________ Other __________

Epithelial Cells ______/HPF Parasites __________

Comments:

____

____

____

____

www.ProAdvantagebyNDC.com

3. Did you note any abnormal test results? If so, describe what these results may indicate.

____

____

____

4. Name two results on a urine test-strip analysis that would indicate a bacterial infection and explain why they indicate an infection.

____

____

____

____

____

5. Identify and sketch at least one crystal, cell, and cast seen in the Gram stain of your or a classmate's urine specimen. Also, draw and label the 3 sexually transmitted organisms from the prepared slides. Be sure to use colored pencils in your drawings.

CRYSTAL

Sediment: ______________

Total magnification: 400×

CELL

Sediment: ______________

Total magnification: 400×

CAST

Sediment: ______________

Total magnification: 400×

N. gonorrhoeae

6. Did you or any of your lab partners observe any abnormal crystals, cells, or casts in your urine specimen? If so, describe what the presence of these sediments may indicate.

7. Describe the growth on your eosin methylene blue (EMB) plate, including colony size and color. Based on colony appearance, are you able to exclude any species from your diagnosis? What type of organisms are growing on the EMB plate, Gram positive or negative?

8. What type of organisms are most likely growing on the TSA plate Gram-positive or Gram-negative? How did you determine this?

9. Why does a disease that has asymptomatic hosts problematic for healthcare officials?

10. What are the only ways to 100% prevent sexually transmitted diseases?

11. Discuss the virulence factors of *Neisseria gonorrhoeae*.

12. Discuss the 4 stages of syphilis and determine the antibiotic used to treat syphilis at each stage of the illness.

13. Discuss the types of infections caused by *Chlamydia trachomatis*.

Lab Exercise 13.2

Medical Microbiology: Blood Agar and Hemolysis

Student Objectives

1. Take medical samples, given the appropriate media and collection tools.
2. Understand that microorganisms are everywhere, including in and on your body.
3. Understand the use of specific media for the identification of unknown bacteria.
4. Use differential media to determine hemolysis of blood agar.
5. Differentiate between alpha, beta, and gamma hemolysis.
6. Evaluate the types of bacteria that perform various hemolysis on the blood agar.

Name ______________________________

Pre-Lab Exercise 13.2

1. Describe **beta-hemolysis** on a blood agar plate. Give an example of a Gram-positive bacterium that will show this?

2. Describe **alpha-hemolysis** on a blood agar plate. Give an example of a Gram-positive bacterium that will show this?

3. Briefly describe **two** diseases caused by *Staphylococcus aureus.*

4. Describe why species of *Streptococci* are put into different Lancefield groups.

5. Explain the virulence of exotoxin A and exotoxin B and how these toxins promote an infection from *Streptococcus pyogenes.*

INTRODUCTION

The **blood agar plate** contains 5% sheep blood added to TSA. It is used to determine the type of hemolysis displayed by the bacterium, as well as distinguish between the types of bacteria. We call this type of medium a **differential medium.**

Bacteria cause harm to the body by producing substances which are toxic to the body. These substances are referred to as **toxins.** Many staphylococci and streptococci species produce substances toxic to red blood cells, referred to as hemolysins. **Hemolysins** disrupt the cell membrane of RBCs, leading to lysis of the cell. The lysis of the red blood cell membrane is called "**hemolysis.**"

The most damaging hemolysins are the **alpha-toxin** (a-toxin), produced by *Staphylococcus aureus* and the *Clostridium* genus, and the **streptolysin enzyme**, produced by *Streptococcus pyogenes*. Some streptococci also produce the **beta-toxin**. When these organisms produce these substances on a TSA-B (blood agar) plate, the colonies live off the blood in the agar around where they are growing, rupturing the red blood cell membranes, leaving a characteristic clear zone around the colony. This pattern displayed is known as β-hemolysis. ***β-hemolysis*** is defined as complete lysis of the red blood cell, rupturing the cell membrane, which displays itself as a **totally clear zone** around the colony. The agar appears amber-colored instead of red.

Other bacterial toxins and enzymes do not completely disrupt the red blood cell membrane, and thus will show different characteristics on the blood agar. ***α-hemolysis*** is defined as the reduction (breakdown) of the hemoglobin protein, on the red blood cell membrane, and will show a **"partial clearing"** of RBCs around the colony. This causes the blood in the agar to become discolored and will show a **greenish** discoloration instead of the red color typical of RBCs. In reality, when you look at your plates this will most likely appear a greenish-brown color. This takes you back to learning your color combinations: green + red = brown. This is the same as "bruising of the body": the breakdown of hemoglobin creates green and brown discoloration under your skin. Note: the red blood cells are still intact, so this is not a true lysis.

If **no hemolysis** takes place, the underlying agar remains bright red due to the presence of whole red blood cells. This lack of any hemolysis is known as ***γ-hemolysis*** (gamma).

APPLICATION

Staphylococci

The genus *Staphylococcus* contains both pathogenic and non-pathogenic organisms. They do not produce endospores but are highly resistant to drying, especially when associated with organic matter such as blood, pus, and other tissue fluids. Most staphylococci are found routinely on the surface of the skin. Breaks in skin and mucous membranes allow entrance of these organisms into the body where they may cause disease.

Staphylococci are divided into two major groups based on the coagulase test: coagulase positive (CoPS), which includes just the one species, *Staphylococcus aureus;* and coagulase negative (CoNS), which includes all other species of staphylococci, such as *Staphyloccus epidermidis, S. saprophyticus* , and *S. haemolyticus.* 30% of people carry *S. aureus* as normal microbiota of their nose. Staph species colonize skin, external eye, external ear, and mucous membranes.

All *Staph* colonies are 1–3 mm in diameter. They are opaque, smooth, convex, and circular. They have a butyrous (buttery) appearance. **Colonies of *Staphylococcus aureus*** are usually yellow to orange to white, and are usually **beta hemolytic. Colonies of *Staphylococcus epidermidis*** are usually gray-white and are non-hemolytic (**gamma hemolysis**). **Colonies of *Staphylococcus saprophyticus*** are often yellow to orange and are also non-hemolytic.

Microscopically, *Staphylococcus* is a Gram-positive, non-motile, coccus-shaped bacterium. They are 1 µm in size and occur in grapelike clusters. Sometimes those clusters are disturbed by the staining process, leaving staphylococci in singles, pairs, short chains, and tetrads, along with the clusters.

Staphylococcus aureus is the main cause of staph infections. Infections vary from skin infections to serious progressive, invasive diseases. Symptoms include production of pus, swelling, redness, and pain. If the infection is extensive, or has entered the blood stream, fever is present. Some strains of *S. aureus* can produce toxic-shock syndrome (TSS), sometimes accompanied by rash and diarrhea. It is likely that multiple virulence factors act together to produce a wound infection. Clumping factor, coagulase, and protein A coat the organisms and disguise them from phagocytic attack. Released protein A reacts with immunoglobulin and contributes to inflammation and accumulation of pus.

Most wound infections are due to the patient's own *Stahylococcus* strain. Advanced age, poor general health, immunosuppression, prolonged pre-op hospital stay, and infection at a site (other than the site of surgery) increase infection risk. Treatment is problematic because of developed antibiotic resistance. New antibiotic use is also problematic, but vancomycin is generally effective against methicillin-resistant *Staphylococcus aureus* infections (MRSA).

The most important reservoirs of MRSA are infected or colonized patients. Hospital personnel can serve as reservoirs for MRSA and may harbor the organism for many months. Hospital personnel have been more commonly identified as a link for transmission between colonized or infected patients. The main mode of transmission of MRSA is via hands (especially health care workers' hands) which may become contaminated by contact with colonized or infected patients, colonized or infected body sites of the personnel themselves, or devices, items, or environmental surfaces with body fluids containing MRSA.

Streptococci

Diseases caused by the genus *Streptococcus* are bacterial pneumonia, meningitis, tonsillitis, endocarditis, scarlet fever, erysipelas, and urinary tract infections. *Streptococcus* species are also found normally in the mouth and on the skin surface. The streptococci are classified by two major methods: hemolytic activity and serologic classification of Lancefield.

Rebecca Lancefield specifically studied streptococcal species that displayed **beta hemolytic** patterns on the blood agar plate. She discovered proteins in the cell wall that were unique to certain organisms. These proteins were labeled Group A, Group B, Group C, and so on, through Group M. Currently three Lancefield Groups are of medical importance: Group A, Group B, and Group D.

- Group A Strep—*Streptococcus pyogenes.*
- Group B Strep—*Streptococcus agalactiae.*
- Group D Strep—*Streptococcus bovis, Enterococcus (Streptococcus) faecalis.*

Streptococcus pneumoniae does not possess Lancefield proteins and is not classified in one of the Lancefield groups. Viridans streptococci is the term applied to **alpha hemolytic** *Streptococcus* species that lack Lancefield proteins.

All strep colonies are small, being around 0.5 mm in diameter. They are transparent to opaque, smooth, and circular. All strep colonies are coagulase negative. Colonies of ***Streptococcus pyogenes*** (known as Group A Strep) are convex, usually show β-hemolysis, with large areas of clearing that are two to four times the diameter of the colony. Colonies of ***S. agalactiae*** are flat, have a shiny, mucoid (mucus-like) appearance, and usually show β-hemolysis. Colonies of

S. pneumoniae (known as pneumococci) are flat, shiny, mucoid, and display α-hemolysis. Colonies of ***S. mutans*** are small, translucent, and whitish-gray and ***S. sanguis*** (known as viridans strep) are small, opaque, and gray to whitish-gray. ***S. mutans*** is non-hemolytic, while ***S. sanguis*** is α-hemolytic.

Microscopically, *Streptococcus* is a Gram-positive, non-motile, coccus-shaped bacterium. They are 1 µm in size and occur in pairs or chains. Often, especially in a young culture, they will appear elongated, ovoid to rodlike in shape.

An optimum balance of nutrients and a suitable growth environment are needed to culture a microorganism in the laboratory. The culture media contains all the materials necessary for the microorganism to move quickly into the log phase of growth. Many growth media can be prepared for the microbiologist to isolate, grow, and identify a particular species in the laboratory. Some media have ingredients that encourage the growth of a suspected pathogen, whereas others include materials that discourage the growth of unwanted contaminants. Special media have been designed to identify a suspected pathogen in preparation for biochemical testing. The proper laboratory diagnosis of an infectious disease helps the patient successfully cope with an infection and provides the health care worker with information that may prevent others from becoming ill. Because the cultivation of a suspected pathogen is vital to successful treatment, all those involved in the process must be aware of what happens in the lab.

Streptococcus pyogenes (Group A Strep, also known as GAS)

Streptococcus pyogenes is not considered part of a person's normal microbiota. Spread by droplet transmission, they are the main cause of strep infections, especially strep throat (streptococcal pharyngitis). An especially virulent strain causes streptococcal toxic shock (STSS), due to pyrogenic exotoxin A and exotoxin B which destroys tissue by breaking down protein. Spread by contact transmission, they cause skin infections. Symptoms include acute pain development at the surgical wound site. Swelling, fever, and confusion quickly develop. The overlying skin is tense and discolored. Without prompt treatment, shock and death follow shortly.

Streptococcus pyogenes multiplies in the wound, releasing enzymes and toxins. Subcutaneous fascia and fatty tissue can be destroyed. This is known as necrotizing fasciitis, or the flesh-eating disease. The toxins enter the bloodstream. There are 10-20 cases per 100,000 infections. They are associated with injected drug abuse, diabetes, liposuction, hysterectomy, and bunion surgery, but most occur with minor injuries. They require urgent surgery to release pressure and remove dead tissue, sometimes including amputation. Penicillin works with early infection, but has no effect on organisms in necrotic tissue and no effect on toxins. Contact precautions are needed.

Blood Agar Activity

Materials

Laboratory materials: Blood agar plate (BAP), sterile swab, marker, masking tape.

1. Label the bottom of the blood agar plate with your initials and table number.
2. Remove the sterile swab from its package.
3. Collect a sample with the swab from the back of your throat.
4. Remove the cover of the blood agar plate. Use the swab as your 1st quadrant. Switch to a sterile loop and streak for isolation using the quadrant method.
5. Place the swab back in its package and put in the biohazard bag.
6. Immediately put the top back on the plate. Secure with two small pieces of masking tape.
7. Invert the plate. Place your plate on the classroom tray for incubation.

Name ______________________________

Exercise 13.2 Lab Report

1. Describe the variety of colony types seen on your **blood agar plate** as to size, shape and color. Use words and/or phrases that "paint a picture" to someone not viewing the plate. Be sure to list where you acquired your sample.

2. **Compare your results** with the results of other students in the lab. What factors would influence the different results you observe?

2. What **conclusions** can you draw based on your sample and the samples from other students?

3. Why is the blood agar called a differential medium?

4. Explain the role of hemolysins during a bacterial infection.

Epidemiology

The Center for Disease Control and Prevention gives the following definition for epidemiology:

*Epidemiology is the **study** of the **distribution** and **determinants** of **health-related states or events** in **specified populations**, and the **application** of this study to the control of health problems.*

Epidemiologists are essentially disease detectives. They collect and compile information on the spread of disease. In order to do so, the epidemiologist must look at the mode of transmission, reservoirs and risk factors of disease. They then compile statistics and data on disease and use this information to design disease control strategies.

An Early Example of Epidemiology

The "father of epidemiology" is a British physician, John Snow. In 1854 several London residents contracted cholera, a waterborne disease caused by the bacterium *Vibrio cholera*. Cholera causes an osmotic imbalance in the intestines of its victims, resulting in profuse, watery, "rice water" diarrhea. The diarrhea can cause severe dehydration and death. Snow interviewed the victims and other London residents and was able to use the information he gathered to put together a map and statistical analysis which pinpointed the source of infection as a single well on Broad Street. Once the source of infection was identified, the well was closed, preventing further incidence of cholera.

Basic Principles of Epidemiology

Diseases can be classified based on whether or not they are transmissible.

- **communicable:** transmissible from person to person
- **contagious:** *easily* transmitted from person to person
- **noncommunicable:** not transmissible from person to person

One-way to determine whether or not a disease is caused by an infectious agent is to use **Koch's Postulates.** Koch's Postulates state that:

1. The causative agent must be present in every case of the disease. *For example, in order to be MRSA, methicillin-resistant S. aureus must be present on every victim*
2. The organism must be grown in culture from the diseased host. *This means a sample of MRSA should be collected from the host and grown in the lab*
3. The same disease results when the organism grown in the lab is introduced to a new host.
4. The same organism must be present in the experimentally infected host. *The experimentally infected host must exhibit symptoms of MRSA and MRSA must be collected from the host.*

While Koch's Postulates are helpful in determining whether a disease is infectious or not, they do not work for all microbes. There are microbes that we know are infectious, but do not meet Koch's Postulates because they have not been successfully cultured in a lab. Examples include *Treponema pallidum*, the causative agent of syphilis, and *Mycobacterium leprae*, the causative agent of leprosy. Problems with Koch's Postulates also arise due to postulate 3. Some microbes are very specific in regard to their hosts, and introducing these microbes to humans experimentally would be unethical. Such is the case with HIV.

Rates of Disease in Populations

Epidemiologists work extensively with statistical data. By collecting and analyzing data on the number of cases of a given disease, they can make predictions about the spread of disease and also institute control measures. These statistics include:

- **morbidity rate:** new cases of disease that appear in a given population during a specific time period
- **mortality rate:** ratio of number of people who died of a particular disease per time period in a given population
- **endemic:** the disease is always present to some degree in the population
- **outbreak/sporadic disease:** a sudden rise in the occurance of disease, limited to *either* a small number of people (or particular group of people) or a small geographic area
- **epidemic:** a disease that occurs at higher than normal frequency in a given geographic location
- **pandemic:** a disease that occurs at higher than normal frequency and has crossed international boundaries.

Epidemiologists also keep track of the **prevalence,** or the total number of cases in a given area and the **incidence,** which is the number of *new* cases of a given disease in a given geographic area. For example, the CDC estimates that end of 2011, the prevalence of HIV in the U.S. was just over 1,200,000 cases. But the incidence of HIV is around 50,000 new cases per year.

Studying the Spread of Disease

In order to study the spread of disease in populations, epidemiologists need to know and understand several factors about the causative agent and disease process. These factors make up the Chain of Infection.

There are four components that make up the infectious disease process. This is known as the **Chain of Infection.**

The Chain of Infection

1. The Reservoir: there must a source of the pathogen. Reservoirs are sites where microbes maintain a source of infection, this is basically where the microbe lives. Types of reservoirs are covered in Chapter 13.
2. The Mode of Transmission. The microbe must be transmitted to the host. Modes of Transmission are covered in Chapter 13.
3. The Portal of Entry: how the microbe enters the host. Entry occurs usually through inhalation, ingestion or injection of the pathogen. This is also covered in more detail in Chapter 13.
4. Susceptible Host: the host must be susceptible

Sometimes, the host may have encountered a particular microbe before and developed immunity to it, or the host may have been vaccinated. A microbe that enters the host by the wrong portal of entry will not successfully colonize the host. For example, *Salmonella* on the skin will not cause illness, but in the GI tract it does, whereas many other microbes are destroyed by acid in the GI tract.

Other factors may affect the susceptibility of the potential host. These include, but are not limited to:

- **Gender:** Women are more susceptible to Toxic Shock Syndrome caused by *S. pyogenes*. TSS can result from the normal skin microbiota, *S. pyogenes* entering the vagina and overgrowing due to tampon use.
- **Age:** elderly people are more susceptible to pneumonia than younger people. Many pneumonia-causing microbes are opportunistic and cause disease in elderly victims when they are fighting other infections, have had surgery or suffered and injury.
- **Genetic Background:** As noted in Chapter 14, people of Northern European ancestry have a small chance of inheriting the Delta 32 mutation, making them less susceptible to HIV infection. Likewise, some Africans are less susceptible to malaria because they lack a necessary receptor on their red blood cells.

5. Portal of Exit: the microbe must leave the host and be transmitted to another host or to the reservoir.

Breaking the Chain of Infection

If we can break the chain of infection, then we can reduce the number of microbe-related illnesses and deaths.

Strategies for breaking the chain of infection include:

- eliminating or controlling the reservoir. This can include water sanitation and controlling rodent populations.
- prevent contact with infectious microbes from exit portals, eliminate the means of transmission and block exposure to portals of entry. This includes the use of hand-hygiene, Standard and Universal Precautions and aseptic techniques.
- reduce or eliminate the susceptibility of the host, this can be done through vaccination, proper nutrition and education on risk factors associated with disease.

Epidemiologic Studies and Infectious Disease Reporting

When an outbreak or epidemic occurs, it is the epidemiologists job to determine the source of the outbreak in order to stop it. In order to do this, the epidemiologist carries out a **descriptive study**.

In a descriptive study, the epidemiologist collects as much data as they can about the disease and its victims. The descriptive study also identifies the **index case**, the first person identified with the disease in an epidemic. Information is gathered on the place and time of the outbreak, and on the personal details of the victims, such as their age, gender, ethnicity, occupation and socioeconomic group. This information helps the epidemiologist to pin point the source of the outbreak by examining what the victims have in common.

Analytical studies determine risk factors associated with particular diseases. For example, people who eat diets high in fish are more likely to contract *Anisakis*, a helminthic disease transmitted by eating contaminated fish.

Reporting and monitoring is an important component of epidemiology and public health. Certain diseases have been designated by the CDC as **notifiable**, which means that cases of those diseases must be reported to local public health departments, which then turn the data over to the CDC. This is important for early detection of outbreaks. A 1993 outbreak of *E. coli* in Washington state was detected early and traced to contaminated beef at Jack-in-the-Box restaurants due to infectious disease reporting.

The CDC monitors these trend in infectious disease and publishes them in the Morbidity and Mortality Weekly Report (MMWR).

The World Health Organization (WHO), is a specialized agency of the United Nations, and their mission is to promote technical cooperation for health among nations, carry out programs to control and eradicate diseases and improve the quality of human life. The WHO is responsible for the world-wide effort that resulted in the eradication of smallpox, the first and only microbial disease that humans have been able to eradicate so far.

Hospital Epidemiology

An entire subset of epidemiology is dedicated to studying the transmission of disease in healthcare settings. To this end, hospitals frequently have epidemiologists and infection control specialists on staff. **Nosocomial**, or Health-care Acquired Infections (HAIs) are a big concern for today's healthcare facilities. An estimated 2,000,000 U.S. residents per year will get a nosocomial infection at a total cost to treat of roughly 30 billion dollars. These tricky infections can be **exongenous**, which means they are caused by a microbe introduced to the patient while in the healthcare setting, or **endogenous**, meaning the infectious arose from the patient's own normal microbiota as an opportunistic infection. This is often the case with elderly people who enter the hospital for one reason (broken hip, heart surgery) and then develop pneumonia. **Iatrogenic infections** are infections directly caused by modern medical procedures. An example would be a bladder infection caused by a urinary catheter. **Superinfections** result from antibiotic use and are particularly hard to treat. One of the most common, troubling superinfections today is *Clostridium difficile*, a bacterium normally found living in low numbers in the intestines of most humans. Taking antibiotics kills the other, less resistant normal microbiota of the intestine which leaves room and nutrients for *C. difficile* to over-grow and create a hard-to-treat and painful infection, pseudomembranous colitis. One of the most successful treatments for C. diff is to replace and repopulate the normal microbiota of the intestine with microbiota harvested from the feces of a donor.

The chain of infection for HAIs is mostly the same as the chain of infection for all other infections, with some notable differences.

- The reservoirs are the hospital environment and may include medical equipment, other patients and medical and non-medical healthcare personnel. A recent example of medical equipment harboring dangerous microbes is the transmission of carbapenam resistant *Enterobacteriacea* on endoscopes at UCLA's Ronald Reagan Medical Center
- Microbes may be transmitted directly or indirectly by visitors, patients and personnel
- Potential hosts may be immunocompromised due to age, illness or medical/surgical treatment making them at greater risk for opportunistic infections
- The microbes present in the healthcare environment are more likely to be drug resistant or multi-drug resistant as a result of the prevalence of antibiotics in this setting

A major contributing factors to HAIs is the failure of healthcare personnel to follow infection control guidelines. The number one way to combat HAIs is to follow **infection control guidelines.** These are measures designed to break the chain of infection and include, but are not limited to: handwashing, use of PPE, use of Standard and Universal Precautions, use of Transmission Based Precautions and aseptic techniques (asepsis).

Lab Exercise 14.1

Disease Transmission and Epidemiology

Student Objectives

1. Understand the modes of disease transmission between reservoirs of infection and a new host.
2. Simulate disease transmission.
3. Apply descriptive epidemiology to determine the index case in a simulation.
4. Describe a nosocomial disease and how they can be prevented.
5. Understand how diseases are classified based on occurrence and geographic region.

Direct Contact Disease Transmission Activity

Materials

Laboratory materials: Glitterbug UV lotion, UV view box.

NOTE: This experiment requires four people.

1. ONE student per lab table will apply the Glitterbug UV lotion to his or her RIGHT HAND. You may use a glove on your left hand to rub the lotion all over your right hand.
2. Form a circle with the group of four people.
3. The student with the Glitterbug lotion on the right hand (student #1) will shake the right hand of the student on his or her left (student #2).
4. Student #2 will shake the right hand of the student to the left (student #3).
5. Student #3 will shake the right hand of the student to the left (student #4).
6. Student #4 will shake his/her right hand to the LEFT hand of student #1.
7. Student #1 will shake his/her left hand to the left hand of student #2.
8. Student #2 will shake his/her left hand to the left hand of student #3.
9. Student #3 will shake his/her left hand to the left hand of student #4.
10. Everyone will observe their hands under the UV view box. Go in order, starting with student #1 right hand and proceeding through student #4 left hand.
11. Describe how cross contamination occurred in this experiment. Answer question #1 in the lab report.

Simulating Disease Transmission

1. Choose one test tube containing an unknown solution, and one empty test tube. Place two-three drops of unknown in the clean test tube, set it aside and do not use it.
2. Select a partner for round 1 and record their name in Data Table 1. Pour the contents of both your test tubes together and mix. Divide the contents back into 2 tubes.
3. Select a partner for round 2 and repeat step 2.
4. Select a partner for round 3 and repeat step 2.
5. Add two-three drops of phenolphthalein to your test tube. If it turns pink you are infected.
6. If you are infected write your name and the names of all your partners on the board.
7. Record the data for the entire class in the Data Table in your lab report.

Name ______________________

Exercise 14.1 Lab Report

Direct Contact Disease Transmission Activity

1. Describe how the Glitterbug transferred during the cross contamination experiment.

____

____

____

____

____

2. What is cross contamination? What are important activities you can perform as a healthcare professional to avoid spreading contamination among patients?

____

____

____

____

____

3. List and describe the three routes of transmission.

____

____

____

____

____

4. Does an epidemic always involve large numbers of people? Why or why not?

____

____

____

____

____

SIMULATING DISEASE TRANSMISSION DATA TABLES

Round Number	Partner's Name

Names of Infected Person's Partners

Infected Person	Round 1	Round 2	Round 3

8. How many people were infected after round 3?

____

9. Trace the source of infection. Cross out the names of the uninfected partners in round 1. There should be only two names left. Collect the unused test tubes from these two individuals and test with the pH indicator. Record your results below:

____

10. Write the name of the index case of the outbreak here:

____

11. Make a diagram (flow chart) of the disease transmission route. Show the index case and each carrier infected.

12. Give several examples of what the reservoir of infection could be in our simulated outbreak:

13. Why did we use Descriptive Epidemiology to analyze our simulated outbreak?

14. What are the incidence and prevalence rates of our simulated outbreak?

15. Give several examples of what the causative agent could be in our simulated outbreak.

16. Describe endemic, epidemic, and pandemic.

17. List the steps of the Chain of Infection.

18. Describe techniques that will *break* the Chain of Infection.

____

____

____

____

____

19. Describe Virulence Factors and list several examples.

____

____

____

____

____

20. Define the following terms:

Prodromal period: ____

Acute-Communicable Period: ____________________________________

Acute infection: ____

Chronic infection: ____

Symptoms vs. signs: ____

Vector transmission: ____

Nosocomial infection: ____

Systemic infection: ____

Carrier: ____

The Immune System

The Immune System Part 1: Innate Immunity

The immune system is the body's defense against microbial disease. There are many components to the immune system, and they all work together to keep us healthy. A helpful analogy when studying the immune system is to think of the body as a fortress and the microbes as foreign invaders. Think of what a villain would have to go through to invade a fortress:

- they would need to overcome physical barriers such as fences, high walls, locked windows and doors
- once inside they would need to evade any sensor/detection systems like alarms and motion sensors
- they may need avoid and evade any guard dogs present
- they would need to overcome or evade the security guards

The human immune system functions in much the same way as a heavily guarded fortress. There are several layers of protection. When studying these various methods of protection it is tempting to compartmentalize them for simplicities sake, but in order to really understand how the immune system functions, you must keep in mind that all of the components work together, and that there is a lot of communication going on between the various parts.

Fig. 15.1 The Body is Like a Fortress
David Hughes/Shutterstock.com

Innate vs. Adaptive Immunity

The two major components of the immune system are:

1. The **innate immune system**, also known as **non-specific immunity**. This is the part of the immune system that we are born with and works right away. This is because the innate immune system is made up of **physical and chemical barriers** that are **not specific** to any particular microbe. The job of the innate immune system is to keep *everything* out.
2. The **adaptive immune system**, also known as **specific immunity**. This mechanism requires an active immune response from specific cells and requires a longer duration to remove the pathogens. This mechanism is used when the innate mechanism fails, become overwhelmed, or is bypassed (vaccine needle bypassing the skin into the blood vessel). During this response illness usually occurs because of the longer duration required to promote specific cells. When the adaptive immunity ends, the immune system usually forms memory for that specific pathogen or antigen. Humans are not born with an active adaptive immune system however this specific response to an antigen begins "learning" as soon as the first lines of defense are breached in newborns and infants. As humans age the adaptive system learns and remembers how to best remove or destroy specific antigens this ability to form memory against an antigen is responsible for a fast and efficient secondary response to an antigen. In later years of life the adaptive immune response begins to work inefficiently which is why elderly individuals tend to be more susceptible to certain diseases.

The immune system can also be viewed as three "lines of defense". The first two lines of defense are part of the innate immune system, while the adaptive immune response makes up the third line of defense. When a microbe tries to invade, hopefully it is dealt with by the first line of defense. If it makes it through the 1st line, we hope that our 2nd line of defense will take care of it. If the second line of defense can't eliminate it, then it is dealt with by the 3rd line of defense.

Innate Immunity		Adaptive Immunity
1st Line Defenses	2nd Line Defenses	3rd Line Defenses

Physical Barriers
 	
☐ Skin
☐ Mucous Membranes
☐ Lacrimal Apparatus
☐ Normal Flora
	Proteins, Cells, Chemicals and Processes
	
	☐ Cells
	☐ Processes and Chemicals:
	   o Complement
	   o Inflammation
	   o Fever
		Cells
		
		☐ Lymphocytes

Fig. 15.2
Courtesy of Authors

The First Line of Defense

1. Skin

The skin is made of two layers:

1. The epidermis: multiple layers of tightly packed cells. Very few pathogens can penetrate these layers, and the constant shedding of dead skin removes microbes from the skin's surface.
2. The dermis: which has collagen to make it elastic and help resist abrasions.

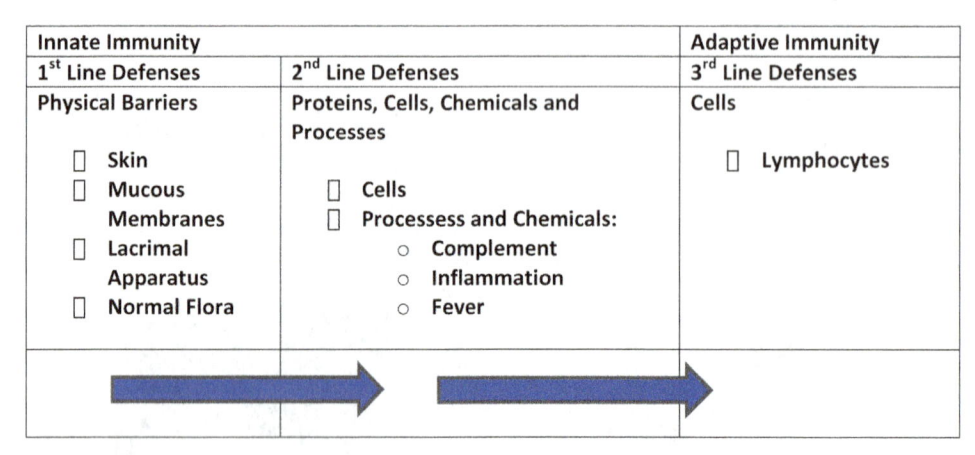

The skin also has chemical components that act as defenses:

1. Perspiration makes the skin salty and contains antimicrobial peptides that kill bacteria
2. Lysozyme, a secreted enzyme that destroys bacterial cell walls by degrading peptioglycan
3. Sebum secreted by the sebaceous glands helps to keep skin pliable in order to resist abrasions, and lowers the pH of skin below the optimum of many bacteria.

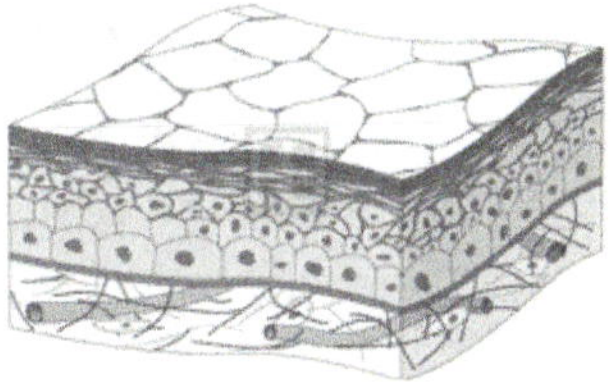

Fig. 15.3 Densly packed layers of skin cells

2. Mucous Membranes

The mucous membranes line all body cavities that are open to the environment and, like the skin, are made of two layers. The upper layer of the mucous membranes, the epithelium, is similar to the epidermis in that the cells are tightly packed to prevent microbial invasion and the cells are continually shed to carry away microbes. Unlike the epidermis, the epithelial layer of the mucous membranes is made up of a thin layer of living cells. A deeper layer of connective tissue supports the epithelium.

3. The Lacrimal Apparatus

The lacrimal apparatus produces tears which wash the surface of the eye during blinking, then are drained away. Tears also contain lysozyme to destroy bacteria.

4. The Normal Microbiota

The normal microbiota help to prevent colonization of the body by pathogenic microbes by competing with the pathogens for space and nutrients. The normal microbiota also produce substances that make the environment unsuitable to colonization by pathogens.

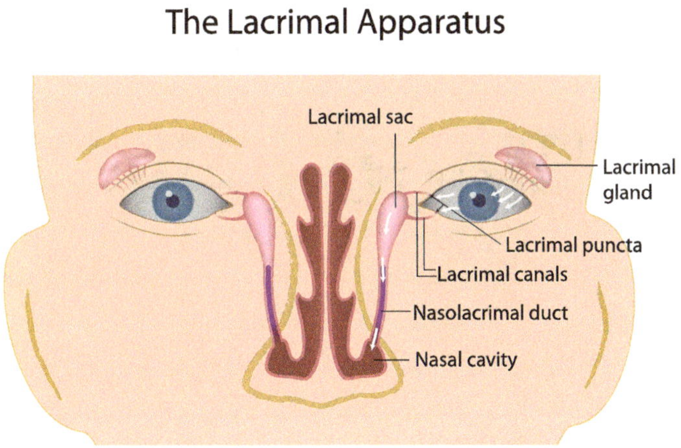

Fig. 15.4 The Lacrimal Apparatus

The Second Line of Defense

If a pathogen succeeds in breaching the first line of defense, the body relies on the second line defenses to protect itself. The second line defenses are composed of a variety of proteins, cells and chemicals.

Cells of the Immune System

All of the cells of the immune system except dendritic cells are blood cells, and are formed in the bone marrow from hematopoietic stem cells. The process are forming blood cells is called hematopoiesis.

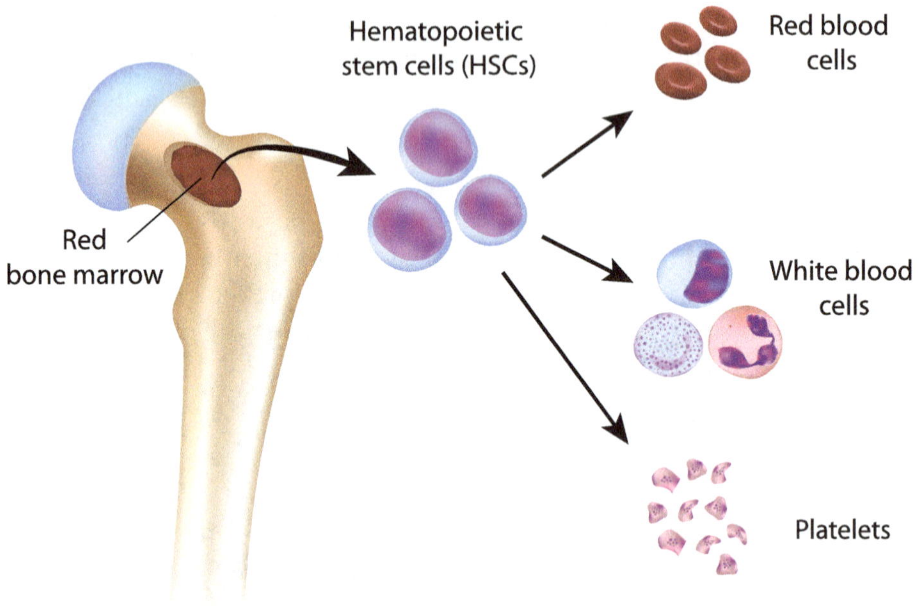

Fig. 15.5 Hematopoiesis
Alila Medial Media/Shutterstock.com

The blood cells that are part of the immune system are the leukocytes, or white blood cells (WBCs).

WBCs are typically divided into two groups, granulocytes and agranulocytes.

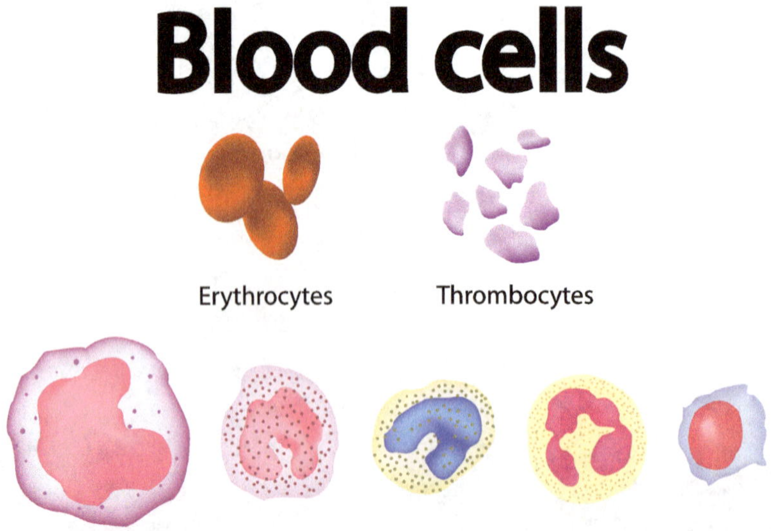

Fig. 15.6 Types of White Blood Cells
MiAdS/Shutterstock.com

Granulocytes

Are named because the large granules in their cytoplasms that are visible using light microscopy.

Neutrophils are the most abundant type of WBC, accounting for 54%–62% of the total blood cell count. They are also known as polymorphonuclear leukocytes (PMNs), because their nuclei can take many forms. They are the first WBCs to arrive at the site of infection, moving from the blood vessels and into the tissue by squeezing out of the capillaries in a process called diapedesis. Once at the site of infection, neutrophils fight microbes in several ways:

- phagocytosis: "phago" means to eat, so the neurophils literally ingest and digest microbial invaders, using digestive enzymes in their lysosomes
- neutrophils use a series of enzymes and chemical reactions to turn oxygen into hyperchlorite, the active ingredient in bleach
- making nitric oxide, which stimulates inflammation
- NETs: neutrophil extracellular traps: the neutrophil kills itself, releasing DNA and histone proteins from the nucleus which combine with proteins from the cytoplasm to create the NET fibers to trap bacterial cells which are then killed by antimicrobial peptides

Neutrophils can be recognized on a blood smear by their dark purple multi-lobed nucleus and lilac-stained cytoplasm. Elevated neutrophil levels may indicate bacterial infection, although stress can also elevate neutrophil levels.

Eosinophils make up 1–3% of the WBC population. They are also phagocytic and their granules contain chemicals related to inflammation. Eosinophils can be phagocytic, but they fight primarily by releasing the antimicrobial chemicals in their granules. The primary targets of eosinophils are parasitic worms. The antimicrobial chemicals released by the cells weakens and kills the worms.

High levels of eosinophils can indicate parasitic worm infections, allergies or autoimmune disorders.

***Recent research suggest that over-reactive eosinophils can cause the symptoms of allergies. The theory is that because of advances in sanitation, parasitic worm infections have become very rare in humans. This has essentially left the eosinophils out of a job, and "bored", which leads them to attack environmental substances as allergens, releasing their inflammation-causing chemicals and initiating the symptoms of allergies. Some allergy sufferers have gone so far as to deliberately infect themselves with worms in order to alleviate their allergy symptoms. ***

Eosinophils can be recognized on a blood smear by their red-orange color and bi-lobed nucleus.

Basophils are the rarest type of WBC, making up only 1% of the total white blood cell population. Basophils are non-phagocytic cells that release several different chemicals that contribute to inflammation, such as anticoagulants, heparin and histamine. They are roughly the same size as eosinophil, appear blue when stained on a blood smear and have fewer, more irregularly shaped granules than eosinophils.

Agranulocytes

Monocytes Account for 3%-9% of the total white blood cell count. They are the largest cells found in the blood. Their nuclei vary in shape, but are usually kidney-shaped or oval. Like neutrophils, monocytes can engulf particles, but at a relatively larger scale, and will engulf numerous bacteria at the same time. These are highly active phagocytes. Monocytes will leave the blood and enter tissue where they will differentiate into one of two special types of white blood cells: macrophages and dendritic cells. Once these special cells are in the tissue, they move through interstitial spaces using a form of self-propulsion called amoeboid motion. These special cells scavenge for "foreigners" of the body. Macrophages also scavenge for damaged or dying host cells that need to be digested. Macrophages and dendritic cells play an important role during the innate (non-specific) mechanism of host immunity; they are found all over the body.

Lymphocytes make up 25–35% of the total WBC count and are the smallest of the WBCs. A typical lymphocyte contains a relatively large, round nucleus surrounded by a thin rim of cytoplasm. They have a relatively long life span that may extend for years. Lymphocytes play an important role in immunity, especially during the adaptive (specific) mechanism of host immunity. When a lymphocyte migrates out of the bloodstream, it typically travels to a lymph node where it waits to be used during an adaptive immune response. There are 3 types of lymphocyte: B cells, T cells and Natural Killer (NK) Cells , all of which are part of the Adaptive Immune Response.

1. *B-cells*. Immature lymphocytes that form in the red bone marrow and differentiate and mature while in the bone marrow. Mature B-cells can be distributed by the blood and constitute 20%–30% of circulating lymphocytes. They settle in the lyphatic system and are abundant in lymph nodes, spleen, bone marrow, secretory glands, and intestinal lining. B-cells act indirectly against antigens by producing and secreting globular proteins called antibodies. Antibodies are carried by body fluids and react in various ways to destroy specific antigens and antigen-bearing particles (pathogens). This type of response is called antibody-mediated immunity (also known as humoral immunity).
2. *T-cells*. Immature lymphocytes that form in the red bone marrow and are released into the blood to reach the thymus to differentiate and mature. Mature T-cells can be distributed by the blood and constitute 70%–80% of circulating lymphocytes. They tend to reside in various organs of the lymphatic system and are abundant in the lymph nodes, thoracic duct, and spleen. T-cells provide an important defense against viral infections, which proliferate inside the host cells where they are somewhat protected. Most viruses, however, cause antigens to be produced on the membranes of the host cells they infect. T-cells can detect these antigens and directly destroy the host cell containing the virus. There are two types of T-cells: helper T-cells (CD4) and cytotoxic T-cells (CD8). Helper T-cells have an important role in the activation of the adaptive (specific) immune response; they act as the director as they release specific chemicals (cytokines) to promote B-cell or cytotoxic T-cell activation and proliferation. Cytotoxic T-cells directly respond to viral infections and destroy the infected host cells using toxic substances.
3. *NK Cells*. Recognize the absence of MHC on self cells that have been infected with a virus or transformed into tumor cells. They also recognize self cells that have been bound by antibody, indicating a viral infection, and destroy those cells.

Processes of the Immune System

Phagocytosis

Phagocytosis is literally cellular eating. There are two primary phagocytes of the immune system: neutrophils and macrophages. The steps of phagocytosis are:

1. Chemotaxis: the phagocytes are attracted to the site of infection by a chemical signal
2. Recognition and adherence: the WBCs "recognize" the microbe and attach to it. Recognition is enhanced by the presence of **opsonins.** Opsonins are proteins that have been made by other cells and bind to bacteria to serve as "eat this" signs. Antibodies and complement proteins serve as opsonins and will be discussed later.

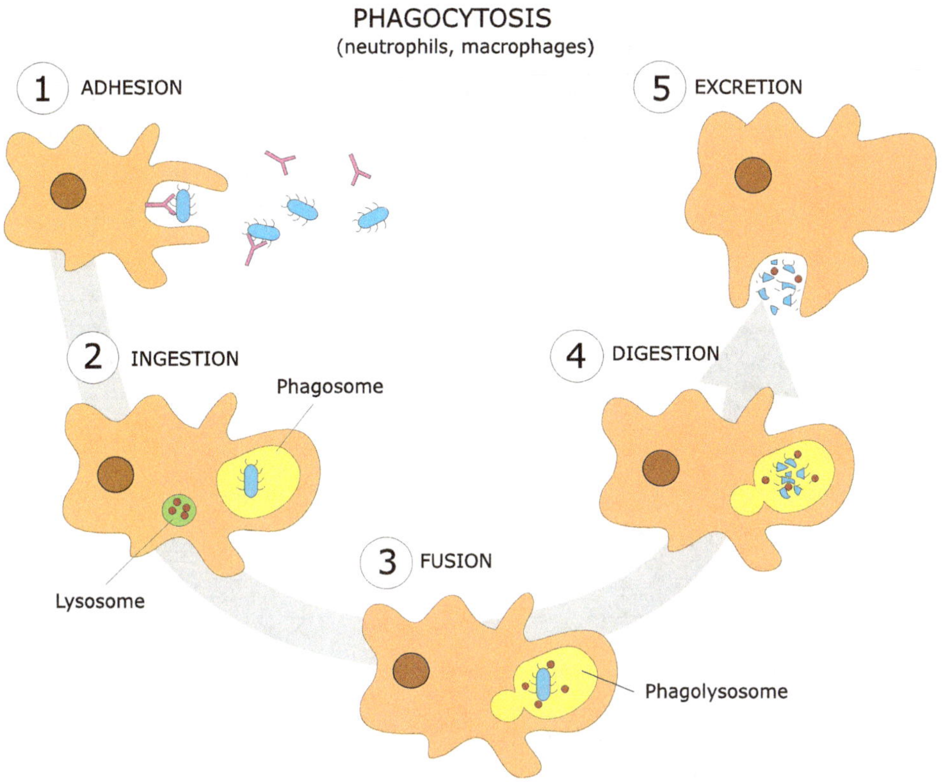

Fig. 15.7 Process of Phagocytosis
Ellepigrafica/Shutterstock.com

3. Ingestion: pseudopodia surround the microbe, eventually forming a vesicle ("bubble") called a phagosome.
4. Digestion: WBC lyosomes fuse with the phagosome forming the phagolysosome. This exposes the microbe to the digestive enzymes and low pH of the lysosome, which will destroy it.
5. Elimination: the digested particles are released from the cell by exocytosis

Iron-binding proteins

Iron is required for both human and bacterial metabolism. Because iron is generally not soluble, it is bound to the protein, *ferritin*, when stored in the liver, and *transferrin* when it is being transported in the bloodstream. While iron is bound to these proteins, it is unavailable for use by bacteria. This is known as sequestering.

Some bacteria, like *S. aureus*, produce "iron-stealing" proteins called siderophores. The siderophores bind more tightly to iron than transferrin, allowing *S. aureus* to steal the iron from body cells. In response, the body produces *lactoferrin*, which can take the iron back.

Chemical Defenses

Sensor and Alarm Systems

Toll-Like Receptors (TLR): are receptors on the surface of phagocytic cells. These are specific to a variety of microbial substances (collectively called PAMPs: pathogen associated molecular patterns), like peptidoglycan, LPS and flagellin. Once a PAMP has bound to a TLR, the phagocyte sends a signal out to the body basically telling it, "we've been invaded!"

NOD Proteins: serve the same function as TLRs, but these receptors are located inside the cell, rather than on the surface. Not much is known about exactly how NODs function.

Cytokines are chemical signals. Each chemical is specific to a particular action. Cytokines are necessary for a coordinated immune response as they allow the cells of the immune sytem to "talk" to each other. A few examples of cytokines are listed below.

Interferons: this word derives from the phrase "*interfere* with viral replication" and that describes what these signals do. Cells that have been infected with a virus release interferons, which signal neighboring cells to stop protein production.
Interleukins (ILs): are signals that generally sent from one leukocyte to another. So far, 35 interleukins have been identified, and each has a unique job. For example, IL-2 is a signal for T cell *proliferation*, but IL-12 is a signal for T cell *differentiation*.
Growth Factors: Signal mitosis in WBCs.
Tumor Necrosis Factor (TNF): is secreted by WBCs to kill tumor cells and regulate inflammation.
Chemokines: are signals for chemotaxis (movement in response to chemical signal). These are the signals that attract WBCs to the site of infection.

***Explain how interferon interferes with viral replication.

***Explain how many of the symptoms of viral infections such as malaise and muscle aches are caused by interferon.

Complement

The complement system is a group of proteins that are catalysts for several events in the immune system. It is often called the "complement cascade" because once started, the product of each reaction serves as the catalyst for *more* reactions downstream, essentially creating a cascade effect. In the simplified diagram below, you can see

- that once activated, the first protein, C3, is cleaved into two active proteins, C3a and C3b.
- C3a initiates inflammation, while C3b serves as an opsonin and activates C5.
- Once activated, C5 is cleaved into C5a and C5b.
- C5a also serves as a signal for inflammation.
- C5b activates C6, C7, C8 and C9 which bind to together to form Membrane Attack Complexes, proteins that punch holes in bacterial cell walls.

Once complement is activated a cascade of reactions occurs, resulting in three outcomes:

- inflammation
- lysis by MACs
- opsonization for phagocytosis. An **opsonin** is a protein that binds to the surface of cell and serves as a signal to phagocytes that they should "eat" the bound cell.

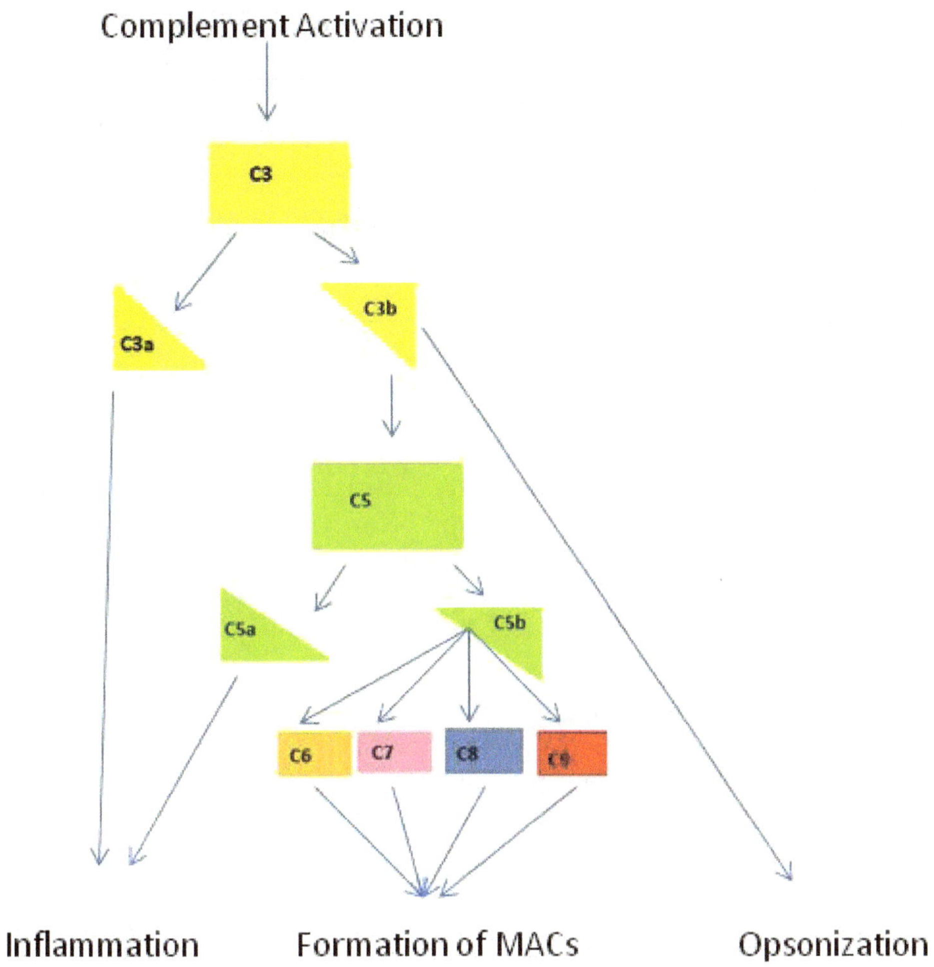

Fig. 15.9 The Complement System
Courtesy of A. Swarthout

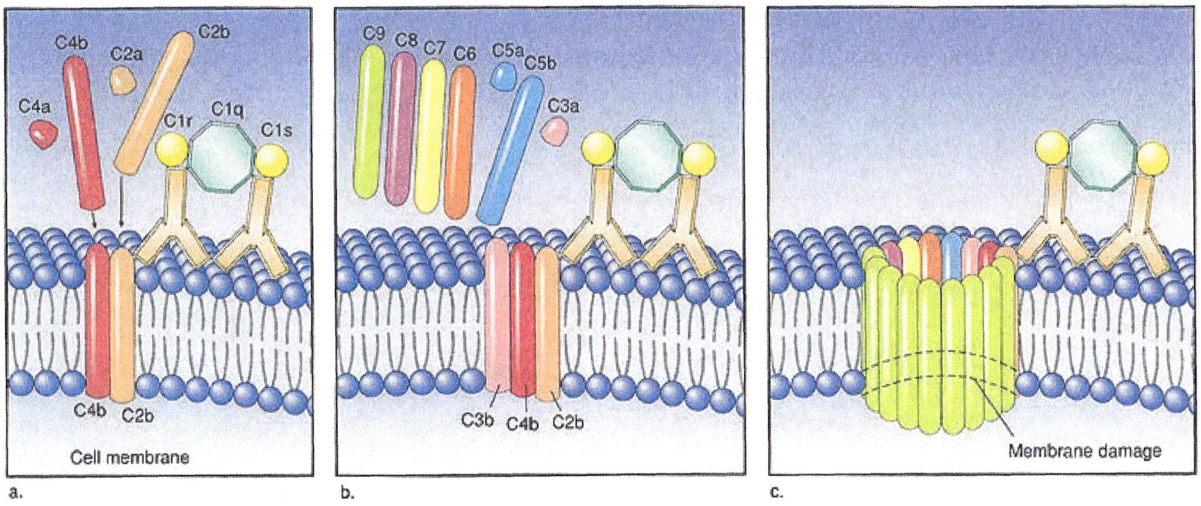

Fig. 15.10 Formation of the membrane attack complex and the pore that results

There are three pathways by which Complement can be activated:

- **The Classical Pathway:** when antibodies bind to antigen, then interact with C3.
- **The Alternative Pathway:** when microbes or their products (ie toxins or glycoproteins) interact with C3.
- **The Lectin Pathway:** when the bacterial polysaccharide, mannose, binds to lectin molecules on our cell surfaces, and this complex interacts with C3.

Inflammation

Inflammation is a nonspecific immune response that can be initiated by a variety of things including: microbes, microbial products and tissue damage. Inflammation is triggered by chemical signals that cause blood vessels to dilate and become leaky in a process called **margination.** This allows the cellular components of the immune system to leave the bloodstream via **diapedesis**. **Chemotaxis** is the movement of white blood cells to site of infection as they follow chemical signals released by cells near the site of infection. Non-cellular components of the blood also "leak" out of circulation to the site of infection. Once at the site of infection neutrophils and macrophages **phagocytose** invading microbes.

Fever

A fever is defined as a body temperature over 37°C. Fevers are triggered by cytokines called **pyrogens.** Macrophages release pyrogens when their TLRs bind to microbial products. The pyrogens travel through the blood and cause the hypothalamus to increase the body's temperature.

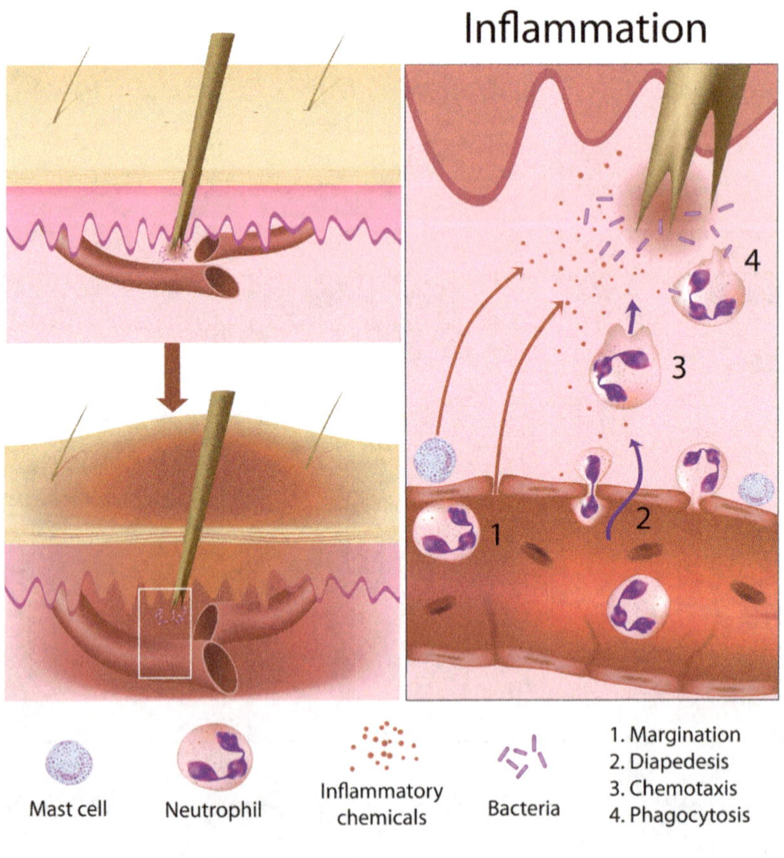

Fig. 15.11 Inflammation
Alila Medical Media/Shutterstock.com

This increase in temperature aids the immune response in two ways:

1. It raises the temperature above the optimum temperature of many microbes
2. It enhances many of the body's responses to microbes. The rate of enzymatic reactions increases, the inflammatory response is enhanced, signaling via cytokines and antibody production are enhanced. In addition, the lymphocytes proliferate more quickly and phagocytes kill more efficiently.

The Immune System part 2: Adaptive Immunity

The third line of defense in the immune system is the adaptive immune response, also known as the specific immune response or acquired immunity. There are five primary attributes of an adaptive immune response:

- Specificity: any adaptive response acts against only one particular molecular shape (antigen) and not others
- Inducibility: the cells of an adaptive immune response are activated in response to a specific antigen
- Clonality: once induced, cells proliferate to form many generations of identical cells
- Unresponsive to self: adaptive immune cells don't act on normal body cells
- Memory: adaptive cells form memory which accounts for a fast secondary response. In fact, the adaptive immune response gets faster with each repeated exposure to a given antigen.

Anatomy of the Lymphoid System

The adaptive immune response takes place in the tissues and organs of the lymphatic system. This system acts as a surveillance system that screens the tissues of the body for foreign antigens and is composed of the lymphatic vessels and the lymphatic cells, tissues and organs.

Primary vs. Secondary Immune Response

A primary immune response occurs the first time a body is exposed to a particular antigen. A primary immune response takes roughly 10–14 days to fully develop. A secondary immune response occurs each time thereafter, and takes 1–3 days to develop. This explains why once someone has had a particular infection, like chicken pox, they are unlikely to get it again.

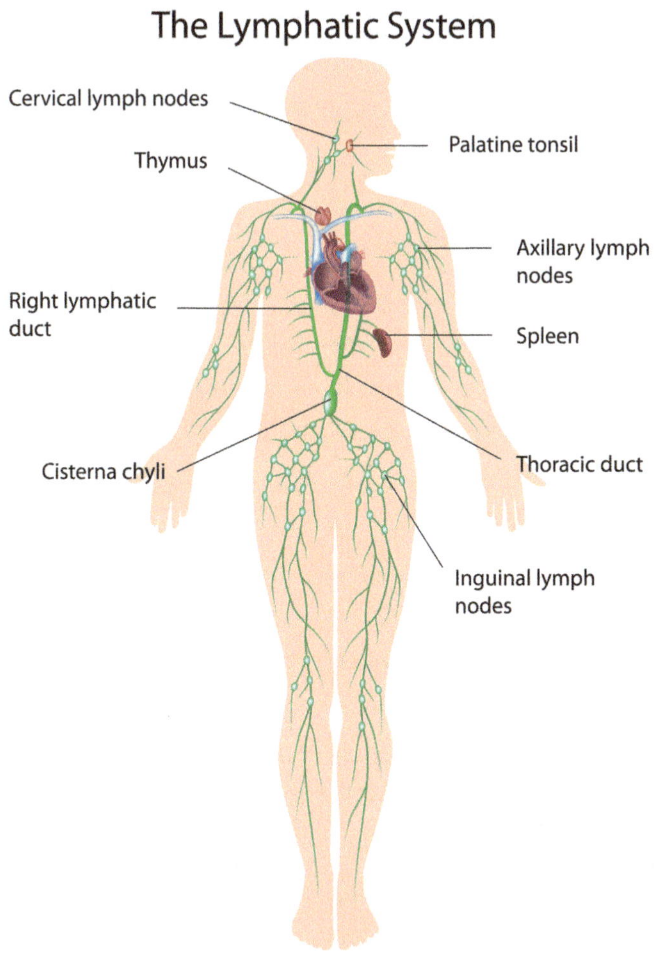

Fig. 15.12 The Lymphatic System
Alila Medical Media/Shutterstock.com

Components of an Adaptive Immune Response

Antigens An antigen is a substance that causes the body to stimulate an adaptive immune response. Various bacterial components as well as the proteins of viruses, protozoa and fungi can serve as antigens, as can particles of food and dust.

Types of antigens:

- Exogenous, these are *extracellular* antigens: toxins and other secretions, or components of the cell wall, cell membrane, flagella or capsid
- Endogenous, *intracellular* antigens that are not accessible to our immune cells, and our immune cells can only respond if the endogenous antigen is incorporated into the body cell's cytoplasmic membrane
- Autoantigens are molecules produced by our own cellular processes that should not, under normal circumstances, produce an immune response.

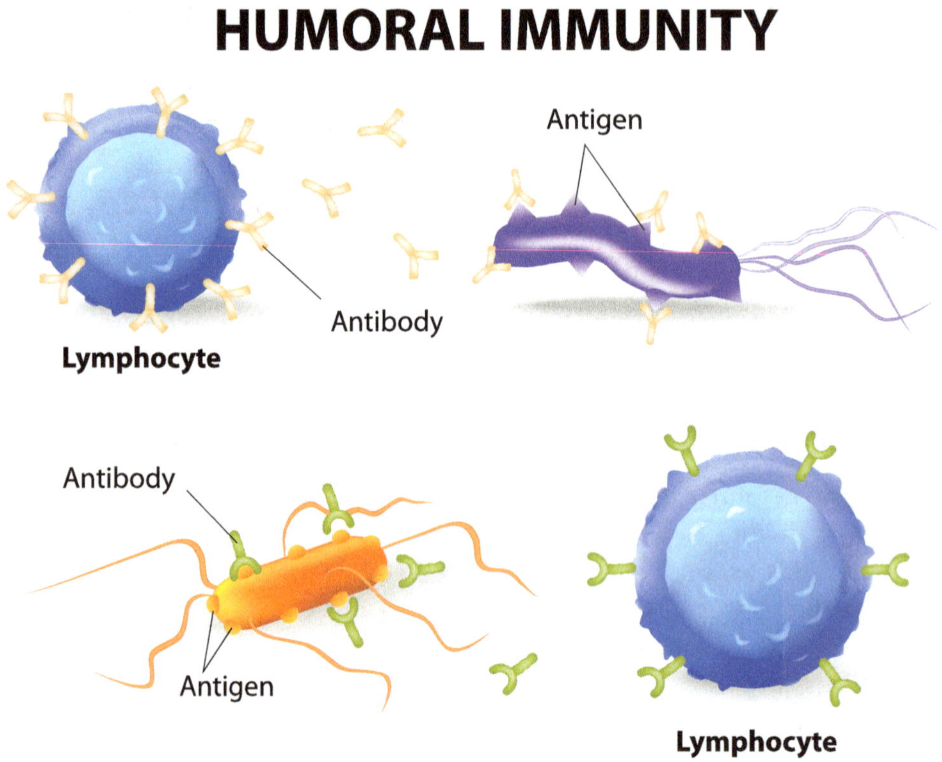

Fig. 15.13 Antibodies Binding to Antigens insert
Designua/Shutterstock.com

Major Histocompatibility Complex (MHC)

MHC proteins hold and position antigens for presentation. This is a way for cells to communicate with each other. The antigen-presenting cell presents ("shows") the antigen to the immune cell, which "looks at" the antigen and initiates an immune response.

There are two classes of MHC:

- MHC class I: found on all normal nucleated body cells
- MHC class II: found only on B cells and Antigen Presenting Cells (APCs). APCs include macrophages and dendritic cells.

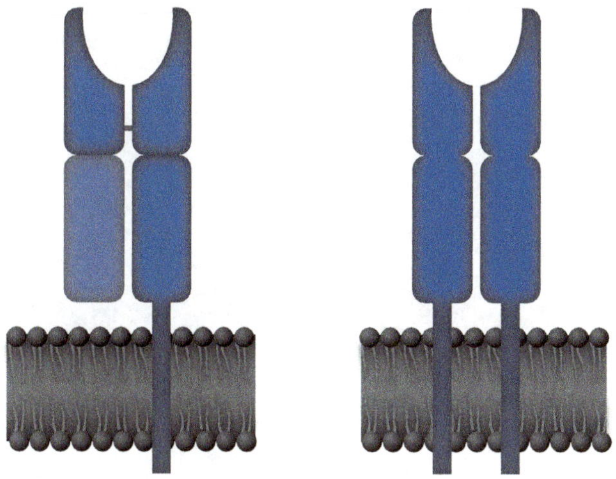

Fig 15.14 MHC Molecules

From Microbiology: A Human Perspective, 7th Edition by Eugene Nester, Denise Andersen and C. Evans Roberts, Jr. Copyright ©2012 The McGraw-Hill Companies, Inc. Reprinted by permission.

Adaptive Immune Cells

Both T cells and B cells develop from stem cells in the red bone marrow. Progenitor cells that migrate to the thymus for maturation become T cells while progenitor cells that become B cells remain in the bone marrow for maturation.

B cells

B cells are found primarily in the spleen and lymph nodes, although a small percentage circulate in the blood. B cells respond to exogenous (extracellular) antigens. A response to extracellular antigens that is mitigated by B cells is called a humoral immune response.

The major of function of B cells is differentiate into plasma cells, which make and secrete antibodies. B cells also differentiate into memory cells, which are involved in secondary immune responses.

Antibodies

Antibodies are special proteins produced by plasma cells. They are “Y”-shaped molecules made up of two light chains and two heavy chains. The job of antibodies is to bind to antigen.

The **Fab region** (arms) is the part of the antibody that actually binds to antigen, allowing the **Fc region** (stem) to stick out away from the bound microbe, serving as red-flag signal to other components of the immune system. (Hint: F-a-b: think Fragment Antigen Binding)

The tips of the antibody are the variable region. This is the part of the antibody that is responsible for antibody-antigen specificity. Antibodies are like enzymes in that the relationship between (lock and key theory) them and what they bind to is very specific. For example, antibodies that bind to antigens of Varicella (chicken pox) virus, will not bind to antigens on the surface of the measles virus and vice versa.

The constant region of the antibody sorts the antibody into one of five classes of antibodies: IgM, IgG, IgA, IgE and IgD. All IgM has the same constant region, but there will be IgM specific to chicken pox, IgM specific to measles, IgM specific to influenza and so on. This is true for all of the classes of antibodies. The “Ig” stands for “immunoglobulin” which is another name for antibody.

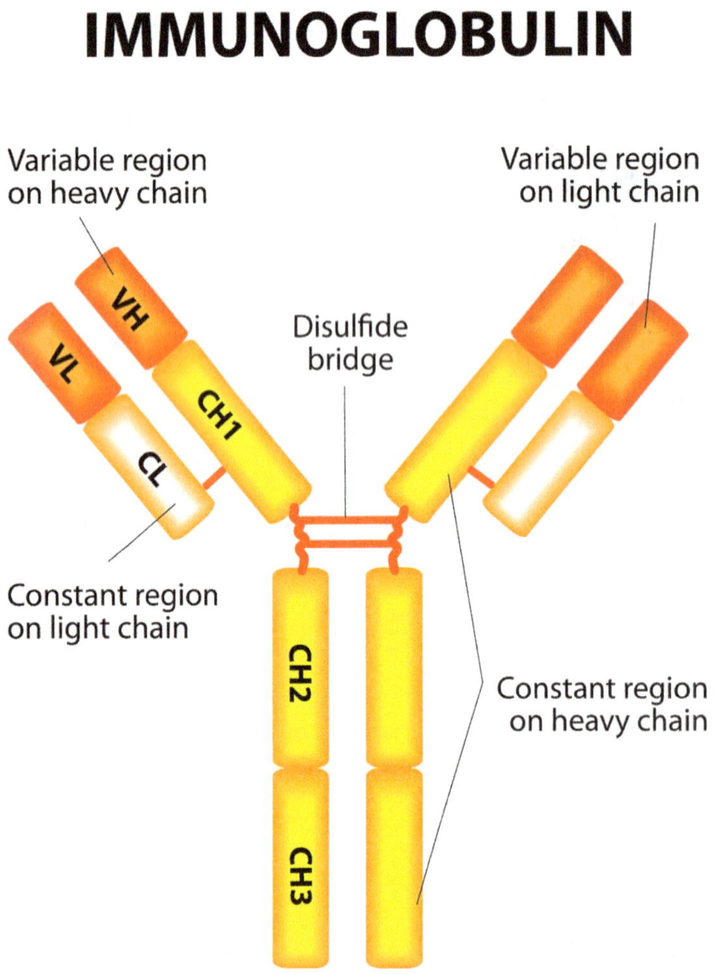

Fig. 15.15 The Structure of Antibody
Designua/Shutterstock.com

Each class of immunoglobulin has unique characteristics.

- IgM is the first antibody produced, forms a pentamer and is found in the mucosa
- IgG is the most abundant antibody (~80% of antibody in serum is IgG), it is the longest lived antibody and it is the antibody that can cross the placenta from pregnant mother to fetus to help protect the baby
- IgA forms a dimer and is associated with bodily secretions like tears, saliva and milk
- IgE is involved in our response to parasitic infections and in the allergic response
- IgD is least common antibody and its function is not well understood by scientists, but it may be involved in coordinating an effective immune response.

Antibodies play a major role in the body's defense against microbes, and people who have immune disorders in which they have no antibodies or their antibodies are immunologically impaired have a much bigger risk of infection.

Outcomes of Antibody—Antigen Binding:

- neutralize bacteria, toxins and viruses by binding to them, thereby preventing the bacteria, etc. from binding to our cells
- serve as opsonins
- agglutinate microbes so that they are easier to phagocytose
- activate complement via the classical pathway

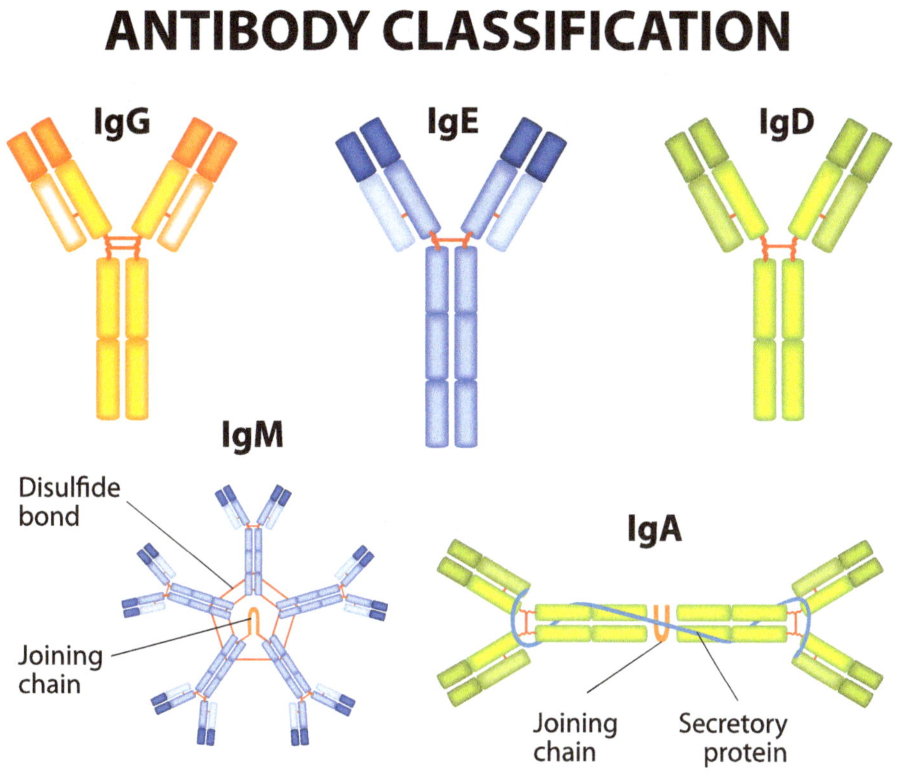

Fig. 15.16 Classes of Antibody
Designua/Shutterstock.com

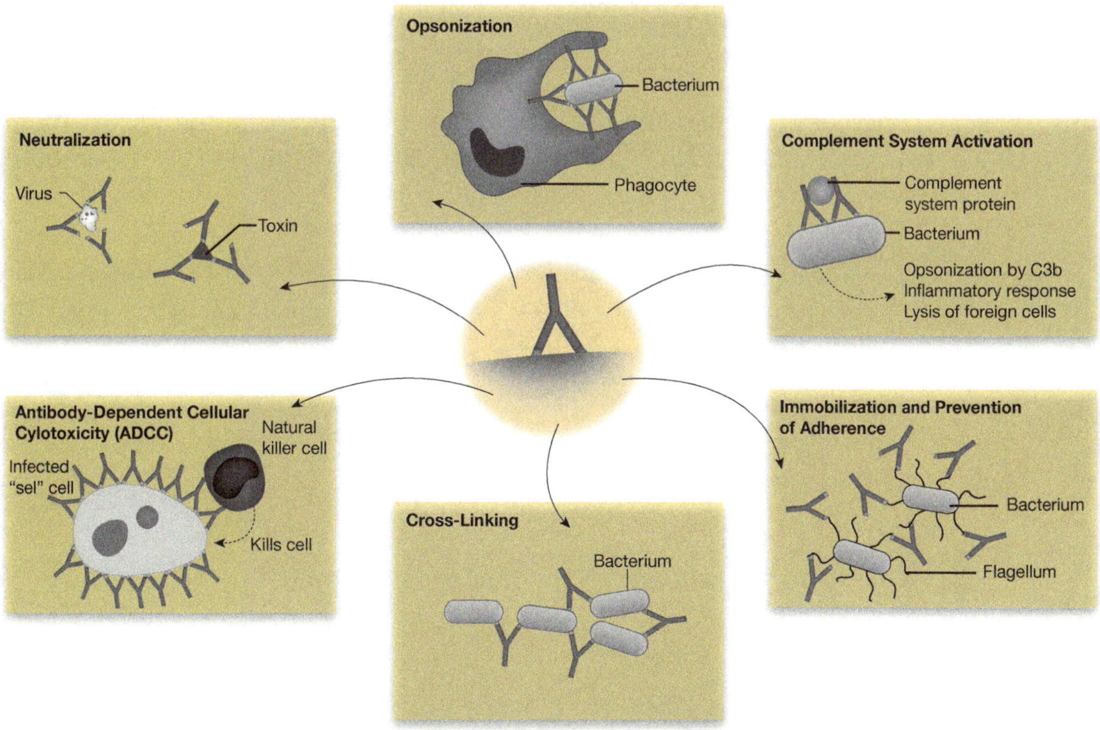

Fig. 15.17 Outcomes of Antibody-Antigen Binding Image showing neutralization, opsonization, etc.
From Microbiology: A Human Perspective, 7/e by Eugene Nester, Denise Andersen and C. Evans Roberts, Jr. Copyright © 2011 McGraw-Hill Companies, Inc. Reprinted by permisison.

- trigger Antibody-dependent cellular cytotoxicity (ADCC). Antibody binds to bacterial cells or virally infected "self" cells. The constant region acts a signal to NK cells that the cell the antibody is attached to needs to be eliminated and the NK cell releases perforins and granzymes, chemicals that trigger apoptosis and lysis of the antibody-bound cell.

B Cell Receptors

B cells have immunoglobulins on their surface that bind and recognize antigen. These B cell receptors (BCRs) are similar in structure to antibodies, but are bound to the cytoplasmic membrane of the cell rather than secreted by it.

Each B cell produces only one type of BCR, but has up to 500,000 copies of it on the surface of the cell. There are billions of different B cells, each with a unique BCR in a single human.

B Cell Responses

B cells can respond to antigen with the help of T helper cells or without the help of T cells. If the response requires the help of T helper cells, it is a **T-dependent response**.

During a B cell T-dependent response:

1. The antigen binds to the BCR of the **naïve** B cell
2. The antigen is internalized by the B cell
3. The B cell **processes** the antigen
4. MHC class II binds to the processed antigen
5. MHC class II **presents** the antigen to a T helper cell

(Steps 1–5: this is clonal selection)

6. The TCR of the T cell interacts with the MHC/antigen complex and CD4 protein ensures antigen is bound to MHC II
7. If the antigen is recognized by the T cell, the T cell sends a cytokine signal, IL-2, to the B cell which **activates** the B cell to undergo several rounds of mitosis, known as **clonal expansion**
8. Some B cell clones differentiate into plasma cells that secrete antibodies, while others become long lived memory B cells. What are the outcomes of antibody-antigen binding?

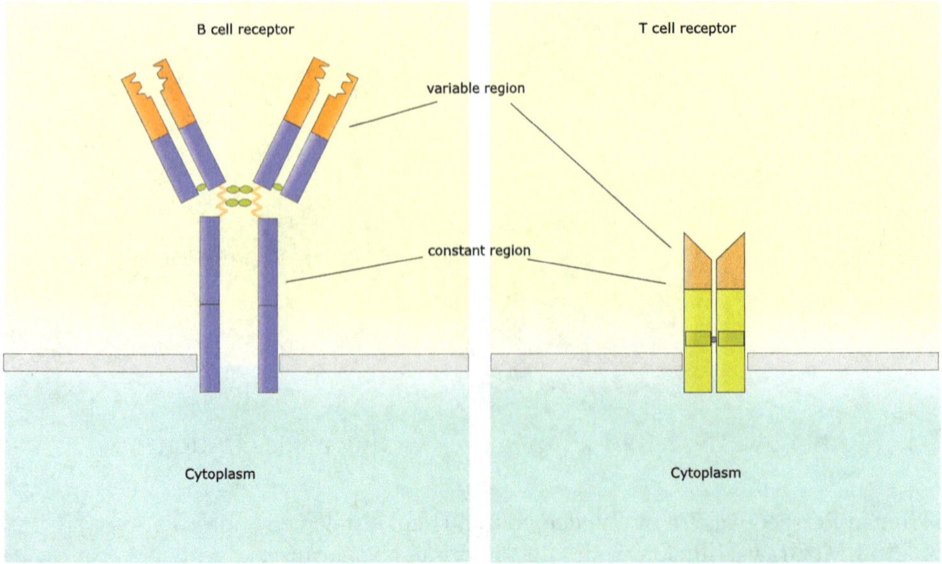

Fig. 15.18 B cell and T cell Receptors
ellepigrafica/ Shutterstock.com

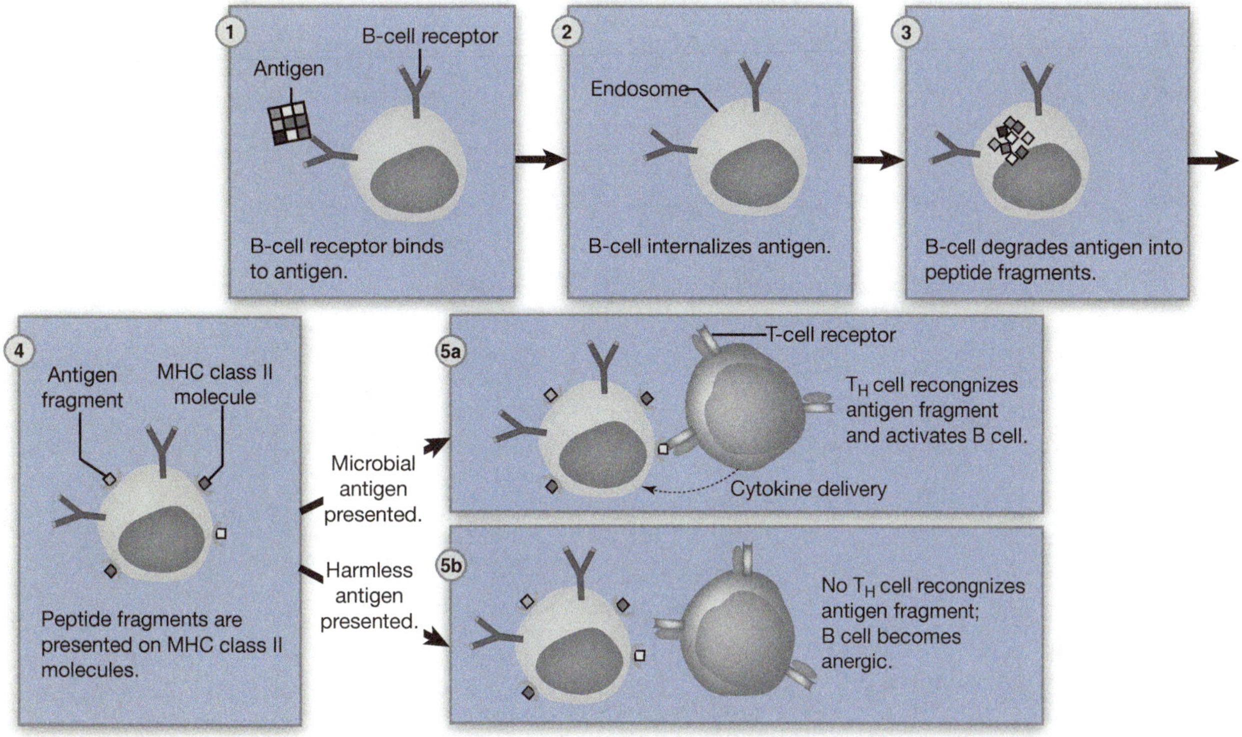

Fig. 15. 19 Naive B cell Activation

Antigen fragment

Peptide fragments are presented on MHC class II molecules.

If an antigen is a large molecule, such as a bacterial polysaccharide, it may bind to several BCRs at once. In this case, the B cell is able to activate itself without the aid of the helper T cell. This is a **T-independent Immune Response**

From Microbiology: A Human Perspective, 7/e by Eugene Nester, Denise Andersen and C. Evans Roberts, Jr. Copyright © 2011 McGraw-Hill Companies, Inc. Reprinted by permisison.

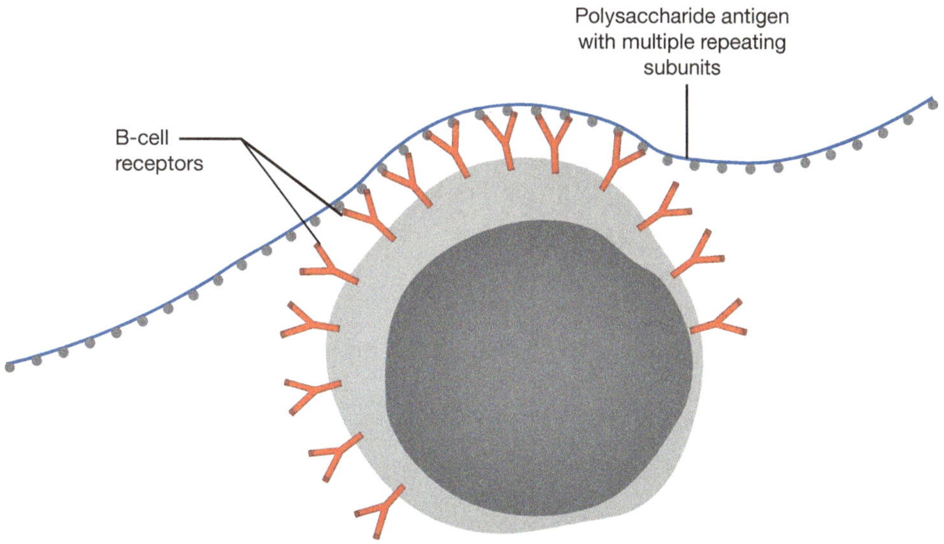

Fig 15.20 B cell T-Independent Response

From Microbiology: A Human Perspective, 7th Edition by Eugene Nester, Denise Andersen and C. Evans Roberts, Jr. Copyright ©2012 The McGraw-Hill Companies, Inc. Reprinted by permission.

The Development of Immunologic Memory

The first time your body encounters a specific antigen you have a **primary immune response.** It takes 10–14 days (2 weeks) for a primary immune response to fully develop. Within that time span, you will typically get sick.

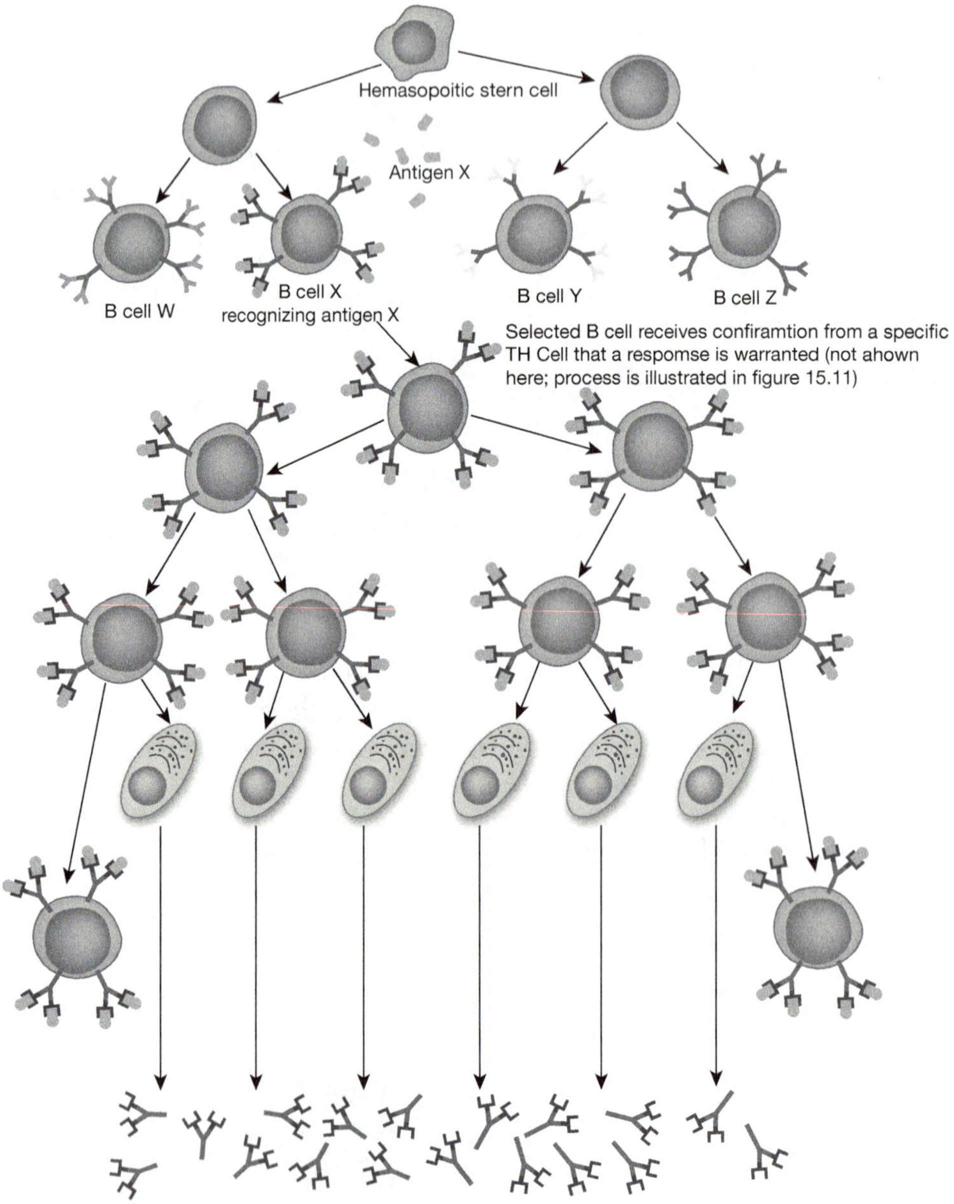

Fig. 15.21 Clonal Selection & Expansion

Selected B cell receives confirmation from a specific TH cell that a response is warranted (not shown here).

Memory B cells:

These long-lived descendants of activated B cells recognize antigen X when it is encountered again.

Anitbodies:

These neutralize the invader and tag it for destruction.

During a **secondary immune response**, which occurs the second (and each subsequent) time the same antigen is encountered, the antigen binds to BCRs on memory cells. The memory cells then become activated B cells, which proliferate and differentiate into plasma cells and memory cells. The secondary immune response is much faster, peaking in roughly 1–3 days instead of 2 weeks. It is also much stronger than the primary immune response. This is why for many microbes, there is no illness on subsequent exposure. The secondary immune response gets stronger and faster on repeated exposure to the antigen.

T cells

T cells are responsible for **cellular immunity**, the part of the immune system dealing with *intracellular* or **endogenous** antigen, such as viruses or bacteria that have successfully invaded our cells.

Unlike B cells, T cells cannot bind to free antigen. T cell receptors (TCRs) only interact with antigen that is bound to **Major Histocompatibility Complex** and presented (shown) to them by other cells.

T helper cells are sometimes called **CD4** cells because they have CD4 co-receptors. These cells help regulate the immune system by activating and coordinating the activities of B cells, macrophages and cytotoxic T cells. The TCRs of T helper cells interact with MHC class II.

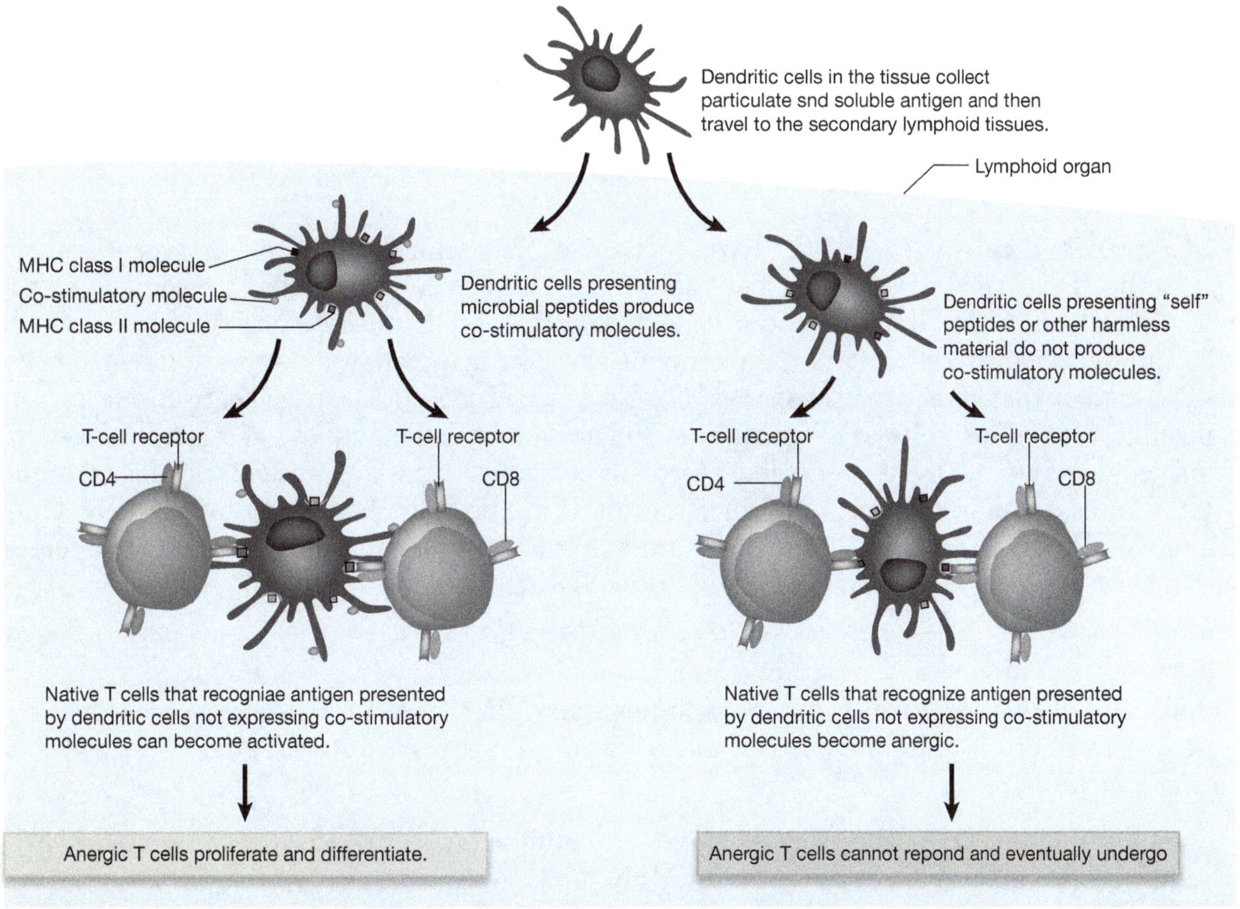

Fig 15.22 MHC/TCR interaction

From Microbiology: A Human Perspective, 7th Edition by Eugene Nester, Denise Andersen and C. Evans Roberts, Jr. Copyright ©2012 The McGraw-Hill Companies, Inc. Reprinted by permission.

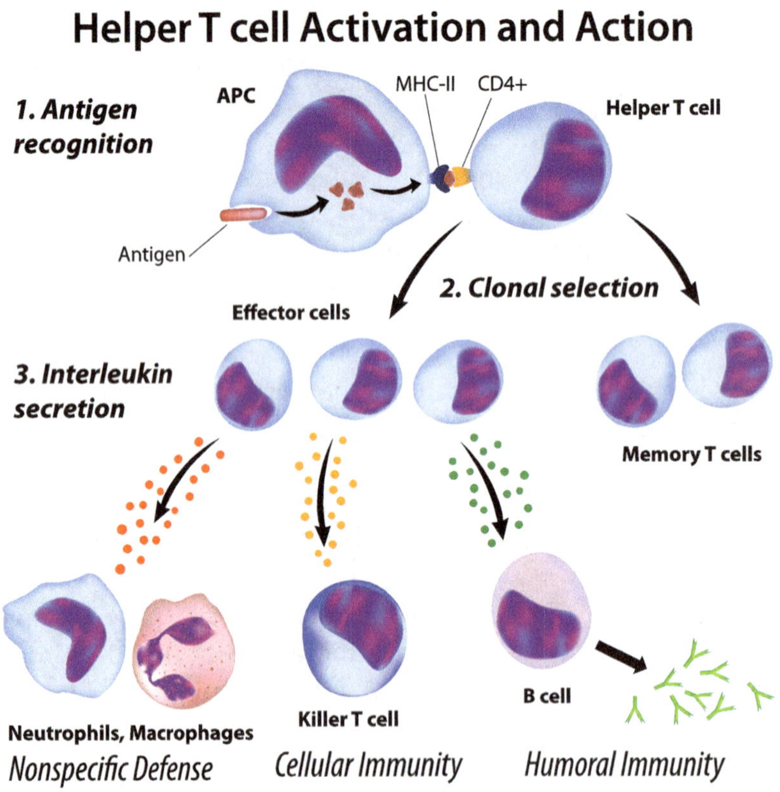

Fig. 15.23 Helper T cell Activation Image

Dendritic cells presenting "self" peptides or other harmless material do not produce co-simulatory molecules

Alila Medical Media/Shutterstock.com

Cytotoxic T cells (Tc or CTL) have TCRs and CD8 which recognize *endogenous* antigen that has been packaged with MHC class I on cells that have been infected by intracellular pathogens, or cells that have become tumor cells. Cytotoxic T cells must be activated by T helper cells, but once activated they produce **perforins** and **granzymes** that directly kill infected cells.

Remember that all *normal*, nucleated cells have MHC class I. As the cell carries out its normal cellular functions, cellular products are produced and small "samples" of these products are displayed on MHC I. CD8 cells act like patrol officers, inspecting the products displayed on the MHC I of each cell to see if it is "normal" or not. If the product displayed is normal, the CD8 cell moves on, leaving the cell alone. If the product is not normal, the cell is most likely infected with a virus or has become a tumor cell and the CD8 cell then kills that cell.

Cell mediated immune response

Cytotoxic T cells recognize endogenous antigen and directly kill cells that have become infected. We will briefly examine one possible mechanism that plays out with virally infected cells during which:

1. Virally infected cells will process antigens by endogenous antigen processing and display them on their cell surface bound to an MHC I protein.
2. Antigen presenting cells such as dendritic cells recognize endogenous antigen associated with the MHC I protein on the infected cells surface.
3. Antigen presentation to a Tc cell will then take place. The dendritic cell will migrate to the local lymph node and play a game of "go-fish" with Tc cells. The Tc cell with the appropriate

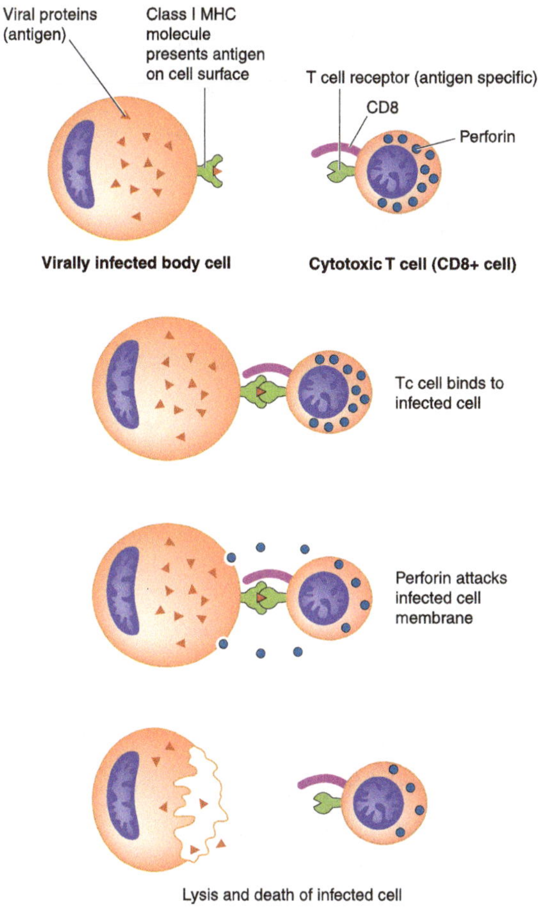

Fig. 15.24 Killing of Infected Cell by Cytotoxic T Cell Image

Anergic T cells cannot respond and eventually undergo

Blamb/Shutterstock.com

TCR will bind to the antigen on the dendritic cell the CD8 "confirms" there is a MHC I molecule associated with the antigen.

4. An activated T helper cell will send a cytokine signal to the Tc cell and cause clonal expansion of the Tc cell. Some of the Tc cells form into long lived memory Tc cells.
5. Activated Tc cells leave the lymph node to seek out virally infected cells displaying the antigen on MHC I.
6. Tc cells bind to the infected cells MHC I/antigen complex with its TCR. Once bound the Tc cell kills the target cell by activating apoptotic factors such as Perforin and Granzyme.

NK Cells

In order to evade the immune system some viruses (and tumor cells) have evolved the ability to cause infected cells to down regulate (stop making) MHC class I. Because cytotoxic T cells can only inspect antigen presented on MHC I, if a cell does not have MHC I, that cell is essentially invisible to the cytotoxic T cell and avoids being killed by it. In cases like this, **NK cells** serve as "back up". One function of NK cells is to inspect body cells to make sure that cells that are supposed to have MHC class I actually do have it. If the NK cell finds a cell without MHC I, it is assumed that the cell is infected and is then killed by the NK cell.

Another function of NK cells is Antibody-Dependent Cellular Cytotoxicity (ADCC). NK cells have Fc receptors on their surfaces that bind to the Fc region of antibodies. Because of these receptors, NK cells can recognize when a cell is bound by antibody, such a bacterial cell or a virally infected self cell. On recognizing that a cell is bound by Ab, the NK cell releases cytokine signals to cause the antibody-bound cell to undergo apoptosis.

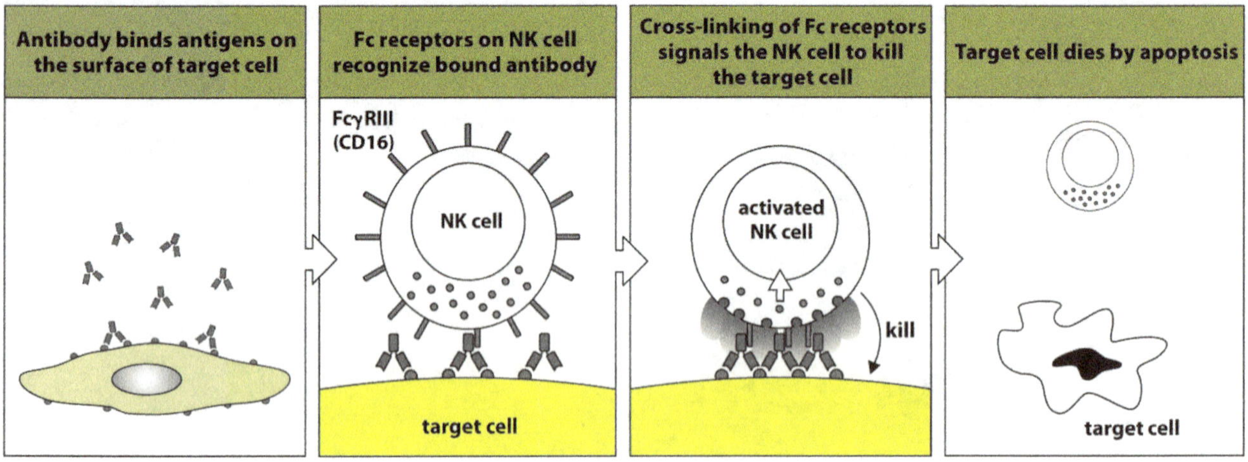

Fig. 15.25 Antibody Dependent Cellular Cytotoxicity

The Immune System Part 3: Vaccines and Immunization

Vaccines are modified or weakened forms of microbes used to induce a primary immune response without causing illness. Upon exposure to the live microbe, a vaccinated individual will have a secondary immune response. Vaccines offer a way to avoid contracting microbes for which there are limited treatment options, such as viruses and drug-resistant bacteria.

History

Variolation, the practice introducing smallpox virus to the body in the hope of causing a mild form of the disease and subsequent immunity, has been documented to have occurred in China as early as the 10th century. Chinese medical practitioners would grind smallpox scabs into a fine powder which was administered by inhalation. Later, people in the Middle East and Europe practiced a form of variolation in which small scratches or punctures were made in the skin and pus from smallpox victims introduced to the wounds.

In 1796, Edward Jenner followed up on an observation that milkmaids did not seem to get smallpox, and therefore, remained free of the facial scarring that many smallpox survivors suffered. Jenner deliberately infected a young boy with cowpox, then after the boy's recovery Jenner exposed him to smallpox. Luckily the boy did not contract the disease, and Edward Jenner is now known as the "Father of Immunology".

Types of Vaccines

Attenuated vaccines are live microbes that have been modified in some way to have reduced virulence. Attenuated vaccines are able to induce a stronger, longer lasting immune response with fewer booster shots because the microbe in the vaccine is alive and able to more closely mimic the actual microbe. Also, because an individual sheds live microbes (just like they would if they were sick), they can "infect" others with the vaccine, contributing to **herd immunity.** Herd immunity is when a critical portion of the population is immune to given disease. When many people have immunity, there are fewer human reservoirs of infection, which limits the transmission of the disease and protects people who are not immune.

The primary drawback to attenuated vaccines is that the microbes are alive and they can mutate back to their disease causing form and actually cause the illness they were created to prevent.

Inactivated vaccines are killed microbes, or parts of killed microbes that induce an immune response. Because these are dead microbes, there is no chance for mutation back to a disease causing form, so they are safer than attenuated vaccines. However, the immunity they induce is not as strong or long lasting, and more boosters are required. There are different types of inactivated vaccines:

whole agent vaccines, which are the whole microbe killed and preserved.

Subunit vaccines: just the antigenic portion of the microbe, these can protein-subunit vaccines (a protein from the microbe) or polysaccharide sub-unit vaccines.

toxoid vaccines: these are inactivated toxins designed to create antibodies to the toxins. An example is the tetanus vaccine, which creates an immune response to the toxin, tetanospasmin, and not to *C. tetani.* The antibodies created by the immune response neutralize the toxin.

Combination Vaccines contain vaccines to more than one microbe in a single dose. Some examples include:

DaPT: diphtheria, pertussis, tetanus. The diphtheria and tetanus vaccines are toxoids, pertussis is a sub-unit vaccine.
MMR: measles, mumps and rubella

Because the response produced by inactivated vaccines is weaker, **adjuvants** are often used to increase the immune response to the antigen. In the U.S., the only adjuvants approved for use by the FDA are aluminum gels and aluminum salts. Adjuvants can be found in the DaPT and Hib vaccines, among others.

Vaccine Schedules

We are most familiar with the idea of childhood vaccination, but recent advances in our understanding of immunity has lead scientists and doctors to believe that adults also need to keep up with their vaccinations. One example is of this is the disease whooping cough (pertussis). This is a nasty bacterial disease that causes uncontrolled coughing. The younger the victim, the more severe the symptoms and the victim can cough so hard they burst blood vessels in their eyes, or worse, suffocate from not being able to catch their breath.

The old wisdom was that this was a childhood illness, and adults did not need to be vaccinated. However, newer research shows that the vaccine wears off in the teen years. Because the severity of symptoms lessons as the person ages, this is not usually harmful in older individuals. However, these older individuals can contract whooping cough and not know that they have it and unwittingly infect others. In fact, epidemiologists have proposed that in cases of a persistent cough lasting two weeks (especially in teens or early 20s) or more, the individual most likely has whooping cough. For this reason, the CDC now advocates adult boosters for whooping cough and doctors frequently combine the pertussis booster with tetanus shots (TDaP).

The CDC's recommended immunization schedule for children and adults can be found on their website www.cdc.gov.

Types of Immunity

Active immunity is when a person's own immune system is generating an immune response. **Passive immunity** occurs when a person benefits from an immune response generated by another person (or animal). Passive immunization is usually done after exposure, when immediate protection is needed, as in the case of rabies or when a person is unable to make their own antibodies, either from illness (cancer affecting B cells) or genetic disorder of the immune system.

Antibodies are collected from blood donors, or from animals genetically engineered to produce specific antibodies and injected into the affected individual. Unfortunately, the protection is limited and the antibodies degrade quickly. Also, repeated injection of antisera can create an allergic reaction called serum sickness.

Sometimes passive and active immunotherapy are combined. This is often done with rabies. When a person is exposed to the rabies virus, the virus kills them before the immune system can respond (with a few recent exceptions).

In this case, the human is given anti-rabies antibodies (passive) and a rabies vaccine (active).

Natural immunity is the result of illness. **Artificial immunity** is the result of vaccination.

	Natural	Artificial
Active	get sick, have an immune response	get vaccinated, have an immune response
Passive	baby gets antibodies from mother across the placenta or in milk	person receives injection of antibodies from blood donors

Immune Testing

Our understanding of how the immune system work can also take us outside the human body. The following are three tests that use antibodies to give us clues to what is going on in the body.

1. **Titer:** antibody titer is measure of how many antibodies to a particular antigen are in the blood. If a person has high Ab titer, that means their immunity to that pathogen is good, if they have low titer, then they have poor titer and need boosters. Hepatitis B titer is routinely checked in healthcare personnel to ensure that they are protected from Hepatitis B virus.
2. **ELISA:** enzyme-linked immunosorbant assay. In this assay, the relationship between antigen and antibody can tell if an individual has antibodies to particular microbe. Having antibodies means that they have (or have had) the disease.

   To conduct an ELISA assay, antibodies to antigen of interest are allowed to attach to plate. The antigen is then added and binds to the antibodies. Next, the enzyme-linked antibodies are added and bind to the antigen. Finally, the substrate of the enzyme linked to the antibodies is added. The enzyme-substrate reaction causes a color change which indicates the presence of the antigen of interest.
3. **Immunofluorescent Assays:** using immunofluorescence, scientists can detect the presence of very small amounts of bacteria in tissue samples, or identify antigens on the surface of eukaryotic cells. Antibodies to the antigens are made and bound to fluorescent tags. The antibodies bind to the antigen, and then the scientist uses a fluorescent microscope to observe the samples. Any cells which have been bound by the antibody will glow due to the fluorescent tag.

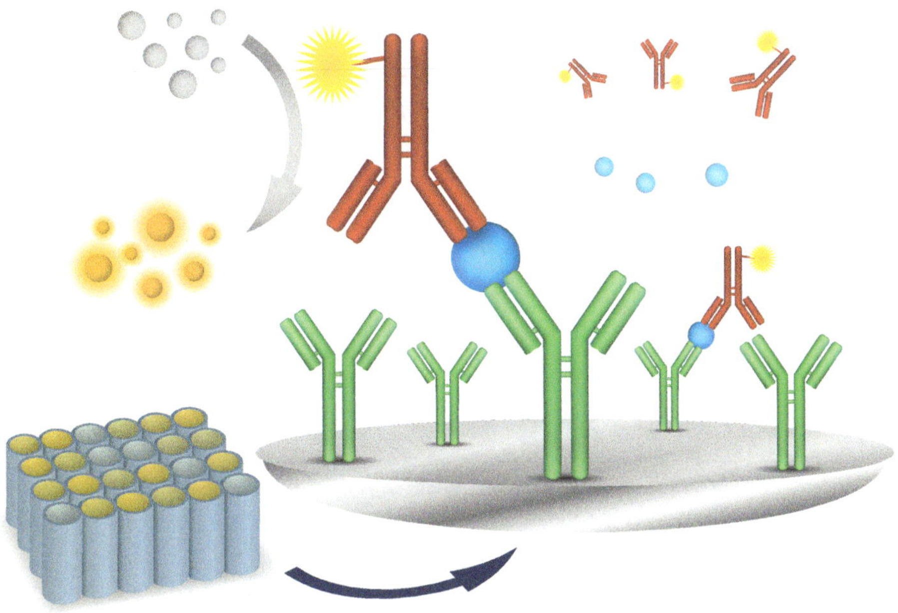

Fig. 15.26 An ELISA Test

Extender_01/Shutterstock.comThe technology used in the ELISA is the same technology used in home pregnancy tests, where antibodies to human chorionic gonadotropin (hCG, a hormone produced by pregnant women) are embedded in the test and bind to hCG if it is present.

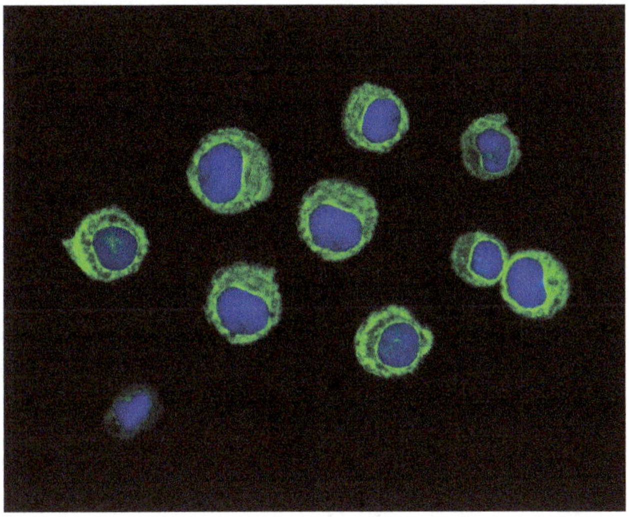

Fig. 15.27 Cells Tagged with Immunofluorescent Antibodies

Dlumen/Shutterstock.com

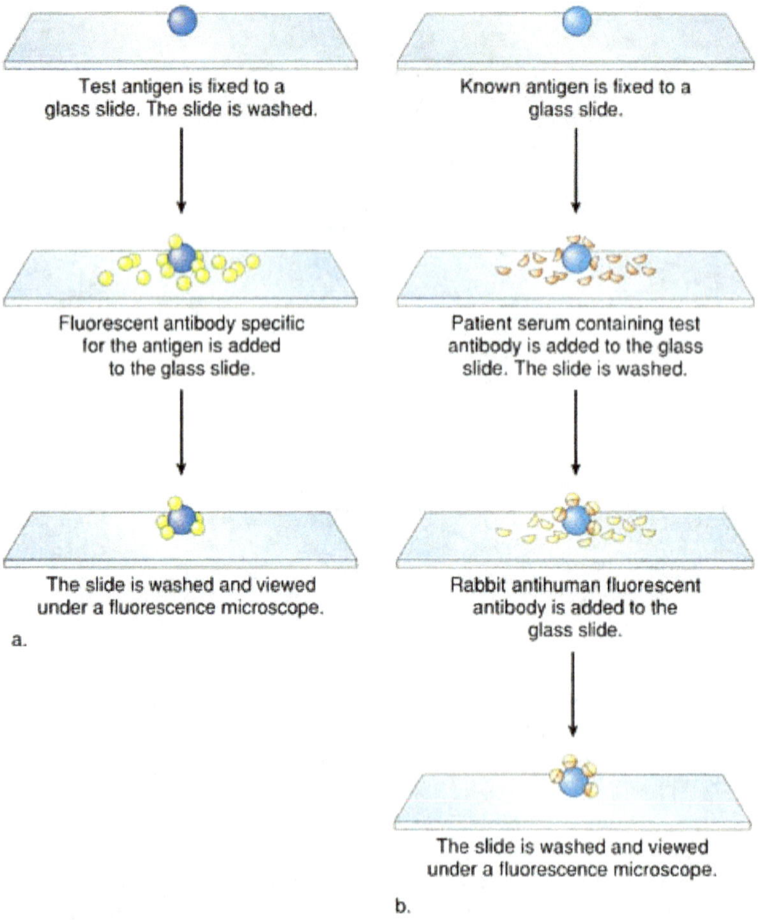

Fig. 15.28 Conduction an Immunofluorescent Assagy
KH 16.17 http://webcom.grtxle.com/customization/uploads/

Lab Exercise 15.1

Medical Microbiology

Innate Immunity: The Role of Normal Microbiota and Antimicrobial Activity of the Skin

Student Objectives

1. Demonstrate and identify the antibacterial characteristics of the skin.
2. Describe the role of normal microbial microbiota in innate immunity.
3. Describe competitive exclusion.
4. Name various microorganisms commonly found on skin of humans and list identify some common infectious diseases each causes.
5. Compare and contrast innate and adaptive immunity.
6. Identify and describe the first line defenses.
7. Apply knowledge on biochemical testing to deduct conclusions on hypothetical scenarios.

Name ______________________________

Pre-Lab Exercise 15.1

1. What is the purpose of this laboratory exercise?

____

____

____

2. Describe the principle of competitive exclusion.

____

____

____

____

3. Explain why mannitol salt agar is a *selective* medium for isolating normal skin microbiota?

____

____

4. List three characteristics of the skin that prevent infection from pathogenic organisms.

____

____

INTRODUCTION

Normal skin microbiota are generally resistant to drying and to relatively high salt concentrations. More bacteria are found in moist areas, such as armpits and the side of the nose, than on dry surfaces of the arms or legs. Transient microbes are present on hands and arms due to contact with the environment but typically find the skin a poor place to reproduce. Also, transient microbes you encounter that can survive on the skin usually do not survive since **competitive exclusion** occurs. Competitive exclusion is a principle that states when two organisms compete for the same critical resources within an environment, one of those organisms will eventually outcompete and displace the other. One of the experiments that will be performed today will demonstrate competitive exclusion.

Most indigenous bacteria on the skin are Gram-positive and salt tolerant. *Propionibacterium* live in hair follicles and feed on sebum from oil glands. The propionic acid they produce via fermentation maintains the pH of the skin between 3 and 5, which suppresses the growth of most other bacteria. Catalase positive, salt tolerant members of Gram-positive cocci (family Micrococcaceae) are the most frequently encountered bacteria on the skin along with members from the genus *Staphylococcus*. The catalase test is a useful test to quickly distinguish between *Streptococcus* which is catalase negative and *Staphylococcus* which are catalase positive.

Staphylococcus aureus is species of bacteria commonly part of the normal microbiota of the skin and is also considered an opportunistic pathogen it is the second most commonly isolated bacteria in hospitals and often associated with nosocomial (hospital acquired) infections. Unlike many other species in the genus *Staphylococcus*, *S. aureus* is able to ferment the sugar mannitol. *Staphylococcus aureus* also produces an enzyme called **coagulase** that clots the fibrin in blood. The production of this enzyme is highly correlated with pathogenicity and distinguishes *S. aureus* from other species of *Staphylococcus*. Testing for mannitol fermentation or for the presence of coagulase are biochemical tests that can be used to identify *S. aureus* from other staph species.

Although many different bacterial genera (and yeasts) live on human skin, we utilize a selective and differential medium to attempt to isolate staphylococci in this exercise. You will also expose two microorganisms to the environment on your skin. Then you will try to determine which of the two is a typical inhabitant based on how they survive the antimicrobial activity of the skin.

Mannitol Salt Agar Plate

Use the swab for about half of the plate then switch to a loop to get isolated colonies. Be sure to flame in between streaks.

Week One—Antimicrobial Activity of the Skin

Materials **per 2 students:** Tryptic Soy Agar (TSA) plate (1), Sterile cotton swabs (5), Sterile saline, Mixed broth culture of Bacillus subtilis and Micrococcus *luteus*

Procedure:

1. Divide the bottom of a TSA plate into 5 sections and label each section with one of the following numbers: 0, 15, 30, 45, and 60.
2. Select four areas on the back of your hand, making sure each area is free from cuts and abrasions. Mark the spots on your hand using a pen or marker with the same numbers (except "0") as above (or use your knuckles as markers).
3. Moisten one sterile cotton swab in the mixed broth culture, then slightly swab each marked area on your hand with the culture. **Do not wash your hands for the duration of this test**. Using the same swab, streak the 0 sector on your plate to show the mix of bacteria at time zero. Biohazard the swab.
4. After fifteen minutes, moisten a fresh swab with **sterile saline** and swab over the area marked 15 on your hand "picking up" the organisms left on your skin. Streak the sector of the agar plate marked 15. Discard swab in the biohazard bag.
5. Repeat the procedure in step 4 above with the area 30 at 30 minutes, 45 at 45 minutes, and 60 at 60 minutes.
6. After the hour long test, wash your hands well with soap and water.
7. Invert and place the labeled plate in the container provided for incubation.

Week One-Isolation of Skin Microbiota

Materials **per student:** Mannitol Salt Agar (MSA) plate (1), Sterile cotton swab (1), Sterile saline

Procedure:

1. Wet the swab with saline, and push against the wall of the test tube to express excess saline. Swab any surface of your skin, the sides of your nose, behind your ear, inside your arm, would be ideal areas for this exercise.
2. Swab one-half of the mannitol salt plate with the swab.
3. Discard the swab in the biohazard bag.
4. Using a sterile loop, streak back and forth into the swabbed area a few times, then streak to obtain isolated colonies according to the diagram on the previous page.
5. *Place your inverted plate in the tub for incubation at 35°C until the next lab period.*

Week Two Antibacterial Activity of the Skin

Examine the TSA plate inoculated last period and estimate the relative amount of growth of Micrococcus luteus (yellow colonies) and *Bacillus* subtilis (off-white colonies). Record your results.

Week Two Isolation of Skin Microbiota

Materials:

Your incubated MSA plate	3% hydrogen peroxide

Procedure:

1. Examine the isolated colonies. Members of the Micrococaceae usually form large opaque colonies. Record the appearance of 3 colonies and any evidence of mannitol fermentation (yellow halos) in your lab manual.
2. Test for catalase production by carefully adding 1 drop of H_2O_2 onto the 3 isolated colonies.

Name ______________________________

Exercise 15.1 Three Lab Report

Antibacterial Activity of the Skin

Sketch your plate and **label** the two organisms in the original culture.

1. **Explain** what happened to the each bacterial species over the hour:

____

____

2. Which organism became "extinct" on your skin?

____

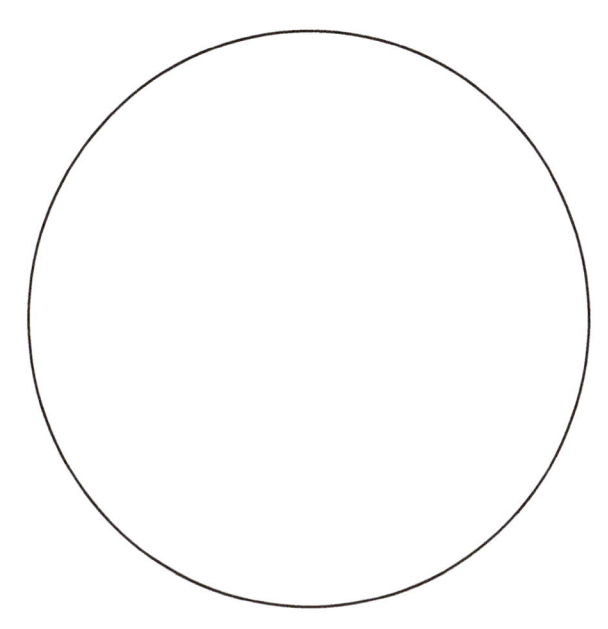

Isolation of Skin Microbiota

Source of inoculum: ________________________________

Mannitol Salt Agar Plate: Describe 3 colony types if possible you may have to look at one of your lab partners plates:

	NUMBER ASSIGNED TO COLONY (COLONY ID)		
	1	2	3
Colony description			
Pigment of colony			
Mannitol fermentation Yes/No			
Catalase reaction +/–			

3. What is a lysozyme? How does it destroy certain bacteria?

____

____

____

4. List at least 3 ***biochemical tests*** that could be performed to positively identify the organisms on your body. **Hint:** you can use tests we have performed in other labs.

____

____

5. An unknown organism is growing in small colonies on an Mannitol Salt agar plate (MSA plate). The media is red to begin with and it is still red following the incubation period. What can you conclude from the results of this growth pattern?

____

____

____

6. You inoculate an MSA plate with an unknown bacterium. The media is red to begin with and it is still red following the incubation period. Following the incubation period there is **no growth**. What can you conclude from the results?

____

____

____

7. An unknown organism is growing in small colonies on an Mannitol Salt agar plate (MSA plate). The media is red to begin with and it is now yellow following the incubation period.

   a. What can you conclude from the results of this growth pattern?

   ____

   ____

   b. You then perform a coagulase test and the result is positive. What can you conclude about the identity of this organism?

   ____

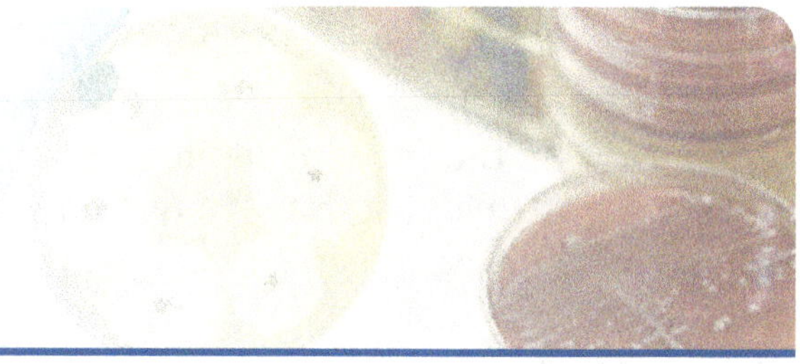

Lab Exercise 15.2

Medical Microbiology:

Blood, Immunology and the Differential White Blood Cell Count

Student Objectives

1. Access, analyze, and use serology information to make predictions based on the evidence discovered.
2. Describe how the immune system enables us to control the transmission of infectious disease-causing microorganisms.
3. Identify and illustrate the innate (non-specific) and adaptive (specific) mechanisms of host immunity.
4. Identify the five major types of leukocytes.
5. Describe the structure and function of red blood cells, white blood cells, and platelets.
6. Illustrate and explain the operation of antibody-mediated and cell-mediated immunity.
7. Describe the various serological tests performed in the microbiology laboratory.

INTRODUCTION

Blood is a type of connective tissue whose cells are suspended in a liquid extracellular matrix. About 55% of the blood is liquid, referred to as plasma. The other 45% is made up of cells. These cells are mainly formed in red bone marrow, and they include red blood cells, white blood cells, and cellular fragments called platelets. Red blood cells transport gases between the body and the lungs. White blood cells defend the body against infections. Platelets play an important role in the stoppage of bleeding (hemostasis). There are major structural and blood-count differences between red blood cells (erythrocytes) and white blood cells (leukocytes). Red blood cells lack a nucleus and look like discs that are biconcave; every milliliter (mL) of blood contains approximately 5 million erythrocytes (see Figure 1). White blood cells are much larger and contain a nucleus; every mL of blood contains approximately 5,000 leukocytes (see Figures 2-6).

A complete blood count (CBC) measures the number of red blood cells and white blood cells in the blood in addition to the total amount of hemoglobin and the amount of hemoglobin per RBC. A white blood cell count (WBC) measures the amount of each type of white cell in the blood. A differential determines the percentage of each type of WBC present compared to normal levels.

Cell Type	Normal Range	Indicate
Neutrophil	54–62%	**High levels:** stress, bacterial infection, eclampsia, rheumatoid arthritis, trauma
		Low levels: aplastic anemia, chemotherapy, radiation exposure/therapy, influenza or other viral infection
Eosinophil	1–3%	**High levels:** parasite infection, allergies
Basophil	1%	**High levels:** infection, cancer, injury
Monocyte	3–9%	**High levels:** leukemia, TB, parasite infection
Lymphocyte	25–33%	**High levels:** bacterial infection, hepatitis, viral infection
		Low levels: chemotherapy, HIV infection, radiation exposure/therapy, sepsis, leukemia

Blood Cells Activity

Materials

Laboratory materials: Microscope, blood-smear with Wright stain slide, pathology blood-smear slides.

1. Obtain a blood-smear slide.
2. Using proper microscope procedures, view the blood slide at the 4X, 10X, and 40X objective lenses.
3. Locate the red blood cells and draw a few red blood cells in the lab report at 400X magnification.
4. Locate each of the five types of white blood cells and draw each cell in the lab report at 400X magnification.

Differential White Blood Cell Count Activity

A differential white blood cell count is performed to determine the percentage of each of the various types of white blood cells present in a blood sample. The test is useful because the relative proportions of white blood cells may change in particular diseases. Neutrophils, for example, usually increase during bacterial infections, whereas eosinophils may increase during certain parasitic infections and allergic reactions.

***NOTE*: Use a partner for this activity.**

1. Using high-power magnification, focus on the cells at one end of a prepared blood slide where the cells are well distributed.
2. Slowly move the entire blood slide back and forth, in a zigzag motion, following a path that avoids passing over the same cells twice.
3. Each time you encounter a white blood cell, identify its type and record it in the lab report by placing a tally mark in its 'Number Observed' box.
4. Continue searching for and identifying white blood cells until you have recorded a total of 100 white blood cells.
5. Total each of the five white blood cells in the 'Total' box.
6. For the 'Percent' of each cell, use the same number in the 'Total' box and add a percent sign. Because percent means "parts of 100," for each type of white blood cell, the total number observed is equal to its percentage in the blood sample.

Pathology of Blood Cells Activity

Pathology is the study of diseases. When a disease occurs, structural alterations of the cells involved may occur. Clinical diagnosis of some diseases can be seen by analyzing the structural changes and quantity of blood cells. The following are diseases we will observe in this section of the lab exercise:

- Mononucleosis = Infection of B-cells from the Epstein Barr virus.
- Malaria = Infection of red blood cells from the protozoan *Plasmodium*.
- Sickle-cell anemia = Genetic mutation. This is an inherited condition which results in some erythrocytes being malformed. The gene for this condition causes the hemoglobin to be incorrectly formed, which in turn causes some erythrocytes to take on a crescent shape. These cells are not able to carry adequate amounts of oxygen to cells.
- Leukemia (Acute and Chronic) = Cancer of the white blood cells; too many leukocytes are made in the red bone marrow.

***NOTE*: The class will work together for this activity; each table will have a different pathology slide.**

1. Obtain a pathology blood-smear slide.
2. Using proper microscope procedures, view the blood slide at the 4X, 10X, and 40X objective lenses.
3. Notice the abnormal shape of the red blood cells or white blood cells.
4. Draw the pathology blood cells in the lab report at 400X total magnification.
5. View the other pathology slides at the other laboratory tables. Draw your observations in the lab report.

Name ____________________________

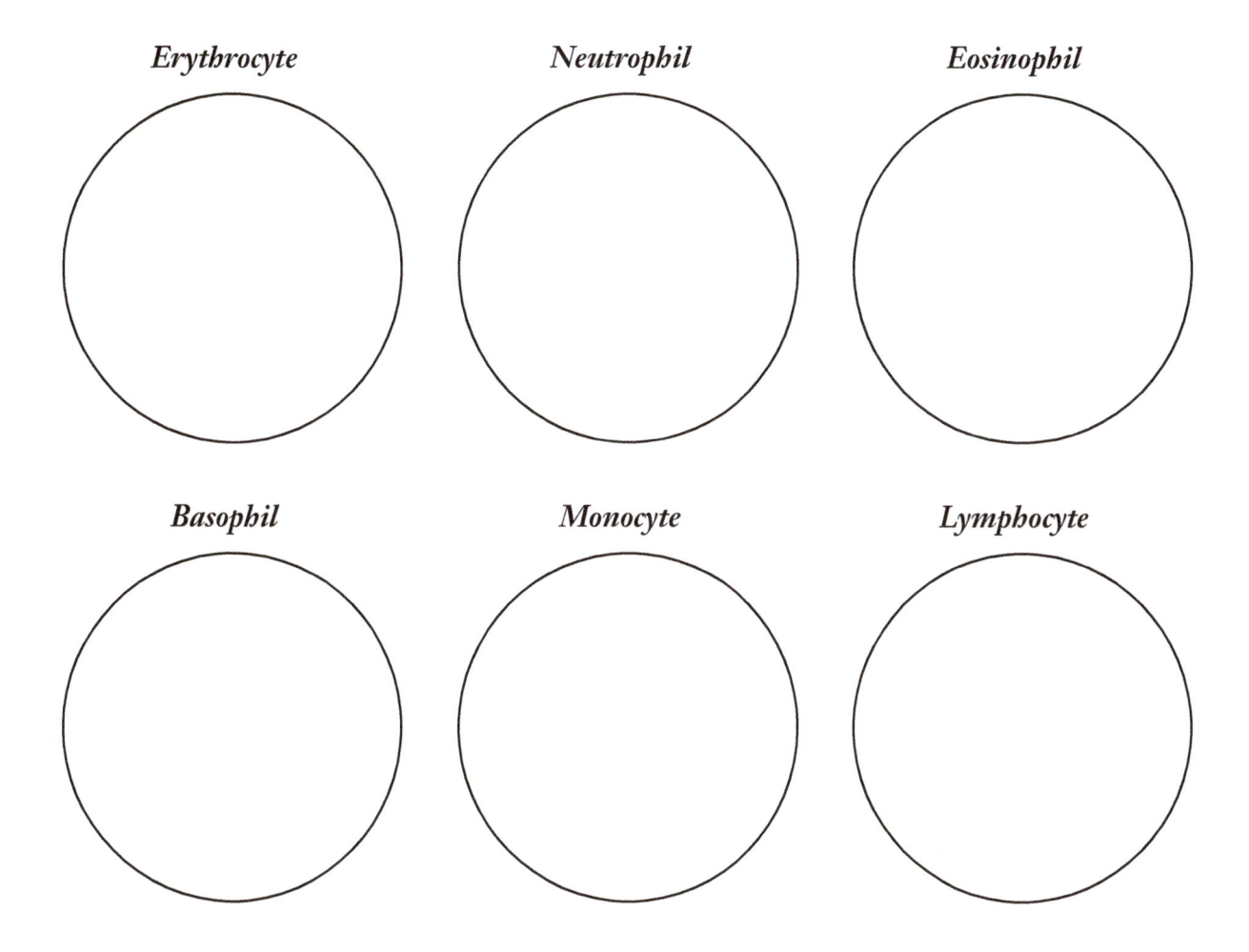

Exercise 15.2 Lab Report

Microscopic Drawings—Blood-Smear Slide

Erythrocyte

Neutrophil

Eosinophil

Basophil

Monocyte

Lymphocyte

1. Describe the shape of a red blood cell.

____

2. Describe the shape and size of a white blood cell.

____

3. What is the largest white blood cell? What is the smallest white blood cell?

____

Differential White Blood Cell Count

WBC Type	Number Observed	Total	Percent
Neutrophil			
Eosinophil			
Basophil			
Monocyte			
Lymphocyte			

4. How do the results of your differential white blood cell count compare with the normal values?

____

____

Microscopic Drawings—Pathology Blood Slides

Mononucleosis *Malaria* *Sickle-Cell Anemia*

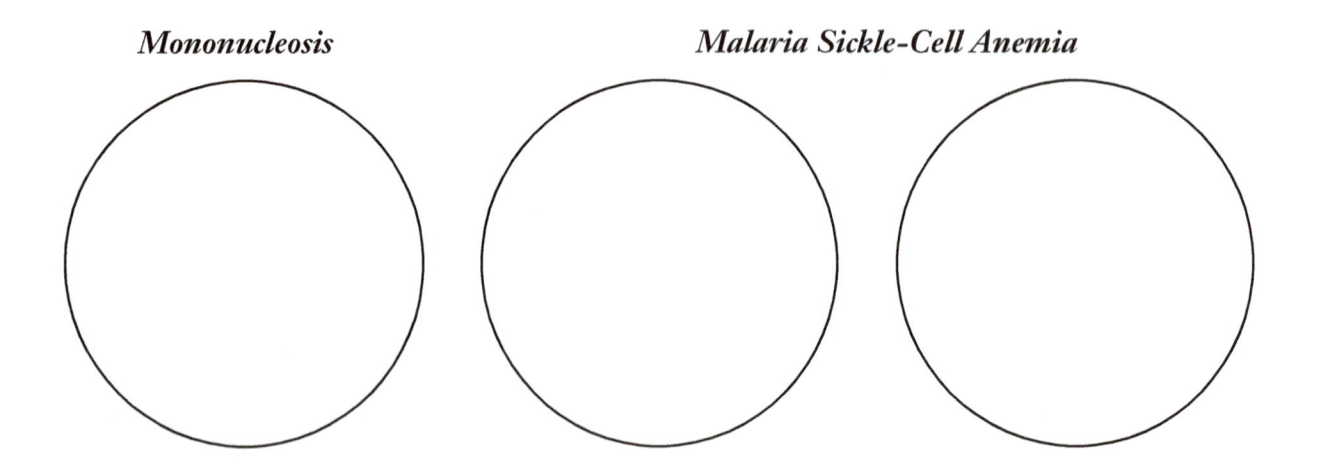

Acute Leukemia

Chronic Leukemia

5. How do the size and shape of the pathological blood cells compare with the normal size and shape of the blood cells?

____

____

Immunology Questions

6. What are the two arms of the immune system?

____

7. List examples of innate (non-specific) immune defense.

____

8. When does adaptive (specific) immune response occur?

____

9. What is the role of red bone marrow?

____

10. What is the role of the thymus?

____

11. What is the role of lymph nodes?

____

12. What is the most numerous white blood cell?

____

13. What is the least numerous white blood cell?

____

14. Describe the role of phagocytes.

____

15. List examples of phagocytic cells.

____

16. What are the differences between B-cells and T-cells?

____

____

17. What are the three types of granular leukocytes?

____

18. Where would you find macrophages?

____

19. What is the function of the helper T-cell?

____

20. What cells will the monocytes differentiate into when they leave the bloodstream?

____

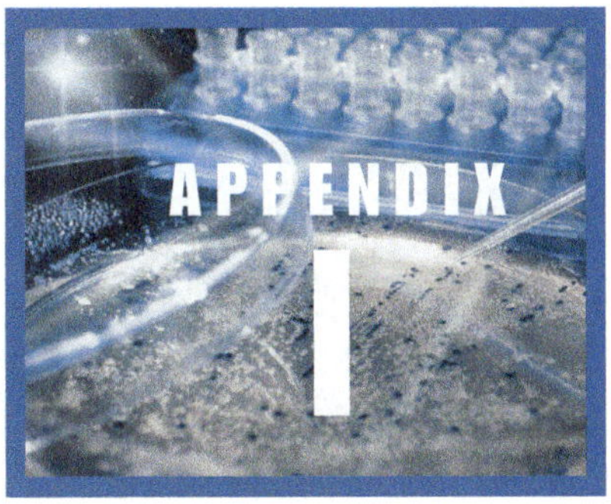

The Unknown Project

Student Objectives

1. Perform proper aseptic techniques while inoculating various types of media.
2. Apply the scientific method and sequence of tests to identify an unknown bacterium.
3. Evaluate the "Flow Chart for the Identification of an Unknown Bacterium" and use the *Bergey's Manual* to determine the sequence of tests required in identifying an unknown bacterium.
4. Perform and evaluate the Gram stain for an unknown bacterium.
5. Perform and evaluate biochemical testing to determine:
   a. sugar fermentation properties
   b. motility
   c. oxygen usage
   d. hemolytic patterns
   e. aerobic metabolism
   f. IMViC test results
   g. oxidase production
   h. eosin utilization
   i. urease production
   j. catalase production
   k. halotolerance

Unknown Procedure—Week One

Materials

Per student: Inoculating loop, inoculating needle, Bunsen burner, biochemical testing media, and a numbered tube with your bacterium.

Per table: One tray labeled "24 hours" for plates; one test-tube rack labeled "24 hours"; one test-tube rack labeled "5 days".

1. Obtain a broth tube with an unknown bacterium species. **Record** your unknown number in the Lab Report.

2. Use the broth culture to perform the **Gram stain** on your unknown. If you are not successful with your Gram staining, repeat the procedure. You cannot progress in the identification of your unknown without obtaining valid results with the Gram reaction. Record your Gram stain results in the Lab Report. Record the shape of your unknown. NOTE: We do not have Gram-negative cocci, only bacilli.

3. Once you know if you have a Gram-positive or a Gram-negative bacterium, you will use the information from **Chapter 4** for your biochemical testing of your unknown. For each test, follow the *Inoculation Procedures*. **Label** each biochemical medium with your initials, table number, and unknown number.

4. Place your inoculated media in your table's tray and test-tube racks. Make sure they are labeled with your table number and section number.

5. Remember, throughout the project you will record your data in this lab. The data sheet lists all the Gram-positive and Gram-negative tests. You will only run the tests and list test results required for your type of bacterium.

Unknown Procedure—Week Two

NOTE: When you have finished with your media, remove the tape from the test tubes, and then discard the test tubes and plates on the autoclave cart.

1. Retrieve your inoculated plates and tubes containing your "unknown" which you inoculated last week. Remember: pick up the test tube by the glass, NOT by the cap.

2. Refer to Chapter 4. Follow along the section named ***Interpretation of Reactions***. This contains directions on how to interpret your tests.

3. Record the results of your work on the *Data Report: Unknown Exercise* (in this lab exercise). This will contain all the data collected from tests you perform to identify your unknown.

4. Use the charts in Appendix 2, the Micro-Atlas, and the *Bergey's Manual* in the lab room. Confirm that you have made a proper determination of your unknown bacterium.

5. Upon determination of your unknown bacterium, record your unknown bacteria name.

Identification of Unknown Project Report

1. **Your instructor will provide you with details on how your report should be constructed.**

Identification of Unknown Project Grade

Your grade on the Unknown Project includes **ALL YOUR WORK**:

1. Your attendance and performance in lab as you complete the identification process.
2. Your utilization of critical thinking skills, demonstrated in the lab as you work your way through the identification process.
3. Correct identification of the unknown microorganism.
4. Thoroughness and accuracy in the data collection process, as demonstrated by the quality of the *Data Sheet*, and use of correct media.
5. Thoroughness and accuracy of the typed written summary of the route and tests you performed to identify your bacterium, as demonstrated by the quality of your flow chart and written summary.

Student Name:

Unknown Number:

Name Of Organism:

DATA SHEET: UNKNOWN EXERCISE

Test	**RESULTS: Circle Choice or Record Results Observed**	**Comments**
1. Gram stain result	Gram-positive Gram-negative	

DATA SHEET: UNKNOWN EXERCISE		
Test	**RESULTS: Circle Choice or Record Results Observed**	**Comments**
Shape	Cocci	
	Bacilli	
	Pleomorphic (various lengths)	
Arrangement	Clusters	
	Chains	
	Branching rods	
	Single rods randomly arranged	
	Palisade layers	
	Other (Specify in comments)	

(*continued*)

DATA SHEET: UNKNOWN EXERCISE		
Test	**RESULTS: Circle Choice or Record Results Observed**	**Comments**
2. Colony Morphology from TSA plate		
a. Size	Small Medium Large	
b. Color	______________________	
Describe the following in the Results column		
c. Opacity		
d. Configuration		
e. Margins		
f. Elevation		

DATA SHEET: UNKNOWN EXERCISE		
Test	RESULTS: Circle Choice or Record Results Observed	Comments
3. Sugar fermentation:		
a. Glucose (G)	Acid (+) or (−); Gas (+) or (−)	
b. Sucrose (S)	Acid (+) or (−)	
c. Lactose (L)	Acid (+) or (−)	
d. Mannitol (M)	Acid (+) or (−)	
4. Motility plate (MT)	Motile Nonmotile	
5. Fluid thioglycollate tube (FT)	Obligate Aerobe Facultative Anaerobe	

(continued)

DATA SHEET: UNKNOWN EXERCISE		
Test	**RESULTS: Circle Choice or Record Results Observed**	**Comments**
6. Oxidase test—TSA plate	Positive for oxidase enzyme Negative for oxidase enzyme	
7. Nitrate broth tube (N)	Positive for Nitrate reduction Negative for Nitrate reduction	
8. Catalase test—TSA plate	Positive for Catalase enzyme Negative for Catalase enzyme	

DATA SHEET: UNKNOWN EXERCISE		
Test	**RESULTS: Circle Choice or Record Results Observed**	**Comments**
9. Mannitol Salt Agar Plate (MSA)	Positive for growth in 7.5% salt Negative for growth in 7.5% salt Positive for mannitol fermentation Unable to determine mannitol fermentation; bacteria could not tolerate 7.5% salt Negative for mannitol fermentation	
10. Blood Agar Plate (BAP)	Positive for beta-hemolysis Positive for alpha-hemolysis Positive for gamma-hemolysis (non-hemolytic)	

(*continued*)

DATA SHEET: UNKNOWN EXERCISE		
Test	**RESULTS: Circle Choice or Record Results Observed**	**Comments**
11. EMB plate	Describe growth characteristics:	
12. Indole test (I)—tryptone broth tube	Positive for indole production Negative for indole production	
13. Methyl Red tube (MR)	Positive for mixed acid production Negative for mixed acid production	
14. Voges-Proskauer test	Positive for acetoin or 2,3-butanediol production Negative for acetoin or 2,3-butanediol production	

DATA SHEET: UNKNOWN EXERCISE		
Test	**RESULTS: Circle Choice or Record Results Observed**	**Comments**
15. Simmon's Citrate slant (SC)	Positive for citrate utilization Negative for citrate utilization	
16. Urea broth tube (U)	Positive for urease enzyme Negative for urease enzyme	

Identification Chart

Gram Negatives

Test	*Alcaligenes faecalis*	*Citrobacter freundii*	*Enterobacter aerogenes*	*E. coli*	*Klebsiella pneumoniae*	*Proteus vulgaris*
Glucose broth	–	+, gas	+, gas	+, gas	+, gas	+
Sucrose broth	–	+	+	–	+	+
Lactose broth	–	+	+	+	+	–
Mannitol broth	–	+	+	+	+	–
Motility	motile	motile	motile	motile	non*	motile
FT	Obligate aerobe	Fac anaerobe	Fac anaerobe	Fac anaerobe	Fac anaerobe	Fac anaerobe
Oxidase	+	–	–	+ or –	–	–
Nitrate broth	–	+	+	+	+	+
EMB	clear	Pink/Red sometimes w/ black center	Purple/red sometimes w/ black center	Dark w/ metallic green sheen	Purple/red sometimes w/ black/ green metallic center	clear
Urea	–	–	–	–	+	+
Indole	–	–	–	+	–	+
MR	–	+	–	+	+	+
VP	–	–	+	–	–	–
Citrate	–	+	+	–	+	–
Notes					*appears motile w/ slimy white center	often darkens agar on TSA plate

Pseudomonas aeruginosa	*Serratia marcescens*	*Alcaligenes viscolactis*	*Citrobacter koseri*	*Enterobacter cloacae*	*Proteus mirabilis*	*Salmonella enteriditis*
+	+	–	+ gas	+ gas	+ gas	+ gas
–	+	–	–	+	–	–
–	–	–	Weak +	+	–	–
–	+	–	+	+	–	+
motile	motile	motile	motile	motile	motile, swarming	motile
Obligate aerobe	Fac anaerobe	Ob aerobe	Fac anaerobe	Fac anaerobe	Fac anaerobe	Fac anaerobe
+ or –	–	+	–	–	–	–
+	+	–	+	+	+	+
clear	Pink–red	No Growth	Translucent pink	Black/dark purple centers some metallic green colonies	translucent pink	translucent pink
–	–*	–	–	–	+	–
–	– *	–	–	–	–	–
–	– *	–	+	–	+	+
–	+	–	–	+	+	–
+	+	–	+	+	–	+
obligate aerobe except in the presence of nitrate	*red pigment makes some tests difficult			slightly yeast–like odor	often darkens agar on TSA plate	

Gram Positives

Test	*Staphylococcus aureus*	*Staphylococcus epidermidis*	*Streptococcus mutans*	*Micrococcus luteus*	*Bacillus subtilis*	*Lactobacillus caseii*
Glucose broth	+	+	+	–	+	–
Sucrose broth	+	+	–	–	+	–
Lactose broth	+	Weak +	–	–	–	–
Mannitol broth	+	–	+	–	Weak +	–
Motility	non	non	non	non	motile	non
FT	Fac anaerobe	Fac anaerobe	Fac anaerobe	Ob aerobe	Obligate aerobe	Fac anaerobe usually w/ little or no aerobic growth
Oxidase	–	–	–	Weak +	+	–
Catalase	+	+	–	+	+	–
Nitrate broth	+	+	–	–	+	–
TSA-B (blood)	β-hemolysis	γ-hemolysis	γ-hemolysis	γ-hemolysis	β	γ-hemolysis
MSA	Salt +, mann +	Salt +, mann –	No growth	Salt +, mann –	No growth	No growth
Special notes			Cells may be slightly oval	Yellow colony color	Cells form chains	Very small colonies

Corynebacterium xerosis	*Mycobacterium smegmatis*	*Bacillus cereus*	*Bacillus megaterium*	*Enterococcus faecalis*	*Streptococcus pyogenes*
+	−	+	−	+	Weak +
+	−	+	−	+	+
−	−	−	−	+	−
−	−	−	−	+	−
non	non	motile	non	Non	Non
Fac anaerobe	Obligate aerobe	Fac anaerobe	Ob Aerobe	Fac anaerobe	Fac anaerobe
−	+	−	−	−	−
+	+	+	+	−	−
+	Weak +	+ no Zn	−	−	−
γ-hemolysis	γ-hemolysis	β-hemolysis	γ-hemolysis	β-hemolysis	β-hemolysis
Salt +, mann −	No growth	No Growth	Weak growth	No Growth	No Growth
Often pleomorphic Grows yellow colonies	Weak + gram stain		Very large cells	Often diplo- or strepto-cell arrangement	

Microbes used in the Lab

*Alcaligenesfaecalis**
Alcaligenesviscolactis
Bacillus cereus
Bacillus megaterium
Bacillus subtilis
Citrobacterfreundii
Citrobacterkoseri
Clostridium acetobutylicum
Corynebacteriumxerosis
Enterobacteraerogenes
Enterobacter cloacae
Enterococcus faecalis
Escherichia coli
Escherichia coli (beta-galac. Def)
Escherichia coli + pBLU
*Klebsiellapneumoniae**
Lactobacillus caseii
Micrococcus luteus
*Mycobacterium smegmatis**
*Proteus mirabilis**
*Proteus vulgaris**
*Pseudomonas aeruginosa**
Pseudomonas fluorescens
Serratiamarcescens
*Salmonella enteriditis**
*Staphylococcus aureus**
Staphylococcus epidermidis
Streptococcus mutans
*Streptococcus pyogenes**
Fungi: Rhizopusstolonifer +/−

*denotes a potential pathogen/BSL-2 microbe
All other microbes are BSL-1.

Index

B

C

D

E

F

G

H

I

N

O

P

Q

R

S